CW00672837

Hydrothermal Chemistry of Zeolites

Hydrothermal Chemistry of Zeolites

R. M. Barrer FRS

Department of Chemistry
Imperial College of Science and Technology
London

1982

ACADEMIC PRESS

A Subsidiary of Harcourt Brace Jovanovich, Publishers

London   New York

Paris   San Diego   San Francisco
São Paulo   Sydney   Tokyo   Toronto

ACADEMIC PRESS INC. (LONDON) LTD
24/28 Oval Road,
London NW1 7DX

United States Edition published by
ACADEMIC PRESS INC.
111 Fifth Avenue,
New York, New York 10003

British Library Cataloguing in Publication Data

Barrer, R.M.
  Hydrothermal chemistry of zeolites.
  1. Zeolites
  I. Title
  549'.68   QE391.Z5
ISBN 0–12–079360–1

Typeset and Printed in Northern Ireland by The Universities Press (Belfast) Ltd.

Preface

One quantitative aspect of inorganic chemistry is geometric in nature, involving the space patterns of crystals. The zeolites provide outstanding examples of this in their numerous and unique structures, each type honey-combed by its individual configuration of windows, cavities and channels of molecular dimensions. It is in particular the permanently porous and rigid nature of the crystals which has rendered them of such interest. In terms of the differing shapes and sizes of molecules and those of the intrazeolitic windows and pores, there can be lock-and-key relationships between guest molecule and host zeolite like those often considered to arise between enzymes and the substrates which they catalyse in specific ways. No apology is needed for presenting as a monograph some of the literature on hydrothermal synthesis with special reference to the formation and transformation of zeolites.

In the foreword to a companion volume† describing the sorbent properties of zeolites and clay minerals I expressed the view that, because of the scale and variety of published work, the time was past for presenting zeolite properties and chemistry in a single volume, but rather that different aspects of zeolite science had now to become the subjects of separate monographs. In accord with this view I have tried in the present volume to survey current knowledge of zeolite synthesis and related topics. This has involved a literature so large and so diversified in its sources that a complete survey has not been possible within the confines of the monograph. Therefore a choice has been made which of necessity omits reference to many useful contributions. Nevertheless, it is hoped that the text is representative and the approach sufficiently systematic and critical to provide background and stimulus for those involved in this particular area of silicate science, in which so much has been achieved, and yet so much remains to be done. The empirical "try it and see" approach, while often successful in the past when the field of synthesis was largely unexplored, is increasingly being replaced by systematic explorations, and explorations in depth, using new and old techniques for

† R.M. Barrer, "Zeolites and Clay Minerals as Sorbents and Molecular Sieves". Academic Press, London and New York, 1978.

examining the course of reaction and the products. Efforts are being made, as indicated in this book, to develop the quantitative physico-chemical basis of zeolite nucleation and crystallization in kinetic, thermo-dynamic and statistical thermodynamic aspects. It can be hoped that as knowledge grows the objective of understanding events occurring at the molecular level will finally be achieved. If so this will result from combining methods of investigating precursor species present in the parent magmas and solutions during the course of reaction with theoreti-cal modelling capable of giving logical interpretation of what is experi-mentally established. Ultimately, improved chemical knowledge of these complex crystallizing systems will, it is hoped, help to design ways of making more and more of the large number of novel zeolites which so far have been constructed only as models. One can at this stage only speculate as to the numbers of these structures which will eventually be synthesized and the uses which will be made of them. Certainly, as an area of chemical research, zeolite synthesis has already proved to be of unusual richness. In conclusion, I wish to thank various authors and publishers for their ready permission to reproduce diagrams.

Spring 1982   *R.M. Barrer*

Contents

Chapter 3
Reactants in Zeolite Synthesis and the Pre-nucleation Stage

Chapter 4
Nucleation, Crystal Growth and Reaction Variables

Chapter 5
Representative Zeolite Syntheses, Crystallization Fields and Transformations

Chapter 6
Isomorphous Replacements in the Frameworks of Zeolites and Other Tectosilicates

Chapter 7
Synthesis and Some Properties of Salt-bearing Tectosilicates

Chapter 1

Occurrence, Classification and Some Properties of Zeolites

1. Introduction

Silicate chemistry, like aliphatic carbon chemistry, is based on the tetrahedron. In silicates each silicon is joined to four oxygens. The SiO_4 unit linked with other SiO_4 tetrahedra through shared oxygens yields numerous island structures, chains, sheets and three-dimensional networks. If in the networks any given SiO_4 tetrahedron shares one of its four apical oxygens with one each of four other tetrahedra then uncharged tectosilicate frameworks are generated, as represented by the crystalline silicas, SiO_2. The SiO_4 tetrahedra may be replaced in tectosilicates by tetrahedra AlO_4, GaO_4 or GeO_4, as described in Chapter 6, to give aluminosilicates, gallosilicates, aluminogermanates and gallogermanates which, like the crystalline silicas, all have three-dimensional frameworks. The frameworks contain cations electrochemically equivalent to the negative charge introduced when Al^{III} or Ga^{III} replaces Si^{IV}. Among these networks the mesh and size of interstice may vary greatly. Thus in felspars ($M^I AlSi_3O_8$ where $M^I = $ Na,K and Rb and $M^{II} Al_2Si_2O_8$ where $M^{II} = $ Ca, Sr and Ba) and in certain felspathoids (e.g.s. nepheline ($NaAlSiO_4$), kalsilite and kaliophilite (both $KAlSiO_4$), leucite ($KAlSi_2O_6$) and eucryptite ($LiAlSiO_4$)) the frameworks are so compact that the cations at normal temperatures are trapped and immobile on interstitial positions.

There are, however, many much more open frameworks so that in addition to the cations needed to neutralize the framework charge there is also room for salt molecules, as in sodalites, cancrinites and scapolites, or for water and often many other molecules as in the zeolites. It is this intracrystalline porosity which confers on zeolites their remarkable properties. Zeolites can be defined as porous tectosilicate crystals in which, as in every aluminosilicate tectosilicate, $O/(Al+Si) = 2$. In all the natural zeolites the intracrystalline pores contain cations and water molecules. With the possible exception of several zeolites of the natrolite group,[1] the water has the characteristic of giving smooth, continuous sorption–desorption isotherms rather than the stepped isotherms observed on formation of salt hydrates. This is a consequence of the comparatively rigid frameworks of zeolites which remain nearly unaltered when emptied of or re-filled by water.

2. Natural Zeolite Formations

The history of zeolites began with the discovery of stilbite in 1756 by the Swedish mineralogist Cronsted.[2] The name zeolite signifies "boiling stone" and refers to the frothy mass which can result when a zeolite is fused in the blowpipe. Volatile zeolitic water forms bubbles within the sinter or melt. Zeolite minerals were first recognized as minor but widespread components in cavities and vugs in basalt. Big crystals can develop in these cavities through the mineralizing action of trapped or circulating solutions on the alkaline matrix, and it is these crystals which, because of their beauty and diversity of form, usually comprise zeolite collections. However, zeolite occurrences in basalt cavities were too scattered to encourage their harvesting on a large scale for industrial exploitation. Despite clear early indications of their remarkable properties no appreciation of, or significant attempts to realize their potential were made until synthesis of crystalline zeolite powders began to be achieved at modest temperatures. Thereafter commenced a surge of activity by chemists and chemical engineers, some results of which are described in subsequent chapters.

Zeolite occurrences are in no way restricted to basalt matrices, although the scale and variety of natural deposits has been appreciated only over the last two decades. Loew[3] in 1875 reported chabazite in a sedimentary tuff bed near Bowie, Arizona; Murray and Renard[4] in 1891 reported the presence of zeolites in marine sediments; in 1914 Johannsen[5] noted that fine-grained zeolites comprised a substantial fraction of Eocene tuffs in Colorado, Wyoming and Utah. This early work was followed by reports of zeolites in saline lake beds,[6,7] vitric tuffs, and in association with bentonites.[9] Several kinds of zeolite were next identified by Coombs[10] in low-grade metamorphic rocks in New Zealand. More studies were soon forthcoming which established the great abundance of zeolites and demonstrated that some deposits contained nearly monomineralic layers in which the original vitreous material was changed to as much as 95% of a single zeolite phase.[11] It has been observed that since the 1950s more than 1000 occurrences of zeolite minerals of sedimentary origin have been reported in more than forty countries.[11] Some of the near-monomineralic deposits are readily mined because they are near or at the surface. They have naturally attracted attention to possible industrial uses mainly complementary to those for which the pure synthetic zeolites have been so successful, for which the abundant but less pure natural zeolites could be more economic.

As a result of many geological explorations zeolite formation is considered to include the following genetic types:[12]

1. Crystals resulting from hydrothermal or hot-spring activity involving reaction between solutions and basaltic lava flows.
2. Deposits formed from volcanic sediments in closed alkaline and saline lake-systems.
3. Similar formations from open freshwater-lake or groundwater systems acting on volcanic sediments.
4. Deposits formed from volcanic materials in alkaline soils.
5. Deposits resulting from hydrothermal or low-temperature alteration of marine sediments.
6. Formations which are the result of low-grade burial metamorphism.

Some zeolite deposits appear to have formed with no direct evidence that the parent material was of volcanic origin. It is of interest to summarize characteristic features of some of the above types of zeolite formations.

Zeolites formed by hydrothermal alteration of basaltic lava flows and hyaloclastic rock sequences in geothermal regions are found in many parts of the world. Ellis[13] considered hydrothermal areas in Wairakei (New Zealand), the Yellowstone National Park and Steamboat Springs (USA), the roman baths at Plombières, (France) and Iceland basalt in hot springs. The non-zeolites opal, kaolin, cristobalite, tridymite and montmorillonite were already present at low temperatures ($\leqslant 100°C$); the zeolites noted, and the temperature ranges involved, were mordenite (50–220°C), heulandite (125–220°C), analcime (150–220°C) and wairakite (200–250°C). At still higher temperatures the non-zeolites albite and adularia were reported. A recent study of Icelandic zeolites in geothermal areas provides more detailed information of the species formed and their temperature zoning, as summarized in Fig. 1.[14] Chabazite, levynite (levyne), thomsonite, mesolite, scolecite, stilbite, phillipsite and analcime were already in evidence at temperatures even below 70°C; gismondine, epistilbite and mordenite began to appear in the zone between 70 and 90°C and laumontite between 90 and 110°C. Finally wairakite was

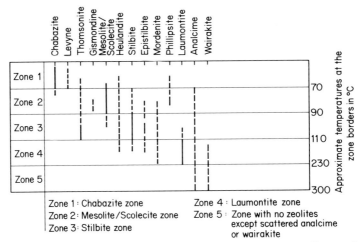

Fig. 1. Temperature zoning of zeolites occurring in Icelandic geothermal areas.[14]

observed in the zone between 110 and 230°C. As the temperature increases, the tendency is for the most highly hydrated zeolites to disappear first (chabazite, levynite, gismondine and phillipsite), while the less highly hydrated species persist to higher temperatures (mordenite, laumontite, analcime and wairakite). Zeolites formed in volcanic rocks of Sicily[15] include analcime, chabazite and its variant herschelite, epistilbite, gmelinite, gonnardite, mesolite, natrolite and thomsonite. Mesozoic basic volcanites which form outcrops from parts of the Italian alps contain many zeolites, often of museum quality, including analcime, chabazite, heulandite, natrolite, scolecite, stilbite and thomsonite.[15] Numerous other locations, world-wide in distribution, provide zeolite-bearing mineral assemblages of similar hydrothermal origin. Well grown zeolite crystals typify this hydrothermal reaction of solutions with alkali-rich rocks such as basalts and some pegmatites.

On the other hand, zeolites formed from sediments usually consist of cemented agglomerates of crystallites of very small dimensions and often with poorly developed crystal faces. It is however these zeolites which form the large deposits and comprise those of technological interest. Iijima[16] has listed twenty three zeolites found in marine and fresh-water non-marine deposits (Table 1). In two instances the "rare" abundances originally given[16] have been revised. Thus at least one massive deposit of erionite is known[11] and sedimentary marine phillipsites appear in substantial amounts.[24] Of all zeolites of sedimentary origins, clinoptilotite so far appears to be among the most abundant.[11,24,25] Alkali and alkaline earth metal cations are those usually present in the mineralizing solutions and consequently are those found in the zeolites. While some clear preferences are shown for particular cations (e.g. Na^+ and/or K^+ in analcime, erionite, clinoptilolite, mordenite, natrolite and phillipsite; or Ca^{2+} in heulandite, laumontite, stilbite, thomsonite and wairakite), these preferences are the result of ion-exchange selectivities, and the cationic compositions could have changed after the period of zeolite formation if those of ambient solutions also changed. The geological ages of the host rocks are upper limits to the ages of the zeolites forming from them, because transformation to zeolites could have occurred more recently and because certain zeolites may have slowly replaced others in reaction sequences.

Schematic representations of spatial zoning for zeolitizing systems of

TABLE 1
Zeolites in marine and fresh water non-marine deposits[16] a

Zeolite	Dominant cations[b]	Geological age of host rock	Abundance
Analcime	**Na**	Quat. ~ Carbon.	Abundant
Chabazite	**Na**, K, **Ca**	Quat. ~ Mio.	Common
Clinoptilolite	**Na, K, Ca**	Quat. ~ Carbon.	Abundant
Epistilbite	**Ca**	Miocene	Rare
Erionite	**K, Na**, Ca	Quat. ~ Eoc.	Common
Faujasite	**Ca**, Na	Quaternary	Rare
Ferrierite	**K**, Na, **Mg**	Miocene	Rare
Garronite	**Ca**, Na	Miocene	Rare
Gismondine	**Ca**, Na, K	Quat. ~ Carbon.	Rare
Gonnardite	**Na**, Ca	Quat. ~ Mio.	Rare
Harmotome	**Ba**	Neogene	Rare
Heulandite	**Ca**, Na	Plio. ~ Carbon.	Abundant
Laumontite	**Ca**	Plio. ~ Devon.	Abundant
Levynite	**Ca**	Pliocene	Rare
Mesolite	**Ca**, Na	Neogene	Rare
Mordenite	**Na, Ca**, K	Quat. ~ Carbon.	Abundant
Natrolite	**Na**	Quat. ~ Perm.	Common
Phillipsite	**K, Na, Ca**	Quat. ~ Carbon.	Abundant
Scolecite	**Ca**	Mio. ~ Perm.	Rare
Stilbite	**Ca**, Na	Quat. ~ Jura.	Common
Thomsonite	**Ca**, Na	Quat. ~ Jura.	Common
Wairakite	**Ca**	Quat. ~ Cret.	Common
Yugawaralite	**Ca**	Miocene	Rare

a Sources of data are Hay[17], Iijima and Utada[18,19], Sheppard[20] Utada et al.[21], Iijima and Harada[22], Harada et al.[23].
b The most characteristic cations are those in bold type.

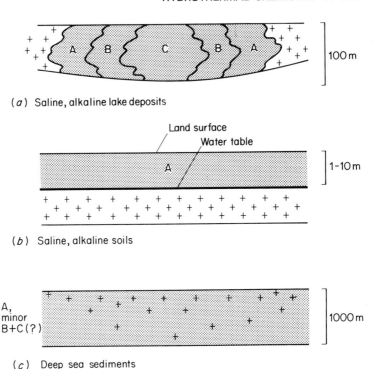

(a) Saline, alkaline lake deposits

(b) Saline, alkaline soils

(c) Deep sea sediments

Fresh glass   Altered glass

Fig. 2. Zoning patterns of authigenic zeolites and felspars in tuffs of saline, alkaline lakes; saline, alkaline soils; and deep-sea sediments.[26] Zone A contains alkali-rich zeolites excluding analcime; zone B contains analcime; and zone C contains felspars.

the genetic kinds referred to earlier in this section are presented in Figs. 2 and 3.[26] In closed, alkaline, saline lakes (Fig. 2(a)) the zones tend to be concentric rings in a horizontal layer, but in saline alkaline soils and in deep sea sediments the zones tend to be horizontal layers in a vertical sequence (Figs. 2(b) and (c)). In the alkaline soils the unconverted parent material is at the bottom, separated at the water table from the zeolitized layer on top; in the deep sea sediments the greatest concentration of fresh material is on top, but with increasing depth the proportion of zeolites increases. In saline, alkaline lake deposits the outer zone is largely unaltered tuff, the next zone is of alkaline zeolites, followed by one of analcime with felspars at the centre.

Figure 3(a) illustrates two characteristic dispositions of zeolites in

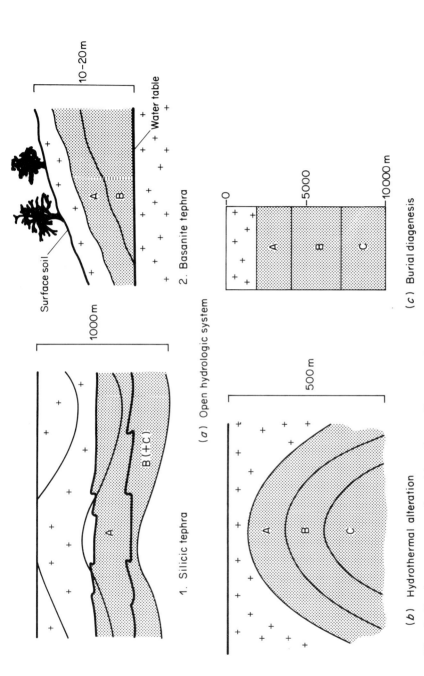

1. Silicic tephra

2. Basanite tephra

(a) Open hydrologic system

(b) Hydrothermal alteration

(c) Burial diagenesis

Fig. 3. Zoning patterns of authigenic zeolites and felspars in tuffs where the zonation is (a) of open system type; (b) hydrothermal; and (c) a result of burial diagenesis.[26]
Zone A is characterized by non-analcime alkali-rich zeolites; zone B by analcime or heulandite; and zone C by K-felspar in (a) and by albite with or without laumontite in (b) and (c). The symbols are the same as in Fig. 2.

relation to water levels and unaltered vitric ash in open hydrologic systems. The zoning is more or less vertical. In hydrothermal zeolitization of the ash the zoning may be complex, but tends in part to imitate, although with vertical zoning of horizontal layers, that developed horizontally in closed, alkaline, saline lakes (Fig. 3(b)). This vertical sequence also tends to be repeated in burial diagenesis, according to Fig. 3(c).

The thicknesses of the zeolitized regions vary greatly, from a number of kilometres in burial diagenesis to a few feet or metres in saline, alkaline soils. The mechanisms of conversion of the sedimentary rocks formed from volcanic ash or other pyroclastic material are not fully understood, but it seems likely that the reaction is between pervading pore waters and aluminosilicate glass with solution and re-precipitation of this glass with or without an intermediate gel phase.[11] The species first formed tend to react further with pore fluids of different composition, often resulting in complex assemblages. The large crystals which form hydrothermally in cavities in basalts must certainly grow by a solution-deposition process because the growing faces often stand well away from the basalt matrices that support them and provide the chemical nutrients for their growth.

3. Zeolite Topologies and Classification

In a companion volume[27] and elsewhere[28] accounts have been given of the topologies of the anionic frameworks of zeolites. It is appropriate here only to summarize aspects of zeolite structures important for illustrating their variety and their classification.

It was observed earlier that the chemistry of tectosilicates is based on the tetrahedron TO_4 where T may be Si or Al. However tectosilicates in general and zeolites in particular are so diverse that secondary structural units, based on small groupings of linked tetrahedra, are needed in describing and systematizing their topologies. Meier[29] in 1967 proposed the secondary building units (SBU) of Fig. 4 as the smallest number of such units from which then known zeolite topologies could be built. In view of the structures of melanophlogite[30] and of ZSM-39,[32] based respectively on those of clathrate hydrates of types I and II,[33] the single 5-ring can be added to the SBU of Fig. 4. In the SBU Al or Si is present at each corner or termination but the oxygens are not shown. They are located near the mid-points of the lines joining each pair of T atoms. It is then found, for example, that the zeolites natrolite, thomsonite and scolecite have tectosilicate frameworks that can be constructed using only the 4-1 units of the figure; that the frameworks of mordenite, dachiardite, epistilbite, ferrierite and bikitaite can be made using only the 5-1 units; while those of **heulandite and** stilbite can be built using only the 4-4-1

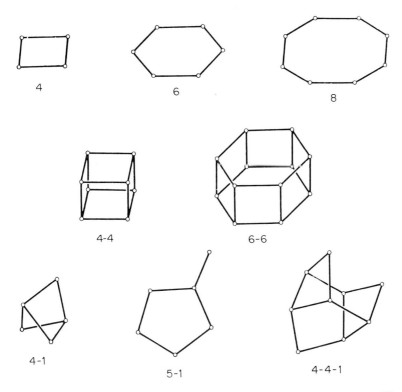

Fig. 4. Secondary building units identifiable in zeolite frameworks.[29]

units. Sometimes more than one SBU is involved. Thus the analcime topology can be made from 4- and 6-ring units; faujasite from 4-, 6- and 8-rings together with the 6-6 hexagonal prisms; Linde zeolite A from 4-, 6- and 8-rings and 4-4 cubic units; and phillipsite, gismondine and paulingite from 4- and 8-rings. In all, twenty five zeolite topologies were so described,[29] two of which are exemplified by Figs 5(a) and (b)[34] for chabazite and Linde zeolite A. In part of each diagram the oxygens are also shown, and in the case of zeolite A so are the Na^+ ions neutralizing the negative charge on the framework. Line diagrams like those in Figs 5(a) and (b) are often shown as stereoscopic pairs which allow ready visualization of the frameworks in three dimensions[28] (see Chapter 7, Fig. 6).

Various other ways of formulating zeolite framework topologies have been developed, involving structural units of greater complexity than the SBU of Fig. 4. These ways include:

1. Stacking of polyhedral cages with other cages, like or unlike the first,

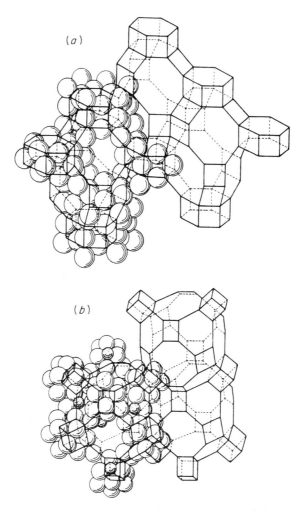

Fig. 5. The crystal structures of (a) chabazite and (b) Linde zeolite
A.[34] Oxygen atoms are shown in parts of each structure, as the larger
spheres, and Na$^+$ ions in part of zeolite A, as the small spheres. Each
corner of the line diagrams is occupied by Al or Si.

with appropriate co-ordination numbers among the cages. The stacked
polyhedra share faces with other polyhedra.
2. Stacking of aluminosilicate layers of a given kind in a number of layer
   sequences and with appropriate T–O–T (T = Al or Si) bonds cross-
   linking each pair of layers. The layers may be of a number of kinds
   often puckered to give different configurations.

3. Cross-linking chains of a given type to other parallel chains of the same kind, by Si–O–T bonds. This may be done in a variety of ways, and the chains can be of a number of different kinds.
4. By the action of an n-fold operator axis upon chains placed alongside the axis. The chains may be puckered to give different configurations. This is easily visualized for the ladder-like chains of 4-rings found in the phillipsite structure or in felspars.
5. Through the so-called "sigma transformation". This is a conceptual device which refers to the expansion of the single tetrahedron, TO_4, or of a tetrahedrally connected structure by imaginary fission of T atoms lying on specified planes or surfaces running through the structure, and creation of new oxygen bridges connecting the pairs of T atoms resulting from the fission.

Each of the above ways of generating tectosilicate frameworks can create, in addition to known topologies, a notable range of so far unknown structures, of which many may in the future by synthesized. Examples of the generation of structures will serve to illustrate each procedure. Thus Fig. 1 in Chapter 7 indicates how 14-hedral sodalite cages, each having 6×4-ring and 8×6-ring faces, are stacked in 8-fold co-ordination by sharing 6-ring faces to yield the sodalite struture. Figures 5(a) and 5(b) of this chapter illustrate the stackings of polyhedra which yield the chabazite and zeolite A structures respectively. In chabazite the polyhedra are hexagonal prisms and 20-hedra (6×8-ring, 2×6-ring and 12×4-ring faces) and in zeolite A the polyhedra are double 4-rings (cubic units of Fig. 4), sodalite cages and 26-hedra (6×8-ring, 8×6-ring and 12×4-ring faces). Melanophlogite, a rare, porous crystalline silica, has a structure made of stacked pentagonal dodecahedra and of tetradecahedra (12×5-ring and 2×6-ring faces), the structure being that shown in Fig. 3 of Chapter 6. The synthetic high-silica zeolite ZSM-39[32] has a structure made by appropriate stacking of pentagonal dodecahedra and hexadecahedra (12×5-ring and 4×6-ring faces). In each case the shared faces are the 5-ring ones. Other structures which can be made by stacking polyhedra of different kinds in different ways include cancrinite, zeolite losod, gmelinite, erionite, offretite, mazzite, faujasite and zeolites RHO and ZK-5.[35] In some instances the stacking of polyhedra also creates continuous wide channels, as found with cancrinite, gmelinite, offretite and mazzite.

Stacking of layers is an equally fruitful way of generating zeolite structures. Thus Table 2 of Chapter 7 summarizes the layer sequences for one important series of tectosilicates including the chabazite–sodalite group. The small letters in that table indicate layers containing single

6-rings and the capital letters denote layers containing hexagonal prisms (double 6-rings). Each layer is linked by Si–O–T bonds to the parallel layer above and to that below. Layers puckered in various ways and cross-linked to identical parallel layers above and below can be identified in and yield the structures of mordenite, dachiardite, epistilbite, ferrierite, bikitaite and zeolite Li-A(BW). These layers and their modes of cross-linking are illustrated in Ref. 27, p. 48. Layers based upon a common structural unit are also found in heulandite, stilbite and brewsterite (Ref. 27, p. 50) and yet another kind of layer structure occurs in laumontite (Ref. 27, p. 51).

Various kinds of chain can be identified in zeolites and other tectosilicates and these can readily serve as the basis for constructing their frameworks, as indicated in the third of the methods listed earlier. The 4-ring can serve as the basis for several kinds of chain. In each 4-ring the vertex of a given tetrahedron may point up (U) or down (D). One then has the configurations UUDD, UDUD, UUUD and UUUU shown in Fig. 6[36] (and the converse ones in which U is replaced by D and D by U). The figure shows three kinds of chain resulting from the first three configurations. The cubic unit resulting from the configuration UUUU is a component of the structure of Linde zeolite A (Fig. 5(b). The double crankshaft chain of Fig. 6 based on the UUDD configuration of the 4-ring can be variously cross-linked to like chains[37] to yield many novel frameworks and also those of felspars and paracelsian and of the zeolites phillipsite, gismondine and merlinoite.[38] The chains in Fig. 6 based upon the UDUD configuration of the 4-ring when cross-linked yield several novel structures and also the non-zeolite banalsite ($BaNa_2Al_4Si_4O_{16}$).[39]

Chains formed by linking the 4-1 units of Fig. 4 are characteristic of the zeolites natrolite, thomsonite and edingtonite. Again the chains are cross-linked differently to their four neighbours in the several zeolite types. They have a period of 6·6 Å and the heights of the free vertices of an isolated chain are multiples of $c/8$. Chains of yet another kind are identifiable in mordenite and dachiardite.[40] These chains are based on 5-rings (Fig. 7) and their cross-linking to like chains can yield in all eight structures comprising six novel zeolite frameworks in addition to the above two zeolites.[41,36]

The ladder-like chain of Fig. 6 based on the UUDD configuration of the 4-ring can serve to illustrate the use of n-fold operator axes in generating zeolite frameworks.[42] These chains can be buckled in various ways of which the double crankshaft form in Fig. 6 is one. If the chain is placed near to and parallel with the operator axis, the centre of each tetrahedron for any configuration may be near (N) or far (F) from the axis, according to the chain configuration. The configuration in Fig. 6 is

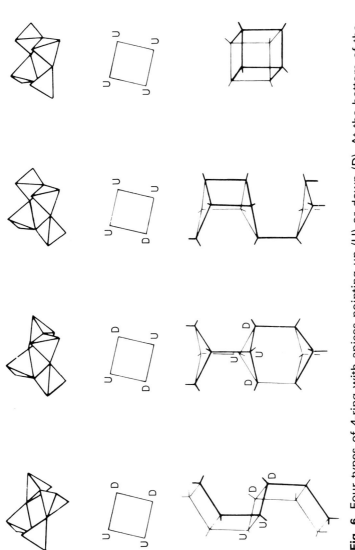

Fig. 6. Four types of 4-ring with apices pointing up (U) or down (D). At the bottom of the diagram three kinds of chain are shown and also the cubic unit which can be formed from two 4-rings of the fourth kind.[36]

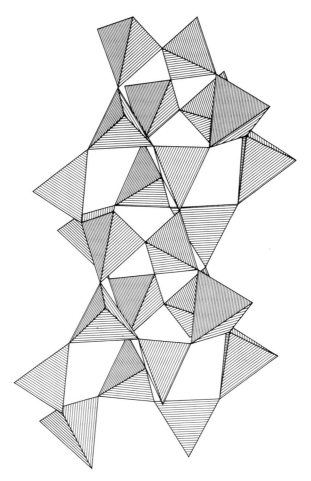

Fig. 7. Chains based on 5-rings which can be identified in mordenite.[40] The chain is shown as linked TO_4 tetrahedra (T = Al or Si).

then designated as NNFF ... in its relation to the axis. Other configurations considered were NF, NNF, FFN, NFFNF, FNNFN, NNFNFF, NFNFNFN, FNFNFNF, NNFFNFF, FFNNFNN, NFNFFNFN, NFNFFNFF and FNFNNFNN. Taking a 3-fold operator as an example the ladders are linked through the unshared oxygens near the axis to produce columns of polyhedra as illustrated in Fig. 8(a) and (c).[42] Continuing with the 3-fold operator axis one may next consider three ways of extending the network in the plane normal to the axis, using the oxygens far from the axis as links:

1. The chain considered is shared by two operators.
2. The chain considered is linked to a second chain belonging to another operator in such a way that:
   (a) the two connected chains do not face other directly (Fig. 8(a));
   (b) the two connected chains do face each other (Fig. 8(c)).

The chain configurations giving rise to the columns of polyhedra in Fig. 8(a) and (c) are respectively NF and NNF. The cross-sections normal to the c axis of the structures generated by operations 2(a) and 2(b) above are shown in Figs. 8(b) and (d) respectively. Numerous structures were generated and, among many which represent unknown zeolites, the

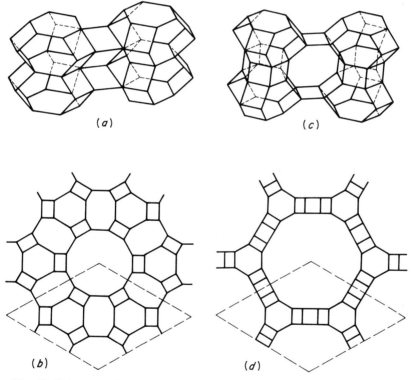

Fig. 8. Columns of polyhedra and their cross-linking produced by a 3-fold operator axis on NF (a) and NFF (c) ladder-like chains. In (a) the connected ladders do not face each other directly and generate the projection of (b) looking along the c-axis. In (c) the ladders do face each other and generate the projection of (d) looking along the c-axis.[42]

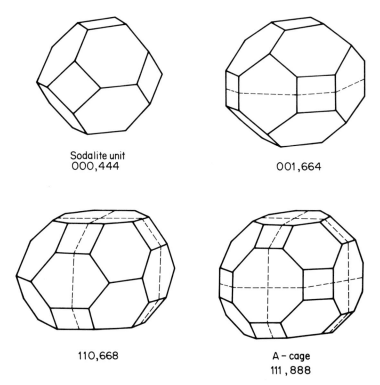

Sodalite unit
000,444

001,664

110,668

A – cage
111 , 888

Fig. 9. An example of the sigma transformation,[43] which converts the 14-hedral sodalite cage into the 26-hedral cage found in zeolite A. Three operations involving planes mutually at right angles and passing through the centre of the cage are involved.

structures of cancrinite, offretite, gmelinite and Linde zeolite L were formed.

The sigma transformation of Shoemaker *et al.*[43] described on p. 11 can be illustrated by the conversion of the 14-hedral sodalite cage to the 26-hedron of zeolite A. To do this three successive transformations are made on mirror planes mutually at right angles and passing through the cage centre (Fig. 9). Similarly, starting with a single T atom one may generate in successive transformations the sequence:

$$T \rightarrow T\text{--}O\text{--}T \rightarrow 4\text{-ring} \rightarrow 6\text{-ring} \rightarrow 8\text{-ring} \rightarrow 10\text{-ring} \rightarrow 12\text{-ring}.$$

From the 4-, 6- and 8-rings one may similarly generate double 4-, 6- and 8-rings (respectively cubic unit, and hexagonal and octagonal prisms found in zeolite A, in chabazite or faujasite, and in zeolite *RHO* or paulingite).

The above ways of generating structures may be regarded as typical. By indicating structural relationships between framework topologies they assist in the placing of zeolites into the sub-divisions or groups shown in Table 2. Sub-divisions, though useful, are however relatively arbitrary. For example, hexagonal prisms are structural units in chabazite, offretite, gmelinite and levynite in one group and in faujasite in another. Similarly sodalite and cancrinite hydrates appear in the chabazite group because their structures are respectively abc... and ab... sequences of layers of single 6-rings analogous to the sequences ABC... and AB... of layers of double 6-rings in chabazite and gmelinite respectively. However sodalite units are important structural components of zeolite A and of faujasite in the faujasite group. It would not be satisfactory to separate sodalite and cancrinite in a classification, but cancrinite would not be well placed in the faujasite group, even though sodalite could be so placed.

Because isomorphous replacements are often found in zeolites, yielding different Si/Al ratios among different specimens of the same zeolite (Chapter 6), the compositions given in Table 2 are idealized and are to be considered only as rough averages which could require revision in the light of new analytical data. In accordance with the recommendation of the IUPAC committee on chemical nomenclature of zeolites[43a] the framework topologies are each represented by three capital italic letters. These letters describe all variants of a framework having the given topology. Thus, stilbite, stellerite and barrerite of Table 2 are all described by the designation *STI*. Likewise all the minerals of the analcime group are believed to have the same framework topology, despite their differences in chemical composition, and so are topologically denoted by *ANA*. The letters are those given by Meier and Olson[28] with the suggested additions of *MTH* for ZSM-3, *MEP* for melanophlogite and *MTN* for ZSM-39. The use of the topological description, following the different listings arbitrarily assigned to the same type of synthetic zeolite by different investigators, has the great advantage of indicating the kind of framework and removing the haphazardness of the listings. Thus zeolite ZK-5[44] and Species P and Q[45,46] have the same framework topology. The description "zeolite ZK-5, *KFI*" and "Species P, *KFI*" at once identifies each.

Where the structural information is insufficient no allocation of the zeolite to a group has been made. This is true of many of the synthetic zeolites reported in the literature (see Chapter 5). It is also true of several recently described natural zeolites, namely cowlesite,[47] svetlozarite[48] and parthéite,[49] although the similarity between the unit cell of the latter

Table 2
Classification of some zeolites and tectosilicates of related topology

	Idealized unit cell composition	Pore volume (cm^3 as liquid H$_2$O per cm^3 of crystal)	Suggested designation of topology
Analcime Group			
Analcime	Na$_{16}$[Al$_{16}$Si$_{32}$O$_{96}$]16H$_2$O	0·18	ANA
Wairakite	Ca$_8$[Al$_{16}$Si$_{32}$O$_{96}$]16H$_2$O	0·18	ANA
Leucite (felspathoid)	K$_{16}$[Al$_{16}$Si$_{32}$O$_{96}$]	0	ANA
Rb-analcime (felspathoid)	Rb$_{16}$[Al$_{16}$Si$_{32}$O$_{96}$]	0	ANA
Pollucite (felspathoid)	Cs$_{16}$[Al$_{16}$Si$_{32}$O$_{96}$]	0	ANA
Viseite	Na$_2$Ca$_{10}$[Al$_{20}$P$_{10}$Si$_6$H$_{36}$O$_{96}$]16H$_2$O	–	ANA
Keoheite	Zn$_{5·5}$Ca$_{2·5}$[Al$_{16}$P$_{16}$H$_{48}$O$_{96}$]32H$_2$O	–	ANA
Natrolite Group			
Natrolite	Na$_{16}$[Al$_{16}$Si$_{24}$O$_{80}$]16H$_2$O	0·21	NAT
Scolecite	Ca$_8$[Al$_{16}$Si$_{24}$O$_{80}$]24H$_2$O	0·31	NAT
Mesolite	Na$_{16}$Ca$_{16}$[Al$_{48}$Si$_{72}$O$_{240}$]64H$_2$O	0·25	NAT
Thomsonite	Na$_4$Ca$_8$[Al$_{20}$Si$_{20}$O$_{80}$]24H$_2$O	0·32	THO
Gonnardite	Na$_4$Ca$_2$[Al$_8$Si$_{12}$O$_{40}$]14H$_2$O	0·35	THO
Edingtonite	Ba$_2$[Al$_4$Si$_6$O$_{20}$]6H$_2$O	0·35	EDI
Metanatrolite	Na$_6$[Al$_{16}$Si$_{24}$O$_{80}$]	–	NAT
Heulandite Group			
Heulandite	Ca$_4$[Al$_8$Si$_{28}$O$_{72}$]24H$_2$O	0·35	HEU
Clinoptilolite	Na$_6$[Al$_6$Si$_{30}$O$_{72}$]24H$_2$O	0·34	HEU
Brewsterite	(Sr, Ba, Ca)$_2$[Al$_4$Si$_{12}$O$_{32}$]10H$_2$O	0·32	BRE
Stilbite	Na$_2$Ca$_4$[Al$_{10}$Si$_{26}$O$_{72}$]32H$_2$O	0·38	STI
Stellerite	Ca$_8$[Al$_{16}$Si$_{56}$O$_{144}$]56H$_2$O	0·39	STI
Barrerite	(Ca, Mg)$_2$(Na, K)$_{12}$[Al$_{16}$Si$_{56}$O$_{144}$]52H$_2$O	0·35	STI
Phillipsite Group			
Phillipsite	(K, Na)$_5$[Al$_5$Si$_{11}$O$_{32}$]10H$_2$O	0·30	PHI
Harmotome	Ba$_2$[Al$_4$Si$_{12}$O$_{32}$]12H$_2$O	0·36	PHI
Gismondine	Ca$_4$[Al$_8$Si$_8$O$_{32}$]16H$_2$O	0·47	GIS
Zeolite Na-P	Na$_8$[Al$_8$Si$_8$O$_{32}$]16H$_2$O	0·47	GIS
Amicite	K$_4$Na$_4$[Al$_8$Si$_8$O$_{32}$]10H$_2$O	0·28	GIS
Garronite	NaCa$_{2·5}$[Al$_6$Si$_{10}$O$_{32}$]14H$_2$O	0·41	GIS
Merlinoite	K$_5$Ca$_2$[Al$_9$Si$_{23}$O$_{64}$]24H$_2$O	0·36	MER
Zeolite Li-ABW	Li$_4$[Al$_4$Si$_4$O$_{16}$]4H$_2$O	0·28	ABW
Mordenite Group			
Mordenite	Na$_8$[Al$_8$Si$_{40}$O$_{96}$]24H$_2$O	0·26	MOR
Ferrierite	Na$_{1·5}$Mg$_2$[Al$_{5·5}$Si$_{30·5}$O$_{72}$]18H$_2$O	0·24	FER
Dachiardite	Na$_5$[Al$_5$Si$_{19}$O$_{48}$]12H$_2$O	0·26	DAC
Epistilbite	Ca$_3$[Al$_6$Si$_{18}$O$_{48}$]16H$_2$O	0·34	EPI
Bikitaite	Li$_2$[Al$_2$Si$_4$O$_{12}$]2H$_2$O	0·20	BIK

Table 2 ((continued)

	Idealized unit cell composition	Pore volume (cm^3 as liquid H_2O per cm^3 of crystal)	Suggested designation of topology
Chabazite Group			
Chabazite	$Ca_2[Al_4Si_8O_{24}]13H_2O$	0·48	CHA
Gmelinite	$Na_8[Al_8Si_{16}O_{48}]24H_2O$	0·43	GME
Erionite	$(Ca, Mg, Na_2, K_2)_{4.5}[Al, Si_{27}O_{72}]27H_2O$	0·36	ERI
Offretite	$(K_2, Ca, Mg)_{2.5}[Al_5Si_{13}O_{36}]15H_2O$	0·34	OFF
Levynite	$Ca_3[Al_6Si_{12}O_{36}]18H_2O$	0·42	LEV
Mazzite (Zeolite Ω)	$(Na_2, K_2, Ca, Mg)_5[Al_{10}Si_{26}O_{72}]28H_2O$	0·37	MAZ
Zeolite L	$K_6Na_3[Al_9Si_{27}O_{72}]21H_2O$	0·28	LTL
Sodalite hydrate	$Na_6[Al_6Si_6O_{24}]8H_2O$	0·34	SOD
Cancrinite hydrate	$Na_6Si_6O_{24}]8H_2O$	0·34	CAN
Zeolite Losod	$Na_{12}[Al_{12}Si_{12}O_{48}]19H_2O$	0·37	LOS
Faujasite Group			
Faujasite (zeolites X and Y)	$Na_{12}Ca_{12}Mg_{11}[Al_{59}Si_{133}O_{384}]26OH_2O$	0·53	FAU
Zeolite ZSM-3	$\{(Li, Na)_2[Al_2Si_{3.2}O_{8.4}]8H_2O\}_m$	0·53	MTH
Paulingite	$(K_2, Na_2, Ca, Ba)_{76}[Al_{152}Si_{525}O_{1354}]70OH_2O$	0·48	PAU
Zeolite A	$Na_{12}[Al_{12}Si_{12}O_{48}]27H_2O$ (pseudo cell)	0·47	LTA
Zeolite RHO	$(Na, Cs)_{12}[Al_{12}Si_{36}O_{96}]46H_2O$	0·41	RHO
Zeolite ZK-5	$Na_{30}[Al_{30}Si_{66}O_{192}]98H_2O$	0·45	KFI
Laumontite Group			
Laumontite	$Ca_4[Al_8Si_{16}O_{48}]16H_2O$	0·35	LAU
Yugawaralite	$Ca_4[Al_8Si_{20}O_{56}]16H_2O$	0·30	YUG
Pentasil Group			
Zeolite ZSM-5 ⎫ Zeolite ZSM-11 ⎭	$Na_n[Al_nSi_{96-n}O_{192}] \sim 16H_2O$, with n typically about 3	0·32 / 0·32	MFI / MEL
Clathrate Group			
Melanophlogite	$[Si_{46}O_{92}]$	–	MEP
Zeolite ZSM-39	$(Na, TMA, TEA)_{0.4}[Al_{0.4}Si_{135.6}O_{272}]nH_2O$	–	MTN

and that of gismondine may eventually place parthéite[49] in the phillipsite group. The water contents of edingtonite and zeolite RHO have been adjusted from those given in the previous classification[50] and also yugawaralite and laumontite have been grouped together since the frameworks of each contain 4-rings linked cornerwise. This grouping is recognized as more arbitrary than others in Table 2. The porous tectosilicates of very high silica content for which structures are known (the pentasils[51] and the clathrate group) are extremely rich in 5-rings, which are also found in the naturally occurring siliceous zeolites of the mordenite group. It may thus be that other very high silica zeolites presently of unknown structure will show this feature. Unit cells have not been given in the table, but are recorded in that given in Ref. 50.

4. Channels, Cavities, Windows and Molecule Sieving

Much of the interest in zeolites as as catalysts and selective sorbents depends upon the arrangement and free dimensions of the intracrystalline channel structures. As found in Nature all zeolites contain molecular water as an intracrystalline fluid which can be removed by heat and evacuation and which can normally be re-sorbed by exposing the crystals to water vapour. Water is a small polar molecule well able to penetrate into the intrazeolitic channel and cavity systems. It is therefore a useful measure of the total intracrystalline pore volume to determine the volume of liquid water which can be recovered on thorough outgassing of the fully hydrated zeolite. This assumes that the molecular volume of zeolitic water is about the same as that of liquid water—an assumption which can be an approximation only, but which should give results about as good as estimating total porosity from crystallographic data. Independently of estimates of pore volume accessible to water, one may take as a measure of the framework density or open-ness the number of $(Al + Si)$ atoms per 1000 Å^3. The smaller this number the more pore space should be present, without however taking into account how accessible this space is. In Fig. 10 the water contents of a number of zeolites are plotted against the above number of $(Al + Si)$ atoms per 1000 Å^3. There is a correlation which does not take the form of a line but of a band with approximate upper and lower envelopes. This suggests that the sorption capacity for water is influenced by factors other than the framework density. One such factor is the way in which the pore volume is sub-divided into individual pores and the accessibility of these. For example the *ab* section of mordenite (Fig. 11) shows two kinds of channel of which only the more open is accessible to guest molecules even as small and polar as water.

A second influence can be exerted by the cations, which can vary in size, valence and numbers. For the same relative vapour pressure, p/p_0, of water the uptakes change according to the cation, as shown for ion-exchanged chabazites,[52] faujasites,[53] phillipsites[54] and zeolite *RHO*.[55] The results for *RHO* are shown in Fig. 12, in each case after equilibration

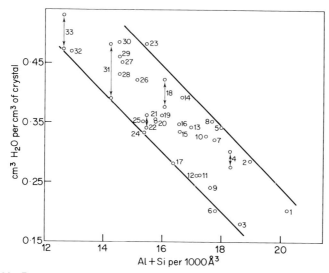

Fig. 10. Reported saturation capacities of zeolites for water plotted against the number of atoms of Al+Si per 1000 Å³ of zeolite framework. See Table 2 for the topological nomenclature used below for synthetic zeolites without natural counterparts.

1. Bikitaite	12. Mordenite	23. Paulingite
2. Zeolite *ABW*	13. Sodalite and heulandite	24. Zeolite *EAB*
3. Analcime	14. Stilbite	25. Erionite
4. Yugawaralite	15. Cancrinite	26. Levynite
5. Epistilbite	16. Edingtonite	27. Zeolite *KFI*
6. Natrolite	17. Zeolite *LTL*	28. Gmelinite
7. Thomsonite	18. Mazzite	29. Gismondine
8. Laumontite	19. Merlinoite	30. Chabazite
9. Ferrierite	20. Phillipsite	31. Zeolite *RHO*
10. Brewsterite	21. Zeolite *LOS*	32. Zeolite *LTA*
11. Dachiardite	22. Offretite	33. Faujasite

over saturated $Ca(NO_3)_2$. The differences in water content were only partially accounted for by the volumes occupied by cations. Changes in the cubic unit cell volume with cation were also insiginficant (a was 15·0 Å for Li^+, Na^+, Rb^+, NH_4^+, Ca^{2+}, Sr^{2+} and Ba^{2+}; 14·8 Å for K^+; and 14.6 Å for Ag^+ and Cs^+). The effects observed must therefore in part result from packing densities of H_2O at the given relative pressure which become smaller the larger the cation. Figure 12 is typical of more porous zeolites. In compact frameworks such as that of analcime the effect of cation exchange can be more dramatic, in that replacement of Na^+ by K^+ or Cs^+ leads to anhydrous structures. For the size of cation to be so

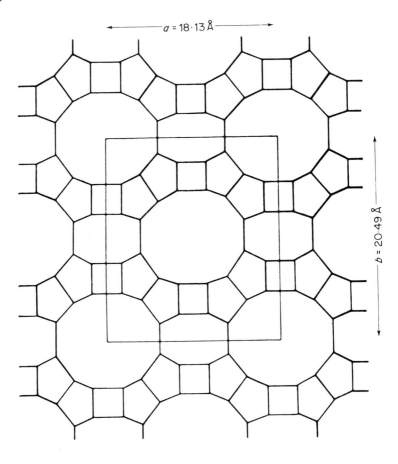

Fig. 11. The *ab* section of the mordenite framework showing the wide channels and also the non-accessible narrow channels.

important, the intracrystalline pore volumes in this case must be parcelled up into many very small interstitial spaces. On the other hand, ferrierite has an accessible intracrystalline free volume not very much greater than that in analcime (see Table 2) but this volume is present as much wider, continuous channels so that ferrierite can sorb permanent and inert gases and also some heterocyclic and aromatic molecules.[56]

There are in principle as many different shapes and dimensions of intracrystalline cavity or channel as there are zeolite topologies. The accessibility of the intracrystalline pores is governed by apertures or windows made by rings of linked tetrahedra. 4-rings and 5-rings are not traversable by guest molecules, and 6-rings can be traversed only by the

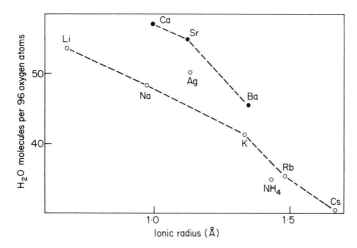

Fig. 12. Relation between water contents and the Pauling ionic radii of ion exchanged forms of zeolite *RHO*.[55]

smallest molecules. The free dimensions of some n-rings ($n = 6$, 8, 10 and 12) are given in Table 3 assuming 2·70 Å as the diameter of the oxygens lining the inner peripheries, first in the case of ideal planar configurations and secondly in the distorted forms (boat-shaped, chair-shaped, puckered, elongated, etc.), observed in representative zeolites. The free dimensions given in the table are very important in determining the molecule-sieving properties of the zeolite. The van der Waals dimensions critical for passage through an n-ring are for several "yardstick" or calibrating molecules estimated to be those in Table 4. Because neither the diffusing molecules nor the oxygens lining the peripheries of the n-ring apertures are hard spheres, and also because of lattice vibrations involving the n-rings, including breathing frequencies, yardstick molecules having critical dimensions somewhat larger than the ring dimensions will nevertheless pass through the rings at appropriate temperatures, in many instances at room temperatures.

However, because the diffusivities of guest molecules tend to depend exponentially upon temperature the molecule sieving property exhibited by a given zeolite can become a strong function of temperature. The lower the temperature the more damped become the lattice vibrations, the more rigid the n-ring, and therefore the more sensitive to the matching of critical dimensions of molecule and n-ring the diffusion becomes. At liquid air temperatures for example one may differentiate between O_2 (critical dimension ~2·8 Å) and N_2 (critical dimension

$\sim 3 \cdot 0$ Å), the N_2 penetrating very much more slowly than O_2 into such zeolites as Na-A, *LTA*,[58] Ca-mordenite (narrow port)[59] and Ca-rich levynite.[60] A second important factor in determining the molecule sieving property is the location of the cations. Thus, in zeolite Na-A the Na^+ ions partially block the 8-ring windows so that n-paraffins cannot penetrate the water-free crystals. However, by exchanging $2Na^+$ by Ca^{2+} the windows are cleared of cations and n-paraffins now penetrate freely. Iso-, neo- and cyclo-paraffins and aromatics are, however, much too large to penetrate and so an important process has been developed for sieving n-paraffins from its mixtures with other hydrocarbons. The n-paraffins

Table 3
Ideal and actual free dimensions of some *n*-rings

		Free dimension (Å)	
n	Ideal planar[a]	Estimated from structures (max. and min.)	
6	$2 \cdot 7 \, (2 \cdot 8)$	Sodalite: $2 \cdot 1$	
		hydrate	
8	$4 \cdot 4 \, (4 \cdot 5)$	Zeolite A, *LTA*: $4 \cdot 1$	
		Chabazite: $4 \cdot 1$ and $3 \cdot 9$	
		Erionite: $5 \cdot 2$ and $3 \cdot 6$	
		Gmelinite: $4 \cdot 1$ and $3 \cdot 4$	
		Levynite: $5 \cdot 1$ and $3 \cdot 2$	
		Offretite: $5 \cdot 2$ and $3 \cdot 6$	
		Zeolite ZK-5, *KFI*: $3 \cdot 9$	
10	$6 \cdot 0 \, (6 \cdot 3)$	Heulandite: $7 \cdot 8$ and $3 \cdot 2$	
		Stilbite: $6 \cdot 2$ and $4 \cdot 1$	
		Dachiardite: $6 \cdot 7$ and $3 \cdot 7$	
		Epistilbite: $5 \cdot 3$ and $3 \cdot 2$	
		Ferrierite: $5 \cdot 5$ and $4 \cdot 3$	
		Zeolite ZSM-5, *MFI*: $5 \cdot 6$ and $5 \cdot 4$	
12	$7 \cdot 7 \, (8 \cdot 0)$	Gmelinite[b]: $6 \cdot 9$	
		Cancrinite[b]	
		hydrate: $6 \cdot 2$	
		Mordenite[b]: $7 \cdot 0$ and $6 \cdot 7$	
		Offretite: $6 \cdot 4$	
		Mazzite: $7 \cdot 1$	
		Zeolite L, *LTL*: $7 \cdot 1$	
		Faujasite: $7 \cdot 4$	

[a] Figures in brackets are estimated assuming $2 \cdot 8$ Å as the diameters of the oxygens.
[b] Either because of stacking faults or because of detrital material deposited in the channels, gmelinite and cancrinite hydrate have not behaved as wide-pore zeolites. Mordenite can also behave as though partially blocked ("narrow-port" mordenites). Cation locations may be important in blocking access.

Table 4

Some van der Waals dimensions in Å critical for passage through n-rings, based on Pauling radii and bond lengths.[51]

Molecule	Dimensions^a	Molecule	Dimensions
He	~2·0	CF_4	5·3 (5·4)
Ne	3·2	$CH(CH_3)_3$	5·6
Ar	3·8$_3$	SF_6	6·0$_6$ (6·0$_6$)
Kr	3·9$_4$	CCl_4	6·9 (7·1)
Xe	4·3$_7$	$C(CH_3)_4$	6·9 (7·1)
CH_4^{b}	4·0 (4·6)	CBr_4	7·5 (7·7)
	4·4$_4$	CI_4	8·2 (8·5)
C_3H_8	4·9		

^{a} Figures in brackets are diameters of circumscribing spheres.
^{b} The value 4·0 Å assumes the Pauling value for CH_4 or the CH_3 group as structureless spheres. The figure 4·4$_4$ is based on Pauling co-valent radii and bond lengths of individual atoms.

serve to produce large tonnages of chlorinated plasticizers and biodegradable detergents.

Molecule sieving, when there are adequate shape and size differences between the molecules to be separated (and even small differences at low temperatures), can be a quantitative, single-step process in which there is a lock and key aspect between the apertures giving access to the interior of the crystal sieve and the potential guest molecule. Guest species of the right shape and size penetrate readily, but molecules of the wrong shape and size can be totally excluded. This was realized and abundantly demonstrated by 1945.[61,62]

Because a given zeolite can often be synthesized with different Si/Al ratios one may vary the number of cations per unit cell. Through ion exchange one may vary the cation size, or by introducing ions of different charge from those originally present the number of ions per unit cell may again be changed. According to the ion and its size and valence the cation locations often alter.[63] Thus, within a given framework much can often be done to modify the sieving behaviour. Zeolite frameworks can also be lightly impregnated by salts of various anions. In this way the wide pore sorbent offretite was converted to a sieve which, like Ca-A, was able to separate n-paraffins from other hydrocarbons.[64] Sieves can also be modified by introducing measured amounts of guest species, such as water, which are immobile at the low temperatures of subsequent sorption measurements and can, by solvating cations or otherwise, partially block the windows of the zeolite framework.[65,66]

The distribution in space of the diffusion pathways for molecules can play a significant role in molecule sieving, counter-diffusion and catalysis. These pathways can be parallel and without cross-connections (e.g. mordenite, zeolite L, *LTL*, cancrinite hydrate); parallel but with cross-connections having smaller openings than those of the main channels (e.g. gmelinite; offretite); inter-connected in two-dimensional patterns (e.g. levynite and heulandite); and inter-connected in three-dimensional channel patterns (e.g. faujasite, chabazite, erionite, zeolite A, *LTA*, zeolite *RHO*, zeolite ZK-5, *KFI*). Zeolites having parallel channels are more susceptible to blocking by detrital material and by deposition of cracking products than are those with two- or three-dimensional channel systems. Also counter-diffusion of reactants to catalytic centres and of resultants away from these can be more difficult in a one-dimensional channel system than in a two- or three-dimensional one which is able to provide alternative pathways around local obstructions.

Before one leaves the topic of sorption of guest molecules two additional aspects deserve attention. Firstly, zeolites are usually very polar sorbents by virtue of the anionic charge on the framework and the charge-compensating cations. There are strong local electrostatic fields within the intracrystalline pore systems which interact with molecules having permanent dipoles, and also local field gradients which interact with molecular quadrupoles. Molecules with such permanent electric moments are therefore sorbed more energetically than molecules of comparable size and polarizability which do not possess electric moments. Polar molecules are accordingly selectively removed from such mixtures all components of which can penetrate the crystals. Thus zeolites prove to be highly effective desiccants of industrial gas streams. The strong sorption of small polar molecules can of course be combined with molecule sieving. For instance K-A, *LTA* sorbs only water from organic mixtures and this property allows its use in intensive drying of organics like ethers used in Grignard reactions, or of transformer oils.

More recently, especially with the development of very high silica porous tectosilicates, in which the content of framework Al and charge balancing cations can be so low that they are little or no more than trace impurities, molecule sieves have become available which have lost much of their polar character. Small polar molecules such as water are sorbed only weakly and are displaceable by larger organic molecules such as n-paraffins. As with well-outgassed graphitic carbons, the sorbents are organophilic rather than hydrophilic. At the same time the full power of molecule sieving is retained. In the case of the H-form of ZSM-5, *MFI* the sieve mesh is such (see Table 3) that C_6 and C_7 paraffins with no or with one $-CH_3$ side group are at 340°C cracked much more readily than those isomers which have two tertiary or one quaternary carbon atom.[67] This suggests that the two n-paraffins and their isomers with one –CH_3 side group can penetrate zeolite ZSM-5 much more readily than the more branched isomers, even at 340°C.

5. Catalysis

The uses of zeolites as catalysts are basically dependent upon the accessibility to reactants of intracrystalline catalytically active centres, and so upon the shape and size of penetrant molecules. The subject has been discussed at length[58,67,68,69,70] and only some illustrative principles and examples need be described here. Specific catalytic actions can depend upon zeolite basicity, upon the generation of acid sites within the crystals, or upon the introduction of particular cations (by ion-exchange) or of elements such as S, Te and Se or metals like Pt into the crystals.

The catalytic oxidation of H_2S by O_2 exemplifies a process in which the best catalysts were the basic Na- and K-forms of zeolites X and Y.[71] As Bronsted acid sites were progressively introduced the oxidation rate declined; the activity was directly proportional to the number of AlO_4 tetrahedra, and was inversely related to the electrostatic potential of the exchange ions introduced (K^+, Na^+, Li^+ and Ca^{2+}).

It is, however, in hydrocarbon chemistry that zeolites have yielded some of the most notable catalysts, especially with H-zeolites providing Bronsted acid sites and reactions proceeding via carbonium ions. The acid centres arise from the reversible reaction

$$H^+ + \left[\begin{array}{c} \diagdown \\ -\overset{\ominus}{Al}-O-Si- \\ \diagup \end{array} \begin{array}{c} \diagup \\ \diagdown \end{array} \right] \rightleftharpoons \left[\begin{array}{cc} \diagdown & HO \diagdown \\ -Al & Si- \\ \diagup & \diagup \end{array} \right]$$

In the immediate vicinity of the trivalent Al the silanol OH is a powerful proton donor exceeding in acid strength the OH groups present in amorphous silica–alumina cracking catalysts. Accordingly ultrastable often partially rare-earth exchanged zeolite Y, *FAU* has found wide acceptance as an industrial cracking, isomerization and reforming catalyst. Some of the framework Al is removed from such catalysts in the form of oxy-aluminium cations during ultrastabilization, and there may be an amorphous part of the catalyst as well as crystalline faujasite in the hydrogen form. These catalysts improved gasoline yields by as much as 25%. Isomerization and molecular sieving can be combined to yield in

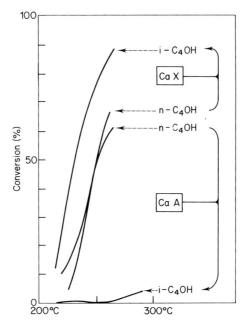

Fig. 13. Dehydration of n- and iso-butanol on Ca-X, which is not shape-selective to either alcohol and on Ca-A which admits only n-butanol and is shape-selective for this alcohol.[67]

principle a gasoline product containing only branched chain and cyclo-paraffins and aromatics. The feed is cracked and isomerized using the catalyst based on H-Y in one unit, while unchanged n-paraffins are then removed from the product in a second unit with zeolite Ca-A as sorbent (iso-, neo- and cyclo-paraffins and aromatics excluded; n-paraffins are sorbed readily). The n-paraffins are then desorbed and returned to the isomerization and cracking unit together with additional feed.

Various shape-selective catalyses have been described which include the following examples. In zeolite Ca-A, chemical dehydration of n-butanol (which penetrates the crystals) but not of iso-butanol (which does not penetrate) was observed. This selectivity was not found with Ca-X which sorbs both alcohols very readily. The % conversion of each alcohol in Ca-A and Ca-X is shown as a function of temperature in Fig. 13.[67] Similarly n-paraffins and n-olefines (which entered the Ca-A) were readily combusted with O_2 while iso-paraffins and iso-olefines, which did not enter Ca-A, did not react appreciably.[72] A Pt-Na-mordenite hydrogenation catalyst was prepared with Pt inside the mordenite. In it H_2, C_2H_6 and CH_3—CH=CH_2 but not $CH_3CH_2CH_3$ were mobile. Accordingly when a $1:1$ mixture of C_2H_4 and C_3H_6 was hydrogenated only the resultant ethane could escape and so only ethylene was able to react in a continuous way.[73]

Shape selectivity was also observed with ZSM-5, *MFI*. This zeolite sorbs and cracks mono-methyl in addition to n-paraffins and is used to dewax heavy oils. A number of shape-sensitive catalytic industrial processes have been developed, as summarized by Weisz[67] in Table 5. The xylene isomerization and ethyl benzene processes, both based on ZSM-5 catalysts, are reported[67,74] to be used in about 90% of United States and 60% of existing western world capacity. The methanol to gasoline (MTG) process converts not only methanol but also other alcohols, aldehydes, ketones, carboxylic acids and esters to hydrocarbons having a limited range of carbon numbers.[75,76] The MTG process appears fully competitive with the SASOL (Fischer–Tropsch) route for production of hydrocarbons from coal especially where gasoline is desired.[77]

Shape-selectivity can be expressed in the following terms, for a pair of molecular species, P and A.[67] Of these P denotes the species whose reaction is favoured, A that whose reaction is hindered by the restricted intracrystalline environment. If the rate constants in absence of structural restraints on molecular motion are respectively k'_P and k'_A then an index

Table 5
Some large-scale processes based on shape-selective zeolite catalysts[67]

Process	Objective	Main process characteristics
Selectoforming	Octane number (ON) increase in gasoline; LPG production	Selective n-paraffin cracking
M-forming	High yield; ON increase in gasoline	Cracking depends on degree of branching; aromatics alkylated by cracked fragments
De-waxing	Light fuel from heavy fuel oil; lube oils with low pour point	Cracking of high molecular weight n- and monomethyl-paraffins
Xylene isomerization	High yield para-xylene production	High throughput, long cycle life; suppression of side reactions
Ethyl benzene	High yield ethyl benzene production; eliminate $AlCl_3$ handling	
Toluene disproportionation	Benzene and xylenes from toluene	
Methanol to gasoline	Methanol (from coal or natural gas) conversion to high grade gasoline	Synthesis of hydrocarbons only, restricted to gasoline range (C_4 to C_{10}) including aromatics

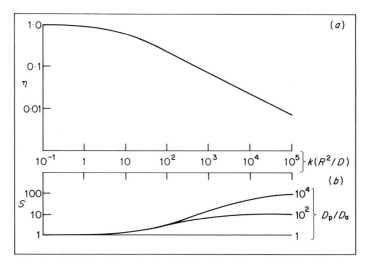

Fig. 14. (a) The effectiveness or utilization factor, η, plotted as a function of $k(R^2/D)$. (b) The selectivity, S, plotted as a function of $k(R^2/D)$ of the antiselective species, A, for several values of the ratio D_P/D_A of the diffusivities of selective (P) and antiselective (A) species.[67]

of shape selectivity, S, can be defined as

$$S = k'_P k_A / k'_A k_P$$

Shape selectivity can arise not only because of restrictions at the point of reaction but also because of different rates of diffusion of P and A within the zeolite to reaction centres. For each reactant the extent to which the reaction rate is slowed by diffusion effects can be expressed in terms of an effectiveness factor, η, related to the ratio $k(R^2/D)$. The top of Fig. 14 shows this relationship, while in the bottom half of the figure S is shown as a function of $k(R^2/D)$ for the species A for three values of the ratio D_P/D_A. R here denotes the radius of the catalyst particle regarded as a sphere and D_P and D_A are the effective diffusivities of P and A within it. S can vary between 1 (no selectivity) and a maximum of $(D_P/D_A)^{1/2}$. For the best selectivity one requires large ratios D_P/D_A and R^2/D_A. The particle radius and therefore R^2/D_A can be varied. Although this treatment is idealized, it gives a semi-quantitative picture of the interplay between diffusion of reactants to catalytic centres and reaction rates at these centres. The reactions involved are considered in the above model to be first-order at the catalytically active centres.

More complicated situations can arise where reaction depends on bimolecular complexes forming at the catalytic centres, as may occur in disproportionation reactions. The accommodation of the large bi-molecular complexes at a catalytic site will be more difficult and more subject to spatial restrictions than for the monomolecular isomerization process. The ratio of effective rate constants for disproportionation and isomerization of *ortho*-xylene was found to increase about 50-fold along the series ZSM-5; ZSM-4 and mordenite; and H-Y, which is the order of increasing open-ness of the zeolite frameworks.[74] In all zeolites the disproportionation rate was less than that for isomerization. Both these experimental features accord with the above qualitative argument, and account for the particular success of ZSM-5 catalysts in suppressing disproportionation and favouring isomerization to *para*-xylene.

This introduction to the catalytic functions of zeolites indicates some perspectives and shows that through the use of porous crystals with lock and key characteristics one has moved a long way towards the specificity of reactions catalysed by enzymes in biochemical environments. This is one reason why zeolites and their synthesis have become of such significance.

6. Cations in Zeolites

The cations are distributed within the same intracrystalline pore systems as the zeolitic water. As a result of the open structures the cations, like the zeolitic water, can be mobile. However, unlike the water, the cations are not free to leave the crystals unless they are replaced by their electrochemical equivalent of other cations because neutralization of the anionic charge of the aluminosilicate framework must be maintained.

The topology of a given zeolite usually provides a set of cation sub-lattices (or site groups), each crystallographically distinct from all the others and each providing a specific number of cation sites. However the total number of these sites may exceed, and often greatly exceed, the number of cations needed for neutralizing the anionic charge. As a consequence the cations and cation vacancies distribute themselves among the sub-lattices. We may consider the following kinds of distribution equilibrium:

1. In a homoionic form of the zeolite, termed A-Z where A is the cation and Z the anionic framework, the cations A and cation vacancies distribute themselves between cation sub-lattices in such a way as to minimize the free energy of the crystal.

2. In a zeolite containing two or more cations, A, B . . . , there is a

competitive equilibrium distribution of A, B, ... and vacancies between each pair of sub-lattices, again minimizing the free energy.
3. In addition, if the zeolite crystals are bathed in a solution containing the cations A, B, ... there will be an equilibrium between the exchange of ions on each sub-lattice and the external solution.

The equilibria in each case will be influenced *inter alia* by the temperature and by the extent of hydration or dehydration of the crystals. For the distributions 1 and 2 the water content is readily changed because no external solution is involved; in this case change in temperature can change the water content so that the cation distributions may alter for this reason and also because the heats of cation interchange between sub-lattices may not be zero. Replacing water by other guest species can also alter the cation distributions.

In formulating distribution equilibria one may not use the method familiar to chemists of assigning chemical potentials to ions on a given sub-lattice.[78] This is because ions A, for example, on a particular sub-lattice are not an independently variable component. The chemical potential is defined as the rate of change of the total free energy of the crystal with the amount of each independently variable component, keeping the amounts of each other such component constant. One cannot vary the cation population, N_i^A, of ions A on sub-lattice i, while keeping constant the populations of ions B, C ... on sub-lattice i and those of A, B, C, ... on all other sub-lattices and at the same time preserve electrical neutrality. Thus the distribution equilibria of ions among sub-lattices cannot be described in terms of chemical potentials.[79] In a mixed hydrated zeolite (A, B)-Z, the independently variable components to which chemical potentials can be assigned are A-Z, B-Z and H_2O. Accordingly, the overall exchange equilibrium between an aqueous solution and the hydrated zeolite can be described in terms of chemical potentials for these three components, as considered in treatments like that of Gaines and Thomas[80] for two exchange ions and of Fletcher and Townsend[81,82,83] for three exchange ions.

However, the inability to assign a chemical potential to ions A or B on any sub-lattice i in no way prevents the formulation of distribution equilibria of A and B between any pair of sub-lattices. The ions A and B on a sub-lattice make a contribution to the free energy of the crystal, and equilibria may be formulated by statistical thermodynamic procedures,[84,85] or by detailed balancing.[78,79] The distribution equilibrium constant, K_{ij}, for the ions A and B between a pair of sub-lattices i and j is given by

$$-\Delta G^{\ominus} = RT \ln K_{ij}, \qquad K_{ij} = \frac{A_i B_j \; \phi_i^A \phi_j^B}{A_j B_i \; \phi_j^A \phi_i^B} \tag{1}$$

A_i, B_j are equivalent cation fractions† of ions A and B on sub-lattices i and j respectively. The ϕ are the equivalent of activity coefficients in bulk systems. They allow for deviations from ideal mixing of A and B on sub-lattices i and j. In energy terms they reflect any variation in the free energy of exchange between sub-lattices other than that attributable to the entropy of mixing.

In a homoionic zeolite, A-Z, the distribution constant for ions A between sub-lattices i and j is

$$-\Delta G^\ominus = RT \ln \frac{A_i \phi_i^A}{A_j \phi_j^A} \tag{2}$$

Calculations of site selectivities have been made for the respective homoionic Li^+, Na^+, K^+, Cs^+, Ag^+, Mn^{2+}, Co^{2+}, Fe^{2+}, Ni^{2+} and Zn^{2+}, forms of dehydrated zeolite A, *LTA* by evaluating cation–lattice interaction energies for each type of ion on each type of site.[86,87,88] 50% ionic and 50% covalent character was assumed for the aluminosilicate framework and coulombic, polarization, dispersion, close-range repulsion and (for the covalent structure) charge transfer contributions were evaluated. The calculations gave site selectivities which agreed rather well with those observed, although energy rather than free energy differences were involved in these calculations.

For the distribution equilibria of ions A and B between an external solution and each of n sub-lattices, detailed balancing[79] or statistical thermodynamics[85] both lead to the relations

$$\frac{a_s^B}{a_s^A} = \frac{a_Z^B}{a_Z^A} \cdot K = \frac{a_1^B}{a_1^A} K_1 = \cdots = \frac{a_i^B}{a_i^A} \cdot K_i = \cdots = \frac{a_n^B}{a_n^A} \cdot K_n \tag{3}$$

where $a_i^B/a_i^A = B_i \phi_i^B / A_i \phi_i^A$. K_{ij} in Eqn (1) is the ratio K_i/K_j from Eqns (3). Included in Eqns (3) is the overall equilibrium constant, K, and the activity ratio a_Z^B/a_Z^A for the whole crystal. It is of interest to show the relation between K and the K_i. If X_i is the fraction of total cationic charge associated with the ith sub-lattice then Eqn (3) gives

$$\frac{a_Z^A}{a_Z^B} (K_1^{X_1} \cdots K_n^{X_n}) = K \left[\left(\frac{a_1^A}{a_1^B} \right)^{X_1} \cdots \left(\frac{a_n^A}{a_n^B} \right)^{X_n} \right] \tag{4}$$

Also, the overall standard state exchange reaction is the sum of the exchange ractions on all the sub-lattices, so that

$$\Delta G^\ominus = \sum_{i=1}^n X_i \, \Delta G_i^\ominus \tag{5}$$

† Defined as $A_i = N_i^A/(N_i^A + N_i^B)$ and $B_j = N_j^B/(N_j^A + N_j^B)$.

where $\Delta G^{\ominus} = -RT \ln K$ and $\Delta G_i^{\ominus} = -RT \ln K_i$. Thus from (4) and (5) one obtains

$$\left.\begin{array}{l} K = K_1^{X_1} \cdots K_n^{X_n} \\[2mm] \dfrac{a_Z^A}{a_Z^B} = \left(\dfrac{a_1^A}{a_1^B}\right)^{X_1} \cdots \left(\dfrac{a_n^A}{a_n^B}\right)^{X_n} \end{array}\right\} \tag{6}$$

as the desired relations.

The expressions (1) to (6) provide the formal basis of ion-exchange of two ions between pairs of sub-lattices or between sub-lattices and solution and show how the latter relate to the overall exchange (Eqn (6)). At present only overall exchange equilibria can be directly measured. Exchange isotherms have been found with at least the four characteristic contours shown in Fig. 15.[79] Figure 15(d) shows a miscibility gap between the end members Na-F, *EDI* and K-F, *EDI*. This behaviour, and also all the other contours, can be represented in terms of a statistical thermodynamic model.[85,92] Miscibility gaps are usually associated with hysteresis loops arising from nucleation difficulties of the new phase on or in a matrix of the parent phase, as discussed in Chapter 4, §2. Isotherms often show changes in selectivity (defined by the mass action quotient) as the extent of exchange increases (Fig. 15(b), (c) and (d)). Selectivities for particular ions can be very great so that zeolites can be very good collectors or scavengers of such ionic species. In this connection zeolite, A, *LTA* is very selective towards Ca^{2+} as compared with Na^+, and to a lesser extent is also selective to Mg^{2+}. The rates of exchange in the small synthetic crystals are also high. Because of environmental damage (eutrophication) associated with phosphate, and hence with phosphate builders in detergents, substitutes for these builders are required in some countries, and zeolite Na-A, *LTL* has proved an adequate if not wholly ideal scavenger of Ca^{2+} in hard water as an alternative to the phosphate.[93,94] Such usage, in which the zeolite is a component of the detergent powder, has resulted in very substantial new production of zeolite A and seems likely to represent a principal industrial application of ion exchange by zeolites. In other areas ion exchange is an important step in the manufacture of zeolite-based catalysts, and in modifying zeolites used as industrial sorbents.

Exchange isotherms or measurements of the maximum extents of exchange under specified conditions have been made for the three zeolites A, *LTA*,[95] X, *FAU*[96] and clinoptilolite[97] having the compositions and pore characteristics given in Table 6. The mean distances between monovalent cations for the compositions in the table are $5 \cdot 3_7$, $5 \cdot 6_8$ and $7 \cdot 0_5$ Å for zeolites A, X and clinoptilolite respectively. The van der Waals

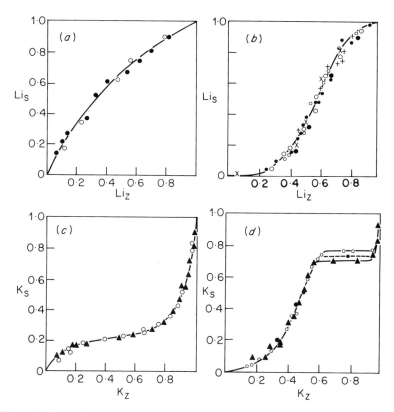

Fig. 15. Four characteristic contours of ion exchange isotherms. (a) Li $\rightleftarrows$ Ni in sodalite hydrate.[89] ○, Na $\rightarrow$ Li; ●, Li $\rightarrow$ Na; 85°C. (b) Li $\rightleftarrows$ Na in cancrinite hydrate.[89] For Na $\rightarrow$ Li: ○, 25°C; o, 85°C; ×, 110°C. For Li $\rightarrow$ Na: ●, 25°C; ●, 85°C; +, 110°C. (c) Na $\rightleftarrows$ K in zeolite Na-P, GIS.[90] For Na $\rightarrow$ K, ○ and for K $\rightarrow$ Na, ▲. (d) Na $\rightleftarrows$ K in zeolite K-F, EDI.[91] For Na $\rightarrow$ K, ○ and for K $\rightarrow$ Na, ▲.

dimensions of some organic cations are summarized in Table 7, measured in three directions at right angles. These dimensions may be compared with the free diameters of the windows giving access to the intracrystalline pore space of the three zeolites (Table 6). The maximum reported uptakes of a number of organic cations are given in Table 8. Penetration of NH_4 into all three zeolites is obviously possible, and so is that of mono-alkylammonium and di-alkylammonium ions with n-alkyl groups as substituents. As with molecule sieving (§3) so also with ion sieving the organic ions and the framework oxygens are somewhat soft and deformable and lattice and ion vibrations also play a part in intracrystalline diffusion. Ions containing quaternary carbon or quarternary nitrogen

Table 6

Zeolite exchangers used and some relevant properties

Zeolite	Unit cell composition (anhydrous)	Pore volume (cm^3 per cm^3 of crystal)	Pore volume per cation (Å^3)	Free dimensions of windows (Å)	Cages in structure
Zeolite A, *LTA**	Na$_{12}$[Al$_{12}$Si$_{12}$O$_{48}$]	~0·47	~73	4·2	Sodalite, 26-hedra type I
Zeolite X, *FAU*	Na$_{82}$[Al$_{82}$Si$_{110}$O$_{384}$]	~0·53	~97	7·4	Hexagonal prisms, sodalite, 26-hedra type II
Clinoptilolite	Na$_6$[Al$_6$Si$_{30}$O$_{72}$]	~0·34	~119	4·0 × 5·5 (8-rings) 4·4 × 7·2 (10-rings) 4·1 × 4·7 (8-rings)	Two-dimensional pattern of channels

*This zeolite A also contained about one NaAlO$_2$ per unit cell.

Table 7
Van der Waals dimensions in Å of some organic cations measured in x, y and z directions[97]

Ion	x direction (length)	y direction (width)	z direction (height)
NH_4	2.8_6	2.8_6	2.8_6
NH_3CH_3	4.9_1	4.0_0	4.0_0
$NH_3C_2H_5$	5.9_0	4.8_8	4.0_0
$NH_2(CH_3)_2$	6.4_2	4.2_8	4.0_0
$NH_3(n\text{-}C_3H_7)$	7.2_6	4.8_8	4.0_0
$NH_3(n\text{-}C_4H_9)$	8.4_2	4.8_8	4.0_0
$NH_3(iso\text{-}C_3H_7)$	6.5_2	5.5_6	4.7_7
$NH_3(tert\text{-}C_4H_9)$	6.5_2	6.1_8	5.6_8
$NH(CH_3)_3$	6.4_2	6.1_0	4.1_7
$N(CH_3)_4$	6.4_2	6.1_0	6.2_2

would not be expected to penetrate zeolite A or clinoptilolite, but smaller quaternary ions such as $N(CH_3)_4^+$ should exchange in zeolite X. On the whole these predictions are supported by the results in Table 8, which serve to illustrate ion sieving as a parallel phenomenon to molecule sieving.

However, ion sieving alone cannot explain the observed maximum extents of exchange. For a series of mono-alkylammonium ions having n-alkyl substituents the maximum exchanges decline rather rapidly with ion volume. There is not enough intracrystalline free volume to accommodate all the organic cations which would be needed to replace all the Na^+ ions. Accordingly a steric factor, not of ion sieve origin, plays an important part in limiting the extent of exchange of larger ions. For example, ~33 $C(CH_3)_4$ molecules fill all the accessible pore space per unit cell in zeolite X, and therefore, since $N(CH_3)_4^+$ has about the same size, a similar number of these ions should do likewise. However, the zeolite X used had 82 Na^+ ions per unit cell.

Even this steric factor plus ion sieving may not adequately cover all the observations. In zeolites A and X for example there are Na^+ ions in sodalite cages, inaccessible to any organic ion, and in zeolite X Na^+ ions are also present in inaccessible hexagonal prisms. Unless the sodium were to vacate all such positions when the organic ions entered the 26-hedra of types I and II this could impose a further limitation upon the maximum observed exchange.

Ion sieving of inorganic cations has also been established for zeolites having narrow windows, such as analcime or sodalite hydrate. Thus Cs^+ cannot exchange with Na^+ in either zeolite under mild conditions. It

Table 8

Maximum exchanges reported for some cations in zeolites A, LTA[95], X, FAU[96] and clinoptilolite[97]

Cation	Solution	Replacement of Na^+(%)
(a) Zeolite A at 100°C for ~4 days		
NH_3CH_3	8 N chloride	77·7
$NH_3C_2H_5$	6 N chloride	59·6
$NH_3(n\text{-}C_3H_7)$	6 N chloride	26·0
$NH_3(n\text{-}C_6H_{13})$	4 N chloride	9·4$_5$
$NH_3(iso\text{-}C_7H_{15})$	4 N chloride	9·1$_4$
$NH(CH_3)_3$	5 N chloride	14·2
$N(CH_3)_4$	6 N chloride	negligible
$N(C_2H_5)_4$	6 N chloride	negligible
(b) Zeolite X at room temperature, after repeated exchanges		
NH_4	2 N chloride	92
NH_3CH_3	2 N chloride	58
$NH_2(CH_3)_2$	1 N chloride	37
$NH(CH_3)_3$	2 N chloride	28
$N(CH_3)_4$	1 N bromide	23
$N(C_2H_5)_4$	1 N iodide	negligible
$NH_3C_2H_5$	1 N chloride	50
(c) Clinoptilolite at 60°C		
NH_4		~100^a
NH_3CH_3		~100^a
$NH_3C_2H_5$		~100^a
$NH_3(n\text{-}C_3H_7)$		~100^a
$NH_2(CH_3)_2$		~100^a
$NH_3(n\text{-}C_4H_9)$	0·0183 N chloride	~40^b
$NH_3(iso\text{-}C_3H_7)$		~27^b
$NH(CH_3)_3$		~46^b
$N(CH_3)_4$		negligible
$NH_3(tert\text{-}C_4H_9)$		negligible

^{a} Extrapolation of measured exchange isotherms; 6 hours allowed per isotherm point

^{b} Maximum observed uptakes; 72 hours allowed per isotherm point.

requires temperatures around 200°C or more for hydrothermal exchange of Cs^+ in analcime. This allows one to effect hydrolysis of CsCl by treatment with Ag-analcime or sodalite:

$$CsCl + H_2O + Ag\text{-}Z \rightarrow CsOH + AgCl \downarrow + H\text{-}Z$$

where Z denotes the zeolite framework. The aqueous solution becomes alkaline and silver chloride is precipitated.[98] One may similarly hydrolyse $N(CH_3)_4Cl$ with Ag-A, *LTA*; or with the silver form of any zeolite into which the $N(CH_3)_4$ ions cannot enter.

7. Conclusion

It has been the intention in this chapter to outline the occurrence and some of the properties of zeolites as a prelude to the account of their synthesis and hydrothermal chemistry which form the main subject of this book. It has been seen that zeolites are widespread in Nature and that they also provide numerous, distinctive framework topologies each producing a different intracrystalline pore and channel configuration. They are without exception non-stoichiometric compounds in which the non-stoichiometry can be of several forms simultaneously. Thus there may be isomorphous replacements through exchanges of cations such as $Na^+ \rightleftarrows K^+$ or $2Na^+ \rightleftarrows Ca^{2+}$ in which cation compositions and, in the latter example, numbers of cations can be changed in a continuous manner. There may also be replacements of the types Al, $Na \rightleftarrows Si$ and Al, $Ca \rightleftarrows Si$, Na so that ratios Si/Al can be altered and are not integers. Replacements of the first type also alter the number of cations present. Zeolites are likewise non-stoichiometric with respect to their water contents, which vary in a continuous manner according to the vapour pressure of water, the temperature and the composition of each given zeolite. Moreover this type of non-stoichiometry is extended when water is replaced wholly or in part by other guest molecules.

Zeolites are also defect structures in the sense that the cations are distributed among a number of sub-lattices which usually provide in all considerably more cation sites than are needed for neutralizing the negative charge on the aluminosilicate framework. Thus one has an equilibrium among the cations and cation vacancies between each pair of sub-lattices. This distribution can itself be influenced not only by the kinds of ions present but also by the temperature, the guest molecules present, the amount of these, and the Si/Al ratio in the crystals.

Finally, because of their open structures and consequent cation mobility zeolites may be considered as solid electrolytes having some of the properties of aqueous electrolyte solutions. They are accordingly ionic semi-conductors[99] with electrochemical properties and uses as potential ion-selective membranes still only very partially explored.[100,101] One may with some confidence predict a continued expansion of the uses of this numerous family of compounds, not only in their present applications as selective sorbents and molecular sieves, ion exchangers and catalysts, but also in novel directions utilizing more of their remarkable properties.

References

1. Cf., W. Eitel, "Silicate Science", Vol. IV. Academic Press, London and New York, 1966, page 471, for dehydration curves of natrolite, mesolite and scolecite, after C.J. Peng, *Amer. Mineral.*, 1955, **40**, 834.

2. A.F. Cronstedt, *Akad. Handl. Stockholm* (1756) **17**, 120.

3. O. Loew, Pt 6 of *U.S. Geog. and Geol. Explor. and Surv. West of 100th Meridian* (1875) **3**, 569.

4. J. Murray and A.F. Renard, "Deep Sea Deposits: Vol. 5, Report on Scientific Results of the Voyage of HMS Challenger during the Years 1873–1876." Eyre and Spottiswoode, London, 1891.

5. A. Johannsen, *Bull. Amer. Mus. Nat. History* (1914) **33**, 209.

6. W.H. Bradley, *Science* (1928) **67**, 73.

7. C.S. Ross, *Amer. Mineral.* (1928) **13**, 195.

8. M.N. Bramlette and E. Posniak, *Amer. Mineral.* (1933) **18**, 167.

9. P.E. Kerr, *Econ. Geol.* (1931) **26**, 153.

10. D.S. Coombs, *Trans. Roy. Soc. New Zealand* (1954) **82**, 65.

11. F.A. Mumpton, *in* "Natural Zeolites. Occurrence, Properties, Use" (Ed. L.B. Sand and F.A. Mumpton), p. 3. Pergamon Press, Oxford, 1978.

12. E.g., R.A. Munson and R.A. Sheppard, *Minerals Sci. Eng.* (1974) **6**, 19.

13. A.J. Ellis, Ph.D. Thesis, University of New Zealand, 1958.

14. H. Kristmannsdottir and J. Tomasson, *in* "Natural Zeolites. Occurrence, Properties, Use" (Ed. L.B. Sand and F.A. Mumpton), p. 277. Pergamon Press, Oxford, 1978.

15. R. Sersale, *in* "Natural Zeolites. Occurrence, Properties, Use" (Ed. L.B. Sand and F.A. Mumpton), p. 285 Pergamon Press, Oxford, 1978.

16. A. Iijima, *in* "Natural Zeolites. Occurrence, Properties, Use" (Ed. L.B. Sand and F.A. Mumpton), p. 175. Pergamon Press, Oxford, 1978.

17. R.L. Hay, *Geol. Soc. Amer. Special Papers* (1966) **85**, 1.

18. A. Iijima and M. Utada, *Sedimentology* (1966) **7**, 327.

19. A. Iijima and M. Utada, *Jap. J. Geol. Geog.* (1972) **42**, 61.

20. R.A. Sheppard, *in* "Molecular Sieve Zeolites—I", *Amer. Chem. Soc. Advances in Chem. Series No. 101* (Ed. R.F. Gould) p. 279. American Chemical Society, 1971.

21. M. Utada, J. Minato, T. Isikawa and Y. Yoshizaki, *Mining Geol.* (Special Issue) (1974) **6**, 291.

22. A. Iijima and K. Harada, *Amer. Mineral.* (1968) **54**, 182.

23. K. Harada, M. Hara and K. Nakao, *Mineral. J.* (1968) **5**, 309.

24. M. Kastner and S.A. Stonecipher, *in* "Natural Zeolites. Occurrence, Properties, Use" (Ed. L.B. Sand and F.A. Mumpton), p. 199. Pergamon Press, Oxford, 1978.

25. R.C. Surdam and R.A. Sheppard, *in* "Natural Zeolites. Occurrence, Properties, Use" (Ed. L.B. Sand and F.A. Mumpton), p. 145. Pergamon Press Oxford, 1978.

26. R.L. Hay, *in* "Natural Zeolites. Occurrence, Properties, Use" (Ed. L.B. Sand and F.A. Mumpton) p. 135. Pergamon Press, Oxford, 1978.

27. R.M. Barrer, "Zeolites and Clay Minerals as Sorbents and Molecular Sieves", Ch. 2. Academic Press, London and New York, 1978.

28. W.M. Meier and D.H. Olson, "Atlas of Zeolite Structure Types". Structure Commission of the International Zeolite Association, 1978.

29. W.M. Meier, *in* "Molecular Sieves", p. 10. Society for Chemical Industry, London, 1968.

30. D.E. Appleman, *Amer. Cryst. Assoc., Mineral. Soc. America, Joint Meeting, Gatlinburg, Tenn.*, 1965, p. 80.
31. B.J. Skinner and D.E. Appleman, *Amer. Mineral.* (1963) 218, 865.
32. J.L. Schlenker, F.G. Dwyer, E.E. Jenkins, W.J. Rohrbaugh, G.T. Kokotailo and W.M. Meier, *Nature* (1981) **294,** 340.
33. R.M. Barrer, *in* "Non-stoichiometric Inclusion Compounds", (Ed. L. Mandelcorn), Ch. 6. Academic Press, London and New York, 1963.
34. D.W. Breck and J.V. Smith, *Scientific American* (1959) **200,** 85.
35. Ref. (27), p. 41.
36. S. Merlino, *Isvj. Jogoslav. Centr. Krist. (Zagreb)* (1976) **11,** 19.
37. J.V. Smith and F. Rinaldi, *Mineral. Mag.* (1962) **33,** 202.
38. E. Galli, G. Gottardi and D. Pongiluppi, *Neues Jb. Mineral. Mh.,* (1979) **1,** 1.
39. N. Haga, *Mineral. Jour.* (1973) **7,** 262.
40. W.M. Meier, *Zeit. Krist.* (1961) **115,** 439.
41. I.S. Kerr, *Nature* (1963) **197,** 1194.
42. R.M. Barrer and H. Villiger, *Zeit. Krist.* (1969) **128,** 352.
43. D.P. Shoemaker, H.E. Robson and L. Broussard, *in* "Proc. of 3rd Internat. Conf. on Molecular Sieves" (Ed. J.B. Uytterhoeven), Zurich, Sept. 3rd, 1973, p. 138.
43a R.M. Barrer, *Pure Appl. Chem.* (1979) **51,** 1091.
44. G.T. Kerr, *Science* (1963) **140,** 1412.
45. R.M. Barrer, *J. Chem. Soc.* (1948) 127.
46. R.M. Barrer, L. Hinds and E.A.D. White, *J. Chem. Soc.* (1953) 1466.
47. W.S. Wise and R.W. Tschnich, *Amer. Mineral.* (1975) **60,** 951.
48. M.N. Maleyev, *Int. Geol. Rev.* (1977) **19,** 993.
49. H. Sarp, J. Deferne, H. Bizouard and B.W. Liebich, *Schweiz. Mineral. Petrogr. Mitt.* (1979) **59,** 5.
50. Ref. (27), pp. 24–27.
51. G.T. Kokotailo and W.M. Meier, *in* "Properties and Applications of Zeolites" (Ed. R.P. Townsend), p. 133. The Chemical Society, 1980.
52. R.M. Barrer, J.A. Davies and L.V.C. Rees, *J. Inorg. Nucl. Chem.* (1969) **31,** 219.
53. R.M. Barrer and G.C. Bratt, *J. Phys. Chem. Solids* (1959) **12,** 130.
54. R.M. Barrer and B.M. Munday, *J. Chem. Soc. A* (1971) 2904.
55. R.M. Barrer, S. Barri and J. Klinowski, *in* "Proc. 5th Internat. Conf. on Zeolites" (Ed. L.V.C. Rees) p. 20. Heyden, London 1980.
56. R.M. Barrer and J.A. Lee, *J. Colloid Interface Sci.* (1969) **30,** 111.
57. R.M. Barrer, *British Chem. Eng.* (1959), **4,** 267.
58. D.W. Breck, "Zeolite Molecular Sieves, Structure, Chemistry and Uses", p. 639. Wiley-Interscience, New York, 1974.
59. R.M. Barrer, *Trans. Faraday Soc.* (1949) **45,** 358.
60. R.M. Barrer, *Nature* (1947) **159,** 508.
61. R.M. Barrer, *J. Soc. Chem. Ind.* (1945) **64,** 130 and 133.
62. R.M. Barrer and L. Belchetz, *J. Soc. Chem. Ind.* (1945) **64,** 131.
63. Ref. 27, p. 78 *et seq.* and p. 82.
64. R.M. Barrer, D.A. Harding and A. Sikand, *J. Chem. Soc. Faraday I,* (1980) **76,** 180.
65. R.M. Barrer and L.V.C. Rees, *Trans. Faraday Soc.* (1954) **50,** 852.
66. L.V.C. Rees and T. Berry, *in* "Molecular Sieves", p. 149. Society for Chemical Industry, London, 1968.

67. P.B. Weisz, *Pure Appl. Chem.* (1980) **52,** 2091.
68. J.A. Rabo (Ed.), "Zeolite Chemistry and Catalysis". American Chemical Society Monograph No. 171, 1976.
69. R.P. Townsend (Ed.), "Properties and Applications of Zeolites". The Chemical Society, 1980.
70. L.V.C. Rees (Ed.), "Proc. 5th Internat. Conf. on Zeolites", Heyden, London, 1980.
71. M. Ziolek and Z. Dudzik, *Zeolites* (1981) **1,** 117.
72. P.B. Weisz, *Erdol u. Kohle* (1965) **18,** 527.
73. N.Y. Chen and P.B. Weisz, *Eng. Prog. Symp. Series* (1967) **63,** 86.
74. W.O. Haag and F.G. Dwyer, *Amer. Inst. Chem. Eng. Ann. Mtg Boston, Mass., Aug.* 1979.
75. C.D. Chang and A.J. Silvestri, *J. Catal.* (1977) **47,** 249.
76. W.W. Kaeding and S.A. Butter, *J. Catal.* (1980) **61,** 155.
77. S.L. Meisel, *Phil. Trans. Roy. Soc., London* (1981) **300A,** 157.
78. R.F. Mueller, S. Ghose and S.K. Saxena, *Geochim. Cosmochim. Acta,* (1970) **34,** 1356.
79. R.M. Barrer, *in* "Proc. 5th Internat. Conf. on Zeolites", (Ed. L.V.C. Rees), p. 273. Heyden, London, 1980.
80. G.L. Gaines and H.C. Thomas, *J. Chem. Phys.* (1953) **21,** 714.
81. P. Fletcher and R.P. Townsend, *J. Chem. Soc. Faraday II,* (1981) **77,** 955.
82. P. Fletcher and R.P. Townsend, *J. Chem. Soc. Faraday II* (1981) **77,** 965.
83. P. Fletcher and R.P. Townsend, *J. Chem. Soc. Faraday II* (1981) **77,** 2077.
84. C. Borghese, *J. Phys. Chem. Solids* (1967) **28,** 2225.
85. R.M. Barrer and J. Klinowski, *Phil. Trans. Roy. Soc., London* (1977) **285A,** 638.
86. M. Nitta, K. Ogawa and K. Aomura, *in* "Proc. 5th Internat. Conf. on Zeolites" (Ed. L.V.C. Rees), p. 291. Heyden, London, 1980.
87. K. Ogara, M. Nitta and K. Aomura, *J. Phys. Chem.* (1978) **82,** 1655.
88. M. Nitta, K. Ogawa and K. Aomura, *Zeolites* (1981) **1,** 30.
89. R.M. Barrer and J.D. Falconer, *Proc. Roy. Soc.* (1956) **236A,** 227.
90. R.M. Barrer and B.M. Munday, *J. Chem. Soc. A* (1971) 2909.
91. R.M. Barrer and B.M. Munday, *J. Chem. Soc. A* (1971) 2914.
92. R.M. Barrer and J. Klinowski, *Geochim. Cosmochim. Acta* (1979) **43,** 755.
93. H.G. Smolka and M.J. Schwuger, to Henkel and Co., Ger. Pat. 2,412,838 (1974).
94. M.J. Schwuger and H.G. Smolka, *in* "Properties and Applications of Zeolites" (Ed. R.P. Townsend) p. 244. The Chemical Society, Special Publication No. 33, 1980.
95. R.M. Barrer and W.M. Meier, *Trans. Faraday Soc.* (1958) **54,** 1074, and (1959) **55,** 130.
96. R.M. Barrer, W. Buser and W.F. Grutter, *Helv. Chim. Acta* (1956) **39,** 518.
97. R.M. Barrer, R. Papadopoulos and L.V.C. Rees, *J. Inorg. Nucl. Chem.* (1967) **29,** 2047.
98. R.M. Barrer and D.C. Sammon, *J. Chem. Soc.* (1956) 675.
99. R.M. Barrer, Proc. *Brit. Ceram. Soc.* (July 1964) No. 1, p. 145.
100. R.M. Barrer and E.A. Saxon-Napier, *Trans. Faraday Soc.* (1962) **58,** 145 and 157.
101. R.M. Barrer and S.D. James, *J. Phys. Chem.* (1960) **64,** 417, 421.

Chapter 2

Introduction to the Hydrothermal Chemistry of Silicates

1. Synthetic Mineralogy

There are two major reasons for studying the synthesis and chemistry of minerals. The first is the need to understand reactions occurring in the lithosphere on or near the surface or in abyssal rock. These reactions are occurring in open systems and may be as vast in scale as in the range of variables such as pressure, temperature and time. The second reason is technical and strategic, a search for cheaper or better replacements for natural minerals and the production of synthetics of better quality having specific built-in properties. In this the chemist is well qualified to achieve the desired end, and in so doing to shed light upon Nature's reaction mechanisms and pathways and to extend synthetic mineralogy into regions of hitherto unexplored chemical and technological interest. It will be seen that zeolite synthesis provides an excellent example of this. Zeolites form in Nature or in the laboratory under conditions in which water is present in considerable amounts often at elevated temperatures and so under hydrothermal conditions. In the present chapter some features of hydrothermal chemistry are outlined, and the versatility of the hydrothermal method is exemplified for the following major groups of silicates: smectites, kandites, micas, felspars, felspathoids and crystalline silicas. The notable flowering in zeolite syntheses, commencing about 40 years ago, will be described in later chapters.

Methods used in mineral synthesis have in part imitated natural processes of change by sintering reactions; by crystal growing from melts or from hydrothermal magmas; or by pneumatolysis in which reactions proceed not in the presence of excess water as in hydrothermal chemistry,

TABLE 1
Methods of growing crystals[1]

Growth from	Conditions	Variants
Vapour	Pressure, P, usually controlled	Deposition from pure vapour. Decomposition at hot wire (van Arkel).
Gaseous solution	High P and temperature T	Isothermal Temperature gradient Pneumatolytic
Aqueous solution or magma	Below 100°C at ambient P; or at higher T (hydrothermal processes)	Isothermal Temperature gradient (regenerative and circulatory) Temperature lowering Precipitation
High temperature mixed melts and solutions	Ambient P and high T	Temperature lowering Evaporation of solvent Composition changes in melt
Pure melt	High T; P depends on substance	Temperature lowering (Bridgman) Crystal pulling (Kyropoulos) Flame fusion (Verneuil) Zone melting (floating zone and crucible techniques)
Solid state	T and P depend on circumstances	Sintering Hot pressing Diamond-forming process.

but at high temperature in the presence of small amounts of reactive volatile components as gas or super-critical fluid (e.g. H_2O, H_2S, CO_2, SO_2, HCl or HF). In Table 1[1] methods of crystal growing have been summarized, to show the considerable number of variations available. In the laboratory some of these have been used to grow large single crystals as near-perfect as possible. However, in the present context it is the hydrothermal method which is of prime interest.

If one were to start with a very large amount of a rock melt of an average composition approaching that of the lithosphere, and to crystallize this mass by very slow cooling, there would be a rather definite sequence in the crystals deposited,[2] as given in Table 2. From the homogeneous, fluid, high-temperature melt silica-free crystals first appear, then orthosilicates such as olivine, and in sequence metasilicates and felspars, quartz as a primary mineral, and mixtures of minerals such as granite. Weathering of these can yield clay minerals and zeolites and eventually aluminas such as diaspore and bauxite. Alternatively at the stage when the melt has cooled and crystallized sufficiently to yield an alkali-rich basalt magma this may be converted in part by saline and

TABLE 2
Crystallization sequence from an initial multicomponent melt[2]

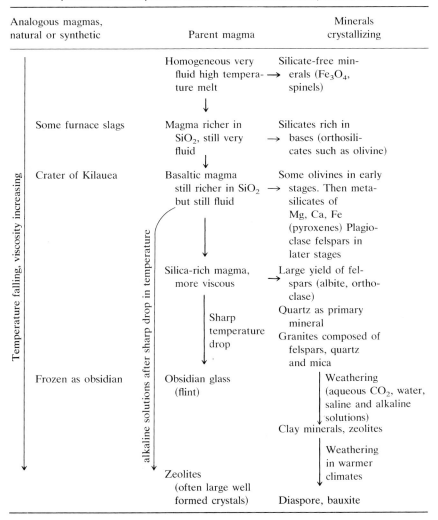

Analogous magmas, natural or synthetic	Parent magma	Minerals crystallizing
	Homogeneous very fluid high temperature melt →	Silicate-free minerals (Fe_3O_4, spinels)
Some furnace slags	Magma richer in SiO_2, still very fluid →	Silicates rich in bases (orthosilicates such as olivine)
Crater of Kilauea	Basaltic magma still richer in SiO_2 but still fluid →	Some olivines in early stages. Then metasilicates of Mg, Ca, Fe (pyroxenes) Plagioclase felspars in later stages
	Silica-rich magma, more viscous →	Large yield of felspars (albite, orthoclase)
	Sharp temperature drop	Quartz as primary mineral. Granites composed of felspars, quartz and mica
Frozen as obsidian	Obsidian glass (flint)	Weathering (aqueous CO_2, water, saline and alkaline solutions)
		Clay minerals, zeolites
	Zeolites (often large well formed crystals)	Weathering in warmer climates. Diaspore, bauxite

Temperature falling, viscosity increasing

alkaline solutions after sharp drop in temperature

alkaline aqueous media to zeolites.[3,4,5,6,7] Zeolitization requires a further big drop in the temperature of a basalt lava.

During crystallization of a natural multi-component magma at depth anhydrous crystals are first deposited (Table 2). This causes volatiles such as water to accumulate in the subterranean crystallization zones, which has been suggested as one cause of vulcanism and of the upwelling of lavas.[8] The accumulation of the volatiles may in turn also lower the viscosity of the remaining magma by dissolving in it and breaking Si–O–Si or Si–O–Al bonds (§5). At the high temperatures involved, very great pressures can develop. This in turn can speed the crystallization processes and so accelerate the build-up of the pressure of the volatiles. When the pressure reaches extreme values explosion can occur, as happened for example in the eruption of Mont Pelée when violent release of red-hot gases caused great devastation and loss of life.[9]

2. Hydrothermal Chemistry

Table 1 indicates that aqueous magmas may react and crystallize either below or above 100°C, in the latter case in closed systems to contain the high pressure of water vapour or super-critical fluid. However there is nothing particular which distinguishes crystal deposition below 100°C from that above. Thus hydrothermal reaction merges into low-temperature reaction, and indeed in zeolite synthesis the trend has been, *inter alia* for greater experimental convenience, to grow zeolites at or below 100°C provided reaction rates are adequate. Table 3[10] gives examples of minerals synthesized by some of the methods of Table 1. The versatility of the hydrothermal procedure is apparent. It yields phyllo- and tectosilicates in considerable variety. Zeolites have been made only by the hydrothermal method, because, as we shall see in §4, open aluminosilicate host frameworks must be stabilized during growth by being filled with guest molecules. On the other hand, the more compact tectosilicates (crystalline silicas and felspars) can be made by a number of the methods of Table 1. With the possible exception of Al-free compounds such as quartz, one cannot synthesize tectosilicates from acid hydrothermal systems. In particular, zeolites never form under acid conditions. The sodalites and cancrinites, like zeolites, have open frameworks but their synthesis can be achieved pyrolytically as well as hydrothermally because non-volatile salts can act like zeolitic water as stabilizers by occupying the pore spaces within the aluminosilicate framework. In hydrothermal syntheses using alkaline salt solutions water and salts may compete as stabilizers so that each may be present within the intracrystalline pores.[53] Indeed sodalite-nosean and cancrinite felspathoids are readily made which contain only zeolitic water plus a small amount of alkali (NaOH),[54] and which are therefore in this form justifiably regarded as zeolites (Chapter 7, §2.2).

The versatility of hydrothermal chemistry owes much to the mineralizing role of water. The factors which promote reactivity in aqueous magmas include:[55]

(a) The stabilizing of porous lattices as zeolites by acting as space fillers, referred to above and in §4.
(b) Through its presence, especially at high pressures, water may be incorporated into otherwise anhydrous glasses, melts and solids. Through chemisorption into siliceous materials Si–O–Si and Al–O–Si bonds hydrolyse and re-form. Chemical reactivity is enhanced and magma viscosity is lowered (§5).
(c) High pressures of water can modify phase equilibrium temperatures (§3).
(d) Water is a good solvent, a property which assists disintegration of solid components of a mixture and facilitates their transport and mixing (§6).

These mineralizing properties can be shown by other substances to varying degrees. As noted already salts may stabilize porous structures by acting as fillers and so catalysing their formation. Silica gel dissolved in sodium metaphosphate crystallized out from the melt over the range 700–950°C as cristobalite only, with $SiO_2.P_2O_5$ as possible intermediate.[56] On the other hand, silica gel dissolved in fused sodium tungstate crystallized only as tridymite in the range 700–850°C. Thus each salt had a different structure-directing effect. In the conversion of amorphous alumina to corundum, an effective catalyst was CaF_2, probably with reversible formation of AlF_3 as an intermediate:[46]

$$3CaF_2 + Al_2O_3 \rightleftarrows 3CaO + 2AlF_3$$

These examples illustrate mineralizing situations not involving water. Water vapour may catalyse sintering reactions such as

$$BaCO_3 + SiO_2 \rightarrow BaSiO_3 + CO_2\uparrow$$

and

$$CaO + SiO_2 \rightarrow CaSiO_3$$

Under specific reaction conditions the first of these was accelerated 22-fold and the second 8.5-fold, the effect being manifest in the term k_0 in the Arrhenius equation $k = k_0 \exp(-E/RT)$ for the rate constant k.[46]

TABLE 3

Examples of mineral syntheses by five methods[2]

Crystal type	Hydrothermal	Pyrolytic	Pneumatolytic	Sintering	Vapour phase
Clay minerals	Kaolinite[11,12,13,14,15] Dickite[14,15] Beidellite[14,15] Sericite[13,15] Nontronite[14] Montmoril- lonite[13,15]				
Micas	Muscovite[12]	Phlogopite[34]			
Zeolites	Analcime[10,16] Mordenite[17] Harmotome[17]				
Sodalite- cancrinite	Sodalite[18,19]	Sodalite[35]			
Minerals	Cancrinite[19] Nosean[19]	Nosean[35] Hauyne[35] Ultramarine[36]			

Crystalline silicas	Quartz[20,21] Cristobalite[22,23]	Quartz[37] Cristobalite[37] Tridymite[37]	Quartz[45]	Quartz[45]	Cristobalite[50]
Felspars	Albite[24] Orthoclase[24,25]	Albite[38] Orthoclase[38]	Albite[45] Orthoclase[45] Anorthite[45]	Albite[46] Anorthite[47]	
Gem-type silicates	Emerald[26,27] Cordierite[28] Zircon[29] Tourmaline[30,31]	Cordierite[39]		Jadeite[48] Zircon[49]	
Non-silicates	Blende[32] Zincite[32]	Fluorite[40] Yttrium Iron Garnet[41,42,43] Apatite[44]			Blende[51,52] Zincite[1]
	Apatite[33]				

3. Pressure, Temperature and Equilibrium Between Phases

In hydrothermal systems there may be a number of co-existing phases, solid, liquid and vapour. Phase equilibria can therefore be relevant and some thermodynamic aspects are considered in this section. We assume three phases, α, β and γ (e.g. solid, liquid and vapour) each containing the same m non-reacting components. The three phases are in equilibrium with each other and every component is distributed between the phases. Therefore one equilibrium condition is

$$\left.\begin{array}{c} \mu_i^\alpha = \mu_i^\beta = \mu_i^\gamma \\[4pt] d\mu_i^\alpha = d\mu_i^\beta = d\mu_i^\gamma \end{array}\right\} \tag{1}$$

for any component i. The phase in which it is present is indicated by the superscript α, β or γ. In addition the Gibbs–Duhem relations are

$$\left.\begin{array}{c} S_t^\alpha \, dT - V_t^\alpha \, dP + \displaystyle\sum_{i=i}^{m} n_i^\alpha \, d\mu_i^\alpha = 0 \\[10pt] S_t^\beta \, dT - V_t^\beta \, dP + \displaystyle\sum_{i=1}^{m} n_i^\beta \, d\mu_i^\beta = 0 \\[10pt] S_t^\gamma \, dT - V_t^\gamma \, dP + \displaystyle\sum_{i=1}^{m} n_i^\gamma \, d\mu_i^\gamma = 0 \end{array}\right\} \tag{2}$$

In these relations S_t and V_t are the total entropy and volume of a given phase; the n_i are the numbers of moles of component i; and P and T are pressure and temperature.

In terms of the phase rule, if $m = 1$ for each of the three phases then the number of degrees of freedom, F, is zero so that P and T are fixed for equilibrium of solid, liquid and vapour—the triple point. When $m = 2$, F is 1 and fixing any one of the four variables (two concentrations, P and T) determines the remaining three. Thus for each P there is a particular T and composition in each of the three phases, so that it is possible to have a three-phase P, T line along which compositions must also be changing. For $m = 3$, F is 2 and so fixing any two of the five variables (three concentrations, P and T) determines the remaining three. One may now have a P, T surface. For any path along it the concentrations must also be changing. When, as a result of altering P and T, the concentrations change there must be transfers of components between phases. Then conservation of mass requires for any component i,

$$dn_i^\alpha + dn_i^\beta + dn_i^\gamma = 0; \quad \text{or}$$

$$dX_i^\alpha \sum_i n_i^\alpha + dX_i^\beta \sum_i n_i^\beta + dX_i^\gamma \sum_i n_i^\gamma = 0 \tag{3}$$

where the X_i are mole fractions.

The three relations (2) may respectively be divided by $\sum_i n_i^\alpha$, $\sum_i n_i^\beta$ and $\sum_i n_i^\gamma$, so that they refer to one mixed mole of each phase, for which entropy and volume are S and V with the appropriate superscript, and the mole fraction of component i is X_i^α, X_i^β and X_i^γ for the three phases. Then

$$
\left.
\begin{aligned}
S^\alpha \, \mathrm{d}T - V^\alpha \, \mathrm{d}P + \sum_i X_i^\alpha \, \mathrm{d}\mu_i^\alpha = 0 \\[2mm]
S^\beta \, \mathrm{d}T - V^\beta \, \mathrm{d}P + \sum_i X_i^\beta \, \mathrm{d}\mu_i^\beta = 0 \\[2mm]
S^\gamma \, \mathrm{d}T - V^\gamma \, \mathrm{d}P + \sum_i X_i^\gamma \, \mathrm{d}\mu_i^\gamma = 0
\end{aligned}
\right\}
\tag{2a}
$$

while for one mixed mole of each phase the second of Eqns (3) becomes

$$
\mathrm{d}X_i^\alpha + \mathrm{d}X_i^\beta + \mathrm{d}X_i^\gamma = 0
\tag{3a}
$$

By subtracting one of each pair of relations (2a) from the other and bearing in mind Eqns (1) there result after a little re-arrangement the expressions:

$$
\left.
\begin{aligned}
\frac{\mathrm{d}P}{\mathrm{d}T} &= \frac{\Delta S^{\alpha\beta} - \sum_i \Delta X_i^{\alpha\beta}(\mathrm{d}\mu_i^\alpha/\mathrm{d}T)}{\Delta V^{\alpha\beta}} \\[3mm]
&= \frac{\Delta S^{\alpha\gamma} - \sum_i \Delta X_i^{\alpha\gamma}(\mathrm{d}\mu_i^\alpha/\mathrm{d}T)}{\Delta V^{\alpha\gamma}} \\[3mm]
&= \frac{\Delta S^{\beta\gamma} - \sum_i \Delta X_i^{\beta\gamma}(\mathrm{d}\mu_i^\alpha/\mathrm{d}T)}{\Delta V^{\beta\gamma}}
\end{aligned}
\right\}
\tag{4}
$$

where $\Delta S^{\alpha\beta} = (S^\alpha - S^\beta)$; $\Delta X_i^{\alpha\beta} = (X_i^\alpha - X_i^\beta)$; and $\Delta V^{\alpha\beta} = (V^\alpha - V^\beta)$, with similar definitions for the other ΔS, ΔX_i and ΔV. The three expressions on the r.h.s. can be equated to each other and then relate the ΔX_i to the ΔV, ΔS and $\mathrm{d}\mu_i/\mathrm{d}T$.

When there are only two phases, α and β, the second and third of Eqns (4) disappear and if the number of components, m, is one the term in $\Delta X_i^{\alpha\beta}$ also vanishes so that

$$
\frac{\mathrm{d}P}{\mathrm{d}T} = \frac{\Delta S^{\alpha\beta}}{\Delta V^{\alpha\beta}} = \frac{\Delta H^{\alpha\beta}}{T \Delta V^{\alpha\beta}}
\tag{5}
$$

where at equilibrium $T \Delta S^{\alpha\beta} = \Delta H^{\alpha\beta} = (H^\alpha - H^\beta)$. H^α and H^β are the enthalpies of one mole of phases α and β respectively. If one phase, α, is liquid or solid and the other, β, is vapour exerting its equilibrium value, then Eqn (5) serves to evaluate $\Delta H^{\alpha\beta}$ since usually $\Delta V^{\alpha\beta} \simeq -V^{\beta}$ ($V^{\beta} \gg V^{\alpha}$). Thus from $P - V^{\beta} - T$ relations for the vapour one may find $\Delta H^{\alpha\beta}$. If the two phases are liquid and solid under hydrostatic pressure P then the reciprocal of Eqn (5) gives the variation of the melting temperature with hydrostatic pressure.

In another situation it is possible to maintain different pressures upon two phases of the same component in equilibrium at constant temperature. Thus by means of a porous membrane piston, through which only the vapour (phase β) of the liquid or solid (phase α) may pass a pressure P^{α} may be maintained upon the condensed phase while the vapour pressure of phase β is P^{β} and is the only pressure to which this phase is subjected. Then

$$\left. \begin{array}{l} d\mu^{\alpha} = V^{\alpha}\, dP^{\alpha} \\ d\mu^{\beta} = V^{\beta}\, dP^{\beta} \end{array} \right\} \tag{6}$$

and since $d\mu^{\alpha} = d\mu^{\beta}$ one finds that

$$\left(\frac{\partial P^{\beta}}{\partial P^{\alpha}} \right)_{T} = \frac{V^{\alpha}}{V^{\beta}} \tag{7}$$

Thus the vapour presure P^{β} over the condensed phase under hydrostatic pressure P^{α} changes with P^{α} and the two changes in pressure must be inversely in the ratios of the two molecular volumes.

In a further equilibrium situation we may consider the distribution at constant pressure of any component i between any pair of multicomponent phases α and β. The equilibrium constant, $K_i^{\alpha\beta}$, is given by

$$K_i^{\alpha\beta} = a_i^{\alpha}/a_i^{\beta}, \qquad \Delta G_i^{\ominus} = -RT \ln K_i^{\alpha\beta} \tag{8}$$

where the a_i are activities and $\Delta G_i^{\ominus}$ is the standard free energy of transfer of i from phase β to phase α. Then

$$\left(\frac{\partial \Delta G_i^{\ominus}}{\partial P} \right)_{T} = -RT \left(\frac{\partial \ln K_i^{\alpha\beta}}{\partial P} \right)_{T} = \Delta V_i^{\ominus} \tag{9}$$

and

$$\left(\frac{\partial \Delta G_i^{\ominus}}{\partial T} \right)_{P} = -\Delta S_i^{\ominus} = (\Delta G_i^{\ominus} - \Delta H_i^{\ominus})/T \tag{10}$$

$\Delta V_i^{\ominus}$, $\Delta S_i^{\ominus}$ and $\Delta H_i^{\ominus}$ are the standard volume change, entropy and heat of transfer of i between phases. When the phase α is a vapour or gas the activity a_i^{α} in it of component i is the relative fugacity $f_i^{\alpha}/(f_i^{\alpha})^{\ominus}$. As required a_i^{α} is unity in the standard state where $f_i^{\alpha} = (f_i^{\alpha})^{\ominus}$. Fugacities have

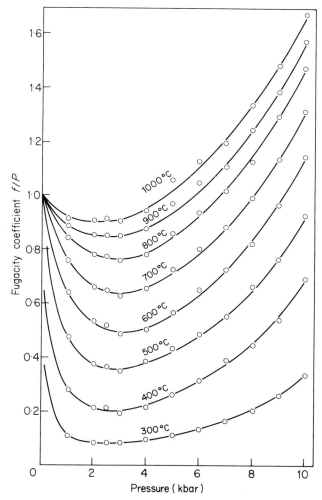

Fig. 1. Plots of the ratio f/P against P for water.[58]

been calculated for water[57,58] and some are shown in the form of plots of the ratio f/P vs P in Fig. 1.[58] Similar calculations have been made of the fugacity of CO_2 between 50 and 1000°C and for pressures from 25 to 1400 bars.[59]

The considerations of this section apply when the components of the several co-existing phases do not react chemically. If one component does so react to give equilibria with any of the other components one may add additional relationships to those in Eqns (1) and (2). If reactants are denoted by A_i and resultants by B_j, and if the stoichiometric numbers are respectively ν_i and ν_j then the reaction may be written as

$$\sum_i \nu_i A_i \rightleftarrows \sum_j \nu_j B_j \qquad (11)$$

and in the phase α for example the additional equilibrium relationship is

$$\sum_i \nu_i \mu^{\alpha}_{A_i} = \sum_j \nu_j \mu^{\alpha}_{B_j} \qquad (12)$$

It will be seen in §5 that water may react in this way when it is dissolved in silicate melts. The relation (12) does not replace Eqns (1) and (2) but supplements them.

4. Stabilization of Porous Crystals by Guest Molecules

When molecules such as water permeate zeolite crystals they lower the chemical potential of the lattice-forming units of the zeolites, just as dissolving water in ethanol lowers the chemical potential of the alcohol. Solution thermodynamics can be used to quantify this behaviour for zeolitic solid solutions[60] just as with clathrate solid solutions[61] or for liquid mixtures.

For one mixed mole of guest molecules, A, and of chosen lattice-forming units, B, of the zeolite the mole fractions are respectively X_A and X_B. The guest molecules are distributed at equilibrium between vapour or liquid solution and crystal phases. If present at all any other component of the gaseous or liquid solution, except A, is assumed to be too large to enter the zeolite lattice. In the zeolite (phase α) the Gibbs–Duhem relation for the mixed mole at constant temperature becomes

$$X_A \, d\mu^{\alpha}_A + (1 - X_A) \, d\mu^{\alpha}_B = V_1 \, dP \qquad (13)$$

where V_1 is the volume of the mixed mole at pressure P and μ^{α}_A and μ^{α}_B are chemical potentials of A and B in it. If μ^{β}_A is the chemical potential of A in the vapour or liquid solution (phase β), or in the pure vapour of A, the equilibrium condition is

$$d\mu^{\alpha}_A = d\mu^{\beta}_A = RT \, d \ln a_A = RT \, d \ln f_A \qquad (14)$$

where a_A and f_A are equilibrium activity and fugacity respectively. Then from Eqns (13) and (14):

$$d\mu^{\alpha}_B = (V_1 \, dP - RTX_A \, d \ln f_A)/(1 - X_A) \qquad (15)$$

From Eqn (15) one may find by integration the change in μ^{α}_B when X_A

changes from 0 to any finite value:

$$\Delta\mu_B = \mu_B^\alpha - (\mu_B^\ominus)^\alpha = \int_{P_0}^{P} \frac{V_1 \, dP}{(1-X_A)} - RT \int_0^{f_A} \frac{X_A}{(1-X_A)} \frac{df_A}{f_A} \quad (16)$$

In this expression P_0 is the hydrostatic pressure when $X_A = 0$ and P the pressure for the finite X_A. Also the fugacity is zero when $X_A = 0$ and is f_A for the finite X_A. If the hydrostatic pressure is held constant then the first integral in Eqn (16) can be omitted. If the vapour or liquid solution is replaced by pure vapour of A then in Eqn (16) P_0 is zero when $X_A = 0$ and is P_A, the equilibrium vapour pressure of A for finite X_A. This latter case is of particular interest and is considered below.

For a porous zeolite crystal which does not swell or shrink appreciably when guest molecules are imbibed, the partial molal volume V_M of the lattice-forming units of the crystal may be taken as a constant. Accordingly for one mixed mole $V_1 = V_M X_B$, or $\dot{V}_M = V_1/(1-X_A) = $ const. Also if M is the molecular weight of a lattice forming unit, and m is that of the guest and if x is the weight of guest molecules per unit weight of host lattice, then $X_A/(1-X_A) = Mx/m$. Accordingly Eqn (16) becomes

$$\Delta\mu_B = V_M p_A - \frac{MRT}{m} \int_0^{f_A} \frac{x}{f_A} \, df_A \quad (17)$$

The relation between fugacity and pressure of the guest species A is

$$\ln f_A/p_A = Bp_A + (C/2)p_A^2 + (D/3)p_A^3 + \cdots \quad (18)$$

where B, C and D are coefficients in the virial equation of state

$$\frac{p_A V_A}{RT} = 1 + Bp_A + Cp_A^2 + Dp_A^3 + \cdots \quad (19)$$

Accordingly from Eqns (18) and (17) one obtains

$$\Delta\mu_B = V_M p_A - \frac{MRT}{m} \int_0^{p_A} x\left(\frac{1}{p_A} + B + Cp_A + Dp_A^2 + \cdots\right) dp_A \quad (20)$$

Thus, if the equation of state is known one may calculate f_A from p_A, and if the sorption isotherm is known (x as a function of p_A) then the isotherm can be given as x in terms of f_A and the integral in Eqn (17) evaluated graphically. Alternatively Eqn (20) with the known coefficients $B, C, D, \ldots$, can also serve with the isotherm to find the integral graphically. Equations (17) or (20) apply even when the pressure is very high. For many isotherms, however, the pressure is low enough to allow one to use the approximation $f_A \sim p_A$. Then

$$\Delta\mu_B = V_M p_A - \frac{MRT}{m} \int_0^{p_A} \frac{x}{p_A} \, dp_A \quad (21)$$

A plot of x/p_A against p_A then serves to evaluate the integral. For strongly sorbed vapours such as water or ammonia the term $V_M p_A$ is normally very small compared with the integral and may safely be neglected. In the Henry's law range and when $f_A \sim p_A$ the ratio x/p_A is a constant, K, and so

$$\Delta\mu_B = \left(V_M - \frac{MRT}{m} K\right) p_A \tag{22}$$

In a tectosilicate one convenient choice of lattice-forming unit is the amount containing two oxygen atoms, e.g. $Na_x Al_x Si_{1-x} O_2$, which approaches SiO_2 as x tends to zero. Alternatively, one may use the gram formula weight (cf. §5), or the unit cell content where this is known. With the first choice Eqn (21) was used to evaluate $\Delta\mu_B$ for faujasite (as zeolite X) when water or ammonia were the guest molecules. According to the chemical analysis of a particular zeolite X the lattice-forming unit was $Na_{0.43} Al_{0.43} Si_{0.57} O_2$. It was not possible, because of the rectangular character of the sorption isotherms, to use $(\mu_B^\ominus)^\alpha$ as the reference state. Instead $\Delta\mu_B$ was determined for the following values of X_A as reference states from which X_A was increased:

NH$_3$-Na-faujasite: $X_A^{ref} = 0.214$ and 0.34
H$_2$O-Na-faujasite: $X_A^{ref} = 0.34$
H$_2$O-Ca-faujasite: $X_A^{ref} = 0.34$ (two samples)
H$_2$O-(Cs, Ca)-faujasite: $X_A^{ref} = 0.34$

The results shown in Fig. 2 indicate a large decrease in chemical potential as X_A is increased from the reference states. The lattice stabilization due to the decrease in chemical potential of the host when the guest is imbibed is effected to some extent by any guest molecule but the more weakly the guest is sorbed the smaller is $\Delta\mu_B$. In addition non-polar liquids have little solvent power for silica or alumina and on account of this would not aid zeolite formation. Liquid ammonia with dissolved alkali metal amides might play a similar solvent role to water and alkali metal hydroxide, but the low temperatures normally involved with liquid ammonia could mean slow nucleation and crystal growth. So far no useful substitute for an aqueous alkaline medium has been found for synthesizing zeolites.

From the free energy of stabilization, $\Delta\mu_B$, determined at two or more temperatures for a given uptake of the guest molecules, the corresponding entropy and enthalpy changes, $\Delta\bar{S}_B$ and $\Delta\bar{H}_B$ respectively, can be determined:

$$\Delta\bar{S}_B = -\frac{\partial}{\partial T}(\Delta\mu_B) \tag{23}$$

$$\Delta\bar{H}_B = \frac{\partial}{\partial(1/T)}\left(\frac{\Delta\mu_B}{T}\right) \tag{24}$$

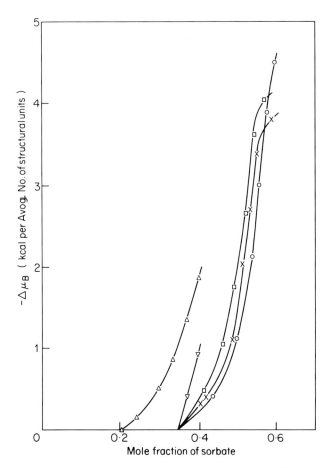

Fig. 2. $\Delta\mu_B$ for NH_3 and H_2O in Na- and Ca-faujasite.[62]
$\triangle$, NH_3-Na-faujasite; mol. fr. $NH_3 = 0.215$ in reference state.

∇, NH_3-Na-faujasite.
$\bigcirc$, H_2O-Ca-faujasite.
$\times$, H_2O-Ca-faujasite.
$\square$, H_2O-(Cs, Ca)-faujasite.⎫ mol. fr. sorbate in reference state $= 0.34$

$\Delta\bar{S}_B$ and $\Delta\bar{H}_B$ for Na-faujasite when CF_4 is sorbed in the zeolite are shown in Figs 3 and 4.[60] They are considerably smaller than the corresponding differential entropy and heat of sorption of the CF_4 in the Na-faujasite.[63] They must allow for any perturbations of the lattice associated with cation shifts or framework alteration when the guest molecules enter the zeolite.

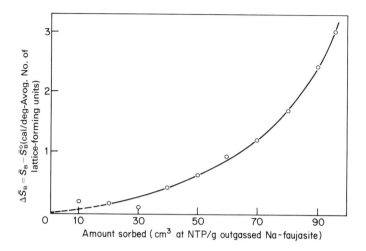

Fig. 3. $\Delta\bar{S}_B$ for CF_4 in Na-faujasite[60] as a function of CF_4 uptake.

Mixtures of guest molecules, all of which can be sorbed by the zeolite, may be volatile or non-volatile or both, for example solutions of salts in water. For guest species A, B, C, . . . the equivalent of Eqn (17) is

$$\Delta\mu_B = V_M(P - P_0) - \sum_{A,B,C,\dots} \frac{MRT}{m_A} \int_0^{a_A} \frac{x_A}{a_A}\,da_A \tag{25}$$

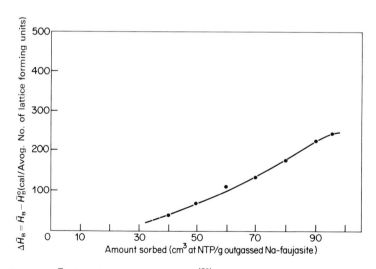

Fig. 4. $\Delta\bar{H}_B$ for CF_4 in Na-faujasite[60] as a function of CF_4 uptake.

where P_0 is the pressure when $X_A, X_B, X_C, \ldots$ are all zero and P is this pressure when sorption equilibrium is established. The fugacity in Eqn (17) is replaced by activity in Eqn (25). The isotherms of each component in the mixture are then required to evaluate $\Delta\mu_B$. The behaviour of mixtures, for example of NaCl and H_2O, is of particular interest for porous tectosilicates of the sodalite and cancrinite types, because these minerals can, as noted earlier, be synthesized with both water and salts included as lattice stabilizers. In the limited intracrystalline pore space the guest molecules compete with each other for the available room so that a displacement equilibrium occurs according to the relation

$$NaCl_o + \nu H_2O_i \rightleftarrows NaCl_i + \nu H_2O_o \qquad (26)$$

where the subscripts o and i denote outside and inside the crystals respectively and ν denotes the number of water molecules displaced from the crystal by each NaCl that enters it. The thermodynamic equilibrium constant K for this competitive distribution is

$$K = \left(\frac{a_i}{a_o}\right)_{NaCl} \left(\frac{a_o}{a_i}\right)_{H_2O}^{\nu} = K_{NaCl}/K_{H_2O}^{\nu} \qquad (27)$$

The overall distribution of Eqn (26) is thus expressible in terms of two distribution constants: K_{NaCl} for NaCl between crystals and solution, and K_{H_2O} for water between crystals and solution. The distribution of salts has been successfully treated in terms of a Donnan membrane equilibrium,[64] but this and a treatment in terms of detailed balancing will be considered in Chapter 7 when salt–zeolite systems are discussed more fully.

Equation (25) shows that $\Delta\mu_B$ when mixtures of guest molecules are imbibed is determined by the sum of terms, one for each guest. While $\Delta\mu_B$ has not yet been evaluated for salt–water mixtures it has been obtained for the sorption of benzene + n-heptane mixtures in the layer silicate montmorillonite made porous by separating the aluminosilicate layers with the large tetramethylammonium ions as the interlayer cations.[60] The total pressure P is $p_{C_6H_6} + p_{n-C_7H_{16}} = p_A + p_B$. The activities a_A and a_B in Eqn (25) can be replaced by the partial pressures p_A and p_B with minimum error. As usual the term $V_M P$ is negligible ($P_o = 0$ when sorption is zero, i.e. in absence of guests with host crystals in a vacuum). The lattice-forming unit of the montmorillonite was taken as the anhydrous unit cell content. The results are shown in Fig. 5 as plots of $a_B/a_B^{\ominus}$ against the mole fraction, X_B of the sorbent. $a_B^{\ominus}$ refers to the starting composition in which $X_B = 0.773$. The three curves in the figure were obtained for three different mole fractions of benzene in the initial benzene + n-heptane mixtures. Again a considerable decrease in $a_B/a_B^{\ominus}$ occurs as the loading of crystals by guest molecules proceeds.

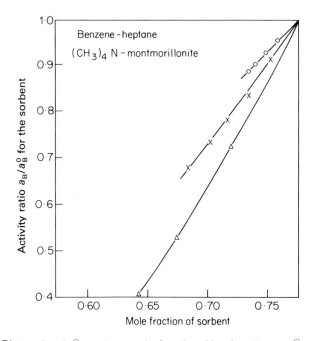

Fig. 5. Plots of $a_B/a_B^\ominus$ against mole fraction X_B of sorbent. $a_B^\ominus$ refers to the starting composition in which $X_B = 0.773$.[60]

The interest in the thermodynamic treatment of the stabilization of porous crystals by uptake of guest molecules lies in its generality, and in the understanding it provides of the conditions necessary to form porous host crystals, whether these are zeolites, smectites, porous felspathoids or clathrates. So far as zeolites are concerned the water can be regarded as a space catalyst in that its presence is essential, through lattice stabilization, to bring about synthesis of the porous tectosilicate frameworks, but once these are formed the water can be removed unchanged and without changing the framework topology.

5. Water in Melts: Pneumatolytic Conditions

Water or other volatile components can act as mineralizers under penumatolytic as well as hydrothermal conditions. In pneumatolysis one has reaction occurring under both high pressures and high temperatures. Burnham and co-workers[65,66] have determined isotherms between water vapour and melts of several compositions. They used a variant of the lattice-forming unit of the previous section in interpreting the results. Thus, in a melt of composition $NaAlSi_3O_8$ the "mole" of silicate was taken as the gram formula weight of $NaAlSi_3O_8$. Whatever the composition of the melt the "mole" was always such as to contain one gram atom of the cation, Na^+ in the above example.

That dissolved water is not primarily in the molecular form but is chemisorbed throughout the melt with dissociation is evidenced by its drastic effect in reducing the viscosity[67] and increasing the electrical conductivity[68] of melts. Evidence that chemisorbed –OH is produced in the reaction comes from infra-red studies of quenched glasses.[69,70] Burnham considered two kinds of reaction to occur, which for the melt $NaAlSi_3O_8$ are:

$$NaAlSi_3O_8 + H_2O \rightarrow AlSi_3O_7OH + NaOH \tag{28}$$

$$AlSi_3O_7OH + nH_2O \rightarrow AlSi_3O_{(7-n)}(OH)_{(2n+1)} \tag{29}$$

In both the water is dissociated to yield chemisorbed –OH groups; in addition caustic soda is formed in Eqn (28). If components of the aqueous fluid phase in equilibrium with components of the melt are denoted by superscript α and those of the melt by superscript β, and if there are n_1^β moles of water incorporated (by reactions (28) and (29)) into each "mole" (i.e. each gram formula weight, $NaAlSi_3O_8$) of original anhydrous melt then

$$F_1^\beta = \frac{n_1^\beta}{1 + n_1^\beta} \tag{30}$$

was taken to be the "mole fraction" of water in the melt. F_1^β is not however the mole fraction of molecular water because little if any free dissolved water exists: it is converted to AlOH, SiOH and OH^- according to Eqns (28) and (29). $F_1^\beta = 0.5$ when $n_1^\beta = 1$. Provided reaction (28) is virtually completed before a significant amount of reaction (29) has occurred the stage $F_1^\beta = 0.5$ corresponds with near completion of reaction (28). For this reaction when n_1^β moles of water ($n_1^\beta < 1$) are incorporated into one mole of the originally anhydrous melt this melt contains

$(1 - n_1^\beta)$ moles $NaAlSiO_8$
n_1^β moles $NaOH$
n_1^β moles $AlSi_3O_7OH$

i.e. an unaltered total of $(1 + n_1^\beta)$ moles of components. Thus F_1^β is unchanged by reaction (28). If, in addition to reaction (28), Δn^β moles of $AlSi_3O_7(OH)$ have reacted with $n\,\Delta n^\beta$ of the n_1^β moles of water according to reaction (29), the melt then contains

$(1 - n_1^\beta + n \, \Delta n^\beta)$ moles $NaAlSi_3O_8$

$\quad n_1^\beta - n \, \Delta n^\beta$ moles $NaOH$

$\quad n_1^\beta - (n + 1) \, \Delta n^\beta$ moles $AlSi_3O_7OH$

$\quad \Delta n^\beta$ moles $AlSi_3O_{6-n}(OH)_{2n+1}$

giving a total of $1 + n_1^\beta - n \, \Delta n^\beta$ moles. Thus $F_1^\beta = n_1^\beta / (1 + n_1^\beta - n \, \Delta n^\beta)$ and so its value has altered.

The equilibrium isotherms for water at 1100°C are shown in Fig. 6[65] for each of a series of melts, the "mole" in each case again being the gram formula weight of melt containing a gram atom of Na^+. For the several melts of Fig. 6 the masses of rock melts taken as equivalent to a gram formula weight of $NaAlSi_3O_8$ are given in Table 4. With these choices of mole the close coincidence of all isotherms is apparent. On the other hand, in a melt of anhydrous composition $NaAlSiO_4$ the isotherm for water uptake agreed with the curve in Fig. 6 only up to about 2 kbar.[65]

The water uptake isotherms of Fig. 6 are all of Type 1 in the Brunauer classification.[73] They represent dissociative chemisorptions of water

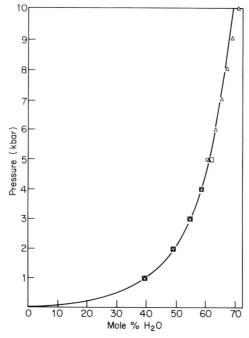

Fig. 6. Equilibrium isotherms for water uptake in melts.[65]
—, Albite; △, Li-pegmatite; ○, andesite; □, basalt. $T = 1100°C$

TABLE 4

Masses of rock melts equivalent to a gram formula weight of $NaAlSi_3O_8$[65]

Melt	Equivalent masses	
	$F_1^\beta < 0.5$	$F_1^\beta > 0.5$
Columbia River basalt[71]	346	290
Mt Hood andesite[71]	273	270
Harding pegmatite[72]	265	263

throughout the melt and it is of interest to consider them further in this light. For reaction (28) the chemisorption sites are Na^+ and $[-\overset{\ominus}{Al}-O-Si<]$:

$$H_2O + Na^+ + \left[\begin{array}{c} \overset{\ominus}{} \\ -Al-O-Si- \\ \end{array} \right] \rightleftharpoons Na^+OH^- + \left[\begin{array}{c} HO \\ -Al \qquad Si- \\ \end{array} \right] \quad (31)$$

$$\underset{1-\theta_1}{} \qquad \underset{1-\theta_1}{} \qquad \underset{\theta_1}{} \qquad \underset{\theta_1}{}$$

In Eqn (31) θ_1 denotes the fraction of complete conversion of the equal numbers of Na^+ and $-Al-O-Si<$ sites to Na^+OH^- and $[-Al\ HO-Si<]$, also produced in equal numbers. Balancing rates of chemisorption and desorption gives for the ideal equilibrium constant and isotherm

$$K_1 = \theta_1^2 / f(1-\theta_1)^2 \quad (32)$$

where f is the equilibrium fugacity of water. For reactions (28) or (31) without any reaction (29), $F_1^\beta = 0.5$ corresponds with $\theta_1 = 1$, i.e. $\theta_1 = 2F_1^\beta$. Equation (32) indicates that f should be proportional to $(F_1^\beta)^2$ only for small values of F_1^β so that it does not fully accord with the apparent proportionality between f and $(F_1^\beta)^2$ up to values of F_1^β approaching 0.5, as reported by Burnham.[65]

For reaction (29) the dissociative chemisorption involves at least $>Si-O-Si<$ sites and the sites $>Al-O-Si<$ with trivalent Al generated in reaction (31). The corresponding equilibria for reaction (29) are

$$H_2O + >Si-O-Si< \rightleftharpoons -SiOH + -SiOH$$
$$\underset{1-\theta_2}{} \qquad \underset{\theta_2}{} \qquad \underset{\theta_2}{}$$

$$H_2O + >Al-O-Si< \rightleftharpoons >AlOH + -SiOH \qquad (33)$$
$$\underset{1-\theta_3}{} \qquad \underset{\theta_3}{} \qquad \underset{\theta_3}{}$$

where θ_2 and θ_3 are the fractions of complete conversion of the two kinds of site to $2[\geqslant\!SiOH]$ and $>\!AlOH+\geqslant\!SiOH$ respectively. Reactions (33) correspond with $n = 1$ in reaction (29). Again by balancing rates of sorption and desorption one obtains

$$\left.\begin{array}{l} K_2 = \theta_2^2/f(1-\theta_2) \\ K_3 = \theta_3^2/f(1-\theta_3) \end{array}\right\} \tag{34}$$

Equations (32) and (34) correct for non-ideality of water in the aqueous fluid but do not allow for non-ideality of the melt solution. This correction could be made by introducing the equivalent of a quotient of activity coefficients into Eqns (32) and (34), there being one such coefficient for each species in Eqns (31) and (33), excluding water. This quotient on present experimental information can only be an extra empirical coefficient and hence it is the idealized eqns (32) and (34) which are considered here.

Information regarding K_2 was obtained by Moulson and Roberts[70] who used infra-red absorption to study the penetration and chemisorption of water in silica glass. θ_2 was small and the uptake of water varied as the square root of the pressure in accord with Eqn (34) for $\theta_2 \ll 1$, confirming the dissociative mechanism of reaction (33) between water and $\geqslant\!Si\text{–}O\text{–}Si\leqslant$. The pressure was varied up to about one atm and at the high experimental temperatures is an adequate measure of fugacity. At a pressure of super-critical water fluid of 70 cm Hg the numbers of OH groups per Si, i.e. θ_2 in Eqn (34), were $\sim\!6\times10^{-3}$ at 600°C and $\sim\!3\times10^{-3}$ at 1200°C. The corresponding values of K_2 are $\sim\!39\times10^{-6}$ and $\sim\!9.8\times10^{-6}$ atm respectively and the heat of the first of reactions (33) in silica glass is about -25 kJ mol^{-1}. For sorption the diffusion coefficient of chemisorbed water between the same two temperatures was given in cm^2 s^{-1} by

$$D = (1\cdot0\pm0\cdot2)\times10^{-6}\exp\left[-(76\,500\pm2100)/RT\right]$$

where the activation energy is in kJ mol^{-1}. The small values of K_2 may support the view that the extent of reaction (29) even with $n = 1$ will be limited, although nothing decisive can be said about the relative values of K_2 and K_3 of Eqns (34). When the fraction θ_2 is calculated for silica glass from the first of Eqns (34) the results are given in Table 5. It is apparent that except at very great pressures the equilibrium of the first of reactions (33) will lie far to the left.

For the case where θ_1, θ_2 and θ_3 are all much less than unity there exist the limiting relations

$$f^{1/2} = \frac{\theta_1}{K_1^{1/2}} = \frac{\theta_2}{K_2^{1/2}} = \frac{\theta_3}{K_3^{1/2}} \tag{35}$$

TABLE 5
θ_2 from Eqn (34) with $K_2 = 39 \times 10^{-6}$ atm^{-1} at
600°C and 9.8×10^{-6} atm^{-1} at 1200°C

Fugacity of water (atm)	θ_2 at 600°C	θ_2 at 1200°C
1	0·0062	0·0031
10	0·0195	0·0098
50	0·043	0·022
100	0·061	0·031
500	0·130	0·068
1000	0·179	0·094
10000	0 46	0·27

Also in a dry melt of composition $NaAlSi_3O_8$ which obeys the $\overset{\ominus}{Al}$–O–$\overset{\ominus}{Al}$ avoidance rule (see Chapter 6) there are three times as many $\geqslant$Si–O–Si$\leqslant$ as of $\geqslant\overset{\ominus}{Al}$–O–Si$\leqslant$ sites while the number of $\geqslant$Al–O–Si$\leqslant$ sites generated by reaction (31) is θ_1 times the initial number of $\geqslant\overset{\ominus}{Al}$–O–Si$\leqslant$. Accordingly θ, the fraction of the maximum possible number of –OH groups which would correspond with $\theta_1 = 1$, $\theta_2 = 1$, $\theta_3 = 1$, is

$$\theta = (\theta_1 + 6\theta_2 + 2\theta_1\theta_3)/9 \tag{36}$$

For large amounts of water the conditions change from pneumatolytic to hydrothermal and one may envisage further reactions such as

$$\tag{37}$$

and other analogous processes, in which the parent aluminosilicate network of the melt or glass is progressively broken with final release of $Al(OH)_3$ and $Si(OH)_4$. The alkali formed by reaction (31) and the fragments released from the network or the partially disintegrated network, are then all available for nucleation and growth of new crystalline phases more stable than the parent melt or glass under the existing, now hydrothermal, conditions.

The extent to which reversible and sometimes irreversible re-arrangements of network atoms is possible, presumably through such processes as are envisaged in reactions (31), (33) and (37), is illustrated by the experiments of O'Neil and Taylor.[74] They heated alkali and alkaline earth felspars with 2 to 3 M aqueous solutions of chlorides at a pressure of 1 kbar. The water was suitably enriched with ^{18}O isotope prior to the treatment of the felspars. It was found that ordered distributions of the Al and Si on lattice sites became disordered even at 350°C; that oxygen isotope partition between crystals and solution was essentially complete; and that cation exchange occurred freely under the conditions used and was nearly complete up to a rather sharp boundary between the ex-changed part and any unexchanged core of the crystals. Although the morphology remained unchanged the ion-exchanged crystals were poorly crystalline and mechanically weak with cleavage cracks into which both cations and anions (Cl$^-$) had penetrated. Occasionally in the high-temperature experiments new euhedral crystals appeared suggesting some solution and re-precipitation. Thus, under hydrothermal conditions, ex-tensive reshuffling of lattice components, including Al, Si and O, can occur while still preserving a defect-rich felspar as the final product.

The ability of water under pneumatolytic and hydrothermal conditions

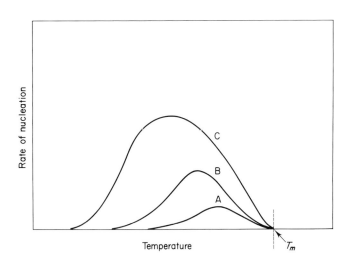

Fig. 7. Rate of nucleation as a function of degree of undercooling and hence also of viscosity,[75] for three magmas, differing only in water content. C refers to the least viscous, most hydrous magma.

to enter aluminosilicate melts, glasses and crystals reversibly as hydroxyl defects, at least transiently and in smaller or greater amounts, must clearly, by breaking Al–O–Si and Si–O–Si bonds, lower the viscosity of melts and glasses[67] and in this way also promote chemical reactivity and change. The rate of nucleation of a new phase from a melt increases with the degree of undercooling of the melt below the melting point of the new crystalline phase. However, as the temperature is lowered by undercooling, the viscosity increases and this results in slower nucleation and crystal growth. Thus the two factors act in opposition, as illustrated formally in Fig. 7.[75] By increasing the water content of the melt its viscosity may be decreased and so rates of nucleation and crystal growth change as illustrated. Curve C represents the behaviour of the most highly hydrated, and so least viscous, melt and A represents that of the least hydrated most viscous melt.

6. Solubility in Water

In §5 the solubility of water in melts was considered. The converse aspect of this is the solubility of various substances in water over the wide temperature and pressure ranges which can be involved in hydrothermal processes. Water is a good solvent for ionic and polar compounds partly because of its high dielectric constant (relative permittivity), which is ~ 78 at 25°C, and partly through such specific factors as hydrogen bonding and chemical reactivity. Other liquids also have high dielectric constants, including glycol (41), glycerol (43), formic acid (57), formamide (84), hydrogen peroxide (93) and hydrocyanic acid (95). For ionic compounds the high dielectric constant of the solvent reduces the coulombic force between cations and anions and so promotes solution. Many hydroxides, carbonates and sulphides are among the more insoluble compounds which may form in hydrothermal reactions. The solubility products, K_{sp}, of a number of these expressed as $(\text{mol dm}^{-3})^n$ are given in Table 6(a), for temperatures around 20 to 25°C. The number of ions formed by the dissociation reaction is denoted by n. Thus for

$$\text{Al(OH)}_3 \rightleftharpoons \text{Al}^{3+} + 3\text{OH}^-$$

n is four. On the assumption of complete ionization of the dissolved species the solubility, S, is expressed as $K_{sp} = S^n$. Thus the concentration of dissolved Al(OH)$_3$ in distilled water is $\sim 1 \cdot 2 \times 10^{-8}$ mol dm^{-3}. There are uncertainties in the numerical values of some solubility products, and also one may query the physical significance of solubility products such as those of HgS, Ag$_2$S or Cu$_2$S because of their very small values. In the

TABLE 6(a)
Examples of solubility products[a] in the range 20–25°C[75a]

Hydroxides		Carbonates		Sulphides	
Compound	Solubility product $(\text{mol dm}^{-3})^n$	Compound	Solubility product $(\text{mol dm}^{-3})^n$	Compound	Solubility product $(\text{mol dm}^{-3})^n$
$Fe(OH)_3$	$6{\cdot}0 \times 10^{-38}$	$PbCO_3$	$1{\cdot}5 \times 10^{-13}$	HgS	$1{\cdot}6 \times 10^{-54}$
$Al(OH)_3$	$5{\cdot}0 \times 10^{-33}$	$SrCO_3$	$7{\cdot}0 \times 10^{-10}$	Ag_2S	$5{\cdot}5 \times 10^{-51}$
$Cr(OH)_3$	$6{\cdot}7 \times 10^{-31}$	$BaCO_3$	$1{\cdot}6 \times 10^{-9}$	Cu_2S	$1{\cdot}2 \times 10^{-49}$
$Cu(OH)_2$	$1{\cdot}6 \times 10^{-19}$	$CaCO_3$	$4{\cdot}7 \times 10^{-9}$	CuS	$8{\cdot}0 \times 10^{-37}$
$Zn(OH)_2$	$4{\cdot}5 \times 10^{-17}$	$MgCO_3$	$4{\cdot}0 \times 10^{-5}$	Fe_2S_3	$1{\cdot}0 \times 10^{-88}$
$Ni(OH)_2$	$1{\cdot}6 \times 10^{-16}$			PbS	$7{\cdot}0 \times 10^{-29}$
$Fe(OH)_2$	$1{\cdot}8 \times 10^{-15}$			CdS	$1{\cdot}0 \times 10^{-28}$
$Pb(OH)_2$	$4{\cdot}2 \times 10^{-15}$			SnS	$1{\cdot}0 \times 10^{-26}$
$Cd(OH)_2$	$2{\cdot}0 \times 10^{-14}$			ZnS	$1{\cdot}6 \times 10^{-23}$
$Mn(OH)_2$	$2{\cdot}0 \times 10^{-13}$			CoS $(\alpha)^b$	$5{\cdot}0 \times 10^{-22}$
$Mg(OH)_2$	$8{\cdot}9 \times 10^{-12}$			NiS $(\alpha)^b$	$3{\cdot}0 \times 10^{-21}$
				FeS	$4{\cdot}0 \times 10^{-19}$
				MnS	$7{\cdot}0 \times 10^{-16}$

^{a} For $M(OH)_3$, $n = 4$; for $M(OH)_2$, $n = 3$; for MCO_3 and MS, $n = 2$; for M_2S, $n = 3$; and for M_2S_3, $n = 5$.
^{b} The most soluble form.

most extreme case, that of HgS, K_{sp} is such that in distilled water there would be one ion only of Hg^{2+} or S^{2-} in $\sim 1{\cdot}3 \times 10^3 \, \text{dm}^3$ of water. However, properly used, solubility products can be useful in various aspects of solution chemistry.[75a] Interfering factors include hydrolysis or complex formation. Also values of K_{sp} will change with temperature and pressure, and with composition for many-component solutions.

In hydrothermal systems, therefore, one may require solubility data at high temperatures, as illustrated in Table 6(b). Both positive and negative temperature coefficients of solubility are recorded. The table includes solubilities reported by Kennedy[80] for quartz and amorphous silica. His results have been re-examined by a number of authors,[87,88,89,90] both under autogenous pressures along the liquid–vapour–solid line up to the critical temperature and in absence of vapour at a fixed hydrostatic pressure. Above the critical temperature of water the density of the supercritical fluid, and hence the solubility of solids in it, depend on the initial degree of filling of the autoclave.

The solubilities of amorphous silica,[91] cristobalite[84] and quartz[80,88] along the three-phase line under autogenous pressure are shown in Fig. 8.

TABLE 6(b)
Solubilities in water[76] (g/100 g)

Substance	100°C	150°C	200°C	250°C	300°C	350°C	374°C	Refs.
ZnS	–	–	5×10^{-5}	–	1×10^{-4}	–	–	(77, 78)
AgI	–	–	–	–	1.9×10^{-3}	5.4×10^{-3}	–	(79)
AgBr	3.7×10^{-4}	$\sim1.9\times10^{-3}$	6.0×10^{-3}	1.1×10^{-2}	1.5×10^{-2}	2.5×10^{-2}	–	(79)
SiO$_2$ (quartz)	–	3.0×10^{-3}	2.4×10^{-2}	4.9×10^{-2}	6.8×10^{-2}	7.0×10^{-2}	2.3×10^{-2}	(80)
AgCl	2.2×10^{-3}	6.4×10^{-3}	1.3×10^{-2}	2.5×10^{-2}	3.1×10^{-2}	5.0×10^{-2}	–	(79)
CaCO$_3$ (1 atm of CO$_2$)	2.2×10^{-2}	9.2×10^{-3}	3.9×10^{-3}	1.5×10^{-3}	6.8×10^{-4}	–	–	(81)
CuI	–	–	12×10^{-2}	4.4×10^{-2}	9.3×10^{-2}	1.9×10^{-1}	–	(77)
SiO$_2$ (amorphous)	3.8×10^{-2}	6.2×10^{-2}	9×10^{-2}	1.26×10^{-1}	1.68×10^{-1}	2.3×10^{-1}	–	(80)
CuBr	–	–	2.2×10^{-1}	5.9×10^{-1}	1.18	1.8	–	(79)
CuCl	–	3.6×10^{-1}	9.5×10^{-1}	1.7	3.5	6.2	–	(79)
NaPO$_3$	6.05×10^{-1}	7.05×10^{-1}	7.5×10^{-1}	7.9×10^{-1}	8.4×10^{-1}	9.2×10^{-1}	9.25×10^{-1}	(82)
La$_2$(SO$_4$)$_3$	9×10^{-1}	$<1\times10^{-1}$	1×10^{-1}	–	–	–	–	(83)
Y$_2$(SO$_4$)$_3$	5.1	2.4	–	–	–	–	–	(83)
ZnSO$_4$	–	–	26.5	10.0	<0.1	–	–	(83)
B$_2$O$_3$	15.6	34.0	77.0	83.5	87.4	91.5	–	(84)
Na$_2$SO$_4$	29.4	29.7	30.7	30.6	19.9	2.3	$<2\times10^{-1}$	(85, 86)
Na$_2$CO$_3$	30.6	28.4	23.3	17.2	8.4	1.9	$<2\times10^{-2}$	(85, 86)
KCl	36.0	40.2	44.5	48.4	53.5	58.2	–	(85, 86)
NaCl	–	–	31.3	34.0	37.4	42.0	44.3	(85, 86)
CdSO$_4$	41	19	1.5	–	–	–	–	(83)
NaBr	54.8	57.2	60.0	63.0	66.1	69.8	–	(85, 88)
Cs$_2$SO$_4$	68.2	71	73.5	74.7	75.5	–	–	(83)
NaI	–	78.3	81.7	84.7	87.4	89.7	90.7	(85, 86)

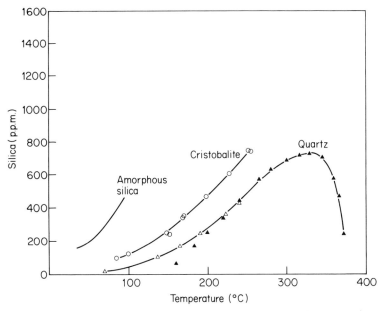

Fig. 8. Solubilities of amorphous silica, cristobalite and quartz along the three-phase line under autogenous pressure.[89]

As expected, the solubility decreases as the free energy of the solid decreases. Figure 9 compares the solubility of quartz along the solid–liquid two-phase line at 1000 atm with that along the three-phase line at autogenous pressure. Anderson and Burnham[90] extended the measurements for quartz into the supercritical region (500–900°C) at a series of constant pressures in the range 1000–10 000 bar. The solubility at constant pressure increases with temperature and at constant temperature increases with pressure.

Below the critical temperature the reaction which dissolves SiO_2 is considered to be

$$SiO_2(s) + 2H_2O(l) \rightleftharpoons Si(OH)_4 \, aq \qquad (38)$$

so that, in terms of activities, a, the equilibrium constant is

$$K = a_{Si(OH)_4}/a_{H_2O}^2 a_{SiO_2} \qquad (39)$$

Standard states for pure liquid water and for quartz could be either the experimental temperature at 1 atm pressure, or 298 K at 1 atm. In the latter case for a pure substance, if the pressure changes from P_0 to P and the temperature from T_0 to T, then the activity changes from a_0 to a

where

$$RT \ln a/a_0 = -\int_{T_0}^{T} S \, dT + \int_{P_0}^{P} V \, dP \qquad (40)$$

where S and V are molar entropy and volume of the substance. When $P_0 = 1$ atm and $T = 298$ K then by definition $a_0 = 1$. Also for a condensed phase like quartz or even liquid water, V will not change very much with pressure and so

$$RT \ln a \simeq -\int_{T_0}^{T} S \, dT + V(P-1) \qquad (41)$$

The integral can in principle be evaluated graphically from plots of S against T.

If the standard state is chosen to be the experimental temperature and 1 atm the integral is not involved and

$$RT \ln a \simeq V(P-1) \qquad (42)$$

For most purposes the term on the r.h.s. of Eqn (42) is small and then

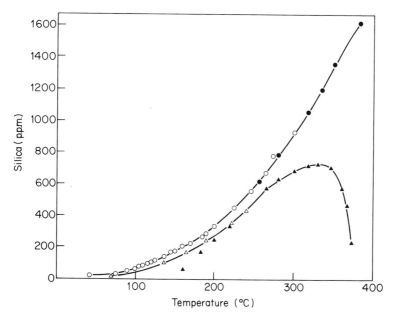

Fig. 9. Solubility of quartz in water as a function of temperature. Upper curve gives the solubility at 1000 atm; lower curve is the solubility along the three-phase line (vapour, liquid water, quartz) at autogenous pressure.[88]

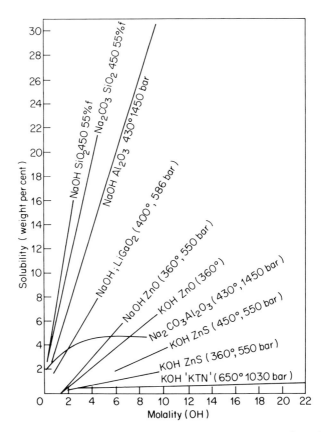

Fig. 10. Solubilities of various refractory compounds as functions of alkali concentration.[94]

$a \sim 1$. Also the solution of $Si(OH)_4$ when reaction (38) occurs is so dilute that $a_{Si(OH)_4} \sim C_{Si(OH)_4}$ where $C_{Si(OH)_4}$ is the concentration (solubility) of dissolved silica (as $Si(OH)_4$), and the water will have virtually the same activity as pure water. Under these conditions, and especially near room temperature

$$K \simeq C_{Si(OH)_4} \qquad (39a)$$

The standard free energy, $\Delta G^{\ominus}$, and enthalpy, $\Delta H^{\ominus}$, for the reaction of Eqn (38) are given in Table 7 at 25°C for quartz and cristobalite. These values give for the free energy of transforming cristobalite to quartz at 25°C one value[89] of $(-3 \cdot 76 \pm 1 \cdot 25) \, \text{kJ mol}^{-1}$ and a second of $\sim -2 \cdot 30 \, \text{kJ mol}^{-1}$. The free energies of formation of cristobalite and quartz from their elements have been reported[89] as $-821\cdot71 \pm 1\cdot25$ kJ mol^{-1} and $-824\cdot09 \pm 1\cdot88$ kJ mol^{-1} so that the free energy of conversion of cristobalite to quartz based on these figures is $-2\cdot38 \pm 1\cdot13$ kJ mol^{-1}.

At constant pressures of 1 atm and 1000 atm the solubilities of quartz at 25°C were estimated as 6 and 12 p.p.m.[88] If the ratio of these concentrations is taken as the ratio of equilibrium constants at the two pressures then from Eqn (8) we may write, in units cm^3 atm,

$$-RT \ln[C_{1000}/C_1] = \int_{1\,\text{atm}}^{1000\,\text{atm}} \Delta V^\ominus \, dP \simeq 999\,\Delta V^\ominus \qquad (43)$$

and thus obtain, for the standard volume change of reaction (38), $\Delta V^\ominus \sim -17$ cm^3.

So far the solubilities of several forms of silica have all referred to distilled water as solvent. When one moves to the alkaline side solubility increases very strongly[92,93] (Table 8). The results indicate an increase in solubility with temperature which is somewhat more rapid with Na_2CO_3 aq than with NaOH aq. This increase is the behaviour required for hydrothermal growth of quartz in a temperature gradient (§8). The greatly increased solubilities show the value of OH$^-$ as a mineralizer; the silica reactants dissolve readily and are therefore freely transportable to growing crystals. The enhanced solubility in alkaline solutions is shared also by other amphoteric oxides such as Al_2O_3, B_2O_3, Ga_2O_3, GeO_2 and ZnO. Some solubilities in alkali are shown in Fig. 10, based on work by Laudise and co-workers.[94] When the amounts dissolved are plotted against the initial concentration of OH$^-$, except for Al_2O_3 in Na_2CO_3, straight lines are obtained. Again alkali is an excellent means of bringing these oxides into solution and so allowing them to mix and react with silica to form

TABLE 7

Solubility of cristobalite and quartz: thermodynamic data

Crystal	Conditions	$\Delta G^\ominus$ for solution at 25°C (kJ mol^{-1})	$\Delta H^\ominus$ for solution^a (kJ mol^{-1})
Cristobalite	Autogenous pressure	$19\cdot0_6$[89]	$19\cdot1_4$[89]
Quartz	Autogenous pressure	(i) $22\cdot8_2$[89,88]	(i) $25\cdot08 \pm 0\cdot63$[89,88]
		(ii) $21\cdot3_6$[87]	(ii) $22\cdot2_4$[87]
	At 1000 atm		$22\cdot49 \pm 0\cdot63$[88]

^{a} $\Delta H^\ominus$ is within error independent of temperature.

TABLE 8
Solubility of quartz (g dm^{-3}) in NaOH and Na$_2$CO$_3$ solutions[92,93]

Temperature (°C)	Na$_2$O wt%	Solubility of quartz in	
		NaOH solutions	Na$_2$CO$_3$ solutions
300	1	20	5
	5	135	28
	15	514	48
350	1	25	13
	5	152	50
	15	515	60
400	1	35	22
	5	155	90
	15	560	105
450	1	35	24
	5	163	124
	15	560	285

aluminosilicates, borosilicates, aluminogermanates, gallosilicates and gal-logermanates.

The solution of an oxide MO in alkali can be written as[94]

$$MO(s) + n\,OH^-(aq) \rightleftarrows MO(OH)_n^{n-}(aq) \rightleftarrows MO(O)_{n/2}^{n-}(aq) + \left(\frac{n}{2}\right)H_2O \tag{44}$$

Activities of dissolved species are not known, but in terms of the mass action quotient, K_m, and with the water in large excess so that its concentration is considered constant, one has

$$K_m = C/C_f^n \tag{45}$$

where C_f is the final concentration of OH^- and C that of the dissolved species. In terms of the initial concentration, C_i, of OH^- one obtains

$$C_f = (C_i - nC) \tag{46}$$

If the equilibrium of Eqn (44) lies well to the right, C_f is small and so

$$C \simeq C_i/n \tag{47}$$

but if the equilibrium is well to the left $C_f \sim C_i$ and so

$$C \simeq K_m C_i^n \tag{48}$$

Thus for the reaction (44) the proportionality between C and C_i (Fig. 10) can be explained either by Eqn (47), or by Eqn (48) with $n = 1$. In the case of Eqn (47), $C/C_i \sim n^{-1}$. The above line of reasoning led to the following tentative equations for solution reactions:

$$3SiO_2 + 2OH^- \rightleftarrows Si_3O_7^{2-} + H_2O \text{ (for NaOH)}$$
$$SiO_2 + 2OH^- \rightleftarrows SiO_3^{2-} + H_2O \text{ (for Na}_2CO_3\text{)}$$
$$Al_2O_3 + 2OH^- \rightleftarrows 2AlO_2^- + H_2O \text{ (for NaOH)}$$
$$ZnO + OH^- \rightleftarrows ZnO.OH^- \rightarrow \tfrac{1}{2}Zn_2O_3^{2-} + \tfrac{1}{2}H_2O$$
$$ZnS + OH^- \rightleftarrows ZnS.OH^-$$

or

$$ZnS + 2OH^- \rightleftarrows \tfrac{1}{2}Zn_2O_3^{2-} + \tfrac{1}{2}H_2O + HS^-$$
$$LiGaO_2 + 2OH^- \rightleftarrows GaO_3^{3-} + Li^+ + H_2O$$
$$KTa_{0.65}Nb_{0.35}O_3 + 2OH^- \rightleftarrows Ta_{0.65}Nb_{0.35}O_4^{3-} + K^+ + H_2O$$

For the last two of these equations and for the reaction involving Al_2O_3 the general form of their mass action quotients is

$$K_m = C^2/C_f^2 \tag{45a}$$

again justifying the linear relations of Fig. 10 for the conditions leading to Eqn (47), or with Eqn (48) replaced, as now required, by

$$C^2 \simeq K_m C_i^2 \tag{48a}$$

When the vapour pressure above saturated salt solutions is plotted as a function of temperature for more soluble salts, including some of those in Table 6(b), the results shown in Fig. 11[76] are obtained, in which the P–T curves pass through maxima. The reason is that, on cooling the salt-

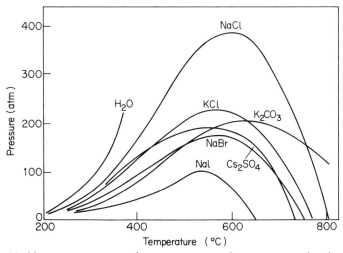

Fig. 11. Vapour pressure of water over various saturated salt solutions.[76]

saturated magmas from the highest temperatures, anhydrous salts crystallize, progressively leaving a residual solution richer and richer in water
with a resultant increase in fugacity of the water (cf. §1). However, when
the temperature is lowered sufficiently the effect of temperature on
fugacity becomes the dominant influence so that as temperatures fall still
more the pressure passes through a maximum and thereafter declines,
being always less than that of pure water. This latter aspect accords with
the lowering of vapour pressure of water by solutes.

The behaviour shown in Fig. 11 merits closer consideration, since it can
be of importance in hydrothermal systems.[95] Figure 12(a) shows projections on planes of the relations between temperature and composition
and between pressure and temperature for a two-component system of A
and B. The triple points A_t and B_t of the two components are joined by
fusion curves, or by curves of co-existing solid, liquid and vapour (A+L+
V or B+L+V), which pass through the eutectic point E. Likewise the
two critical temperatures A_c and B_c are joined by the critical or plait
curve, along which critical phenomena are shown by the various mixtures
of A and B. If decomposition, limited miscibility or transformation occurs
in the system parts of all these curves may lie in a metastable region of
the figure. Since the critical curves lie above the fusion curves and the
curves of co-existent liquid, solid and vapour, only solutions unsaturated
by A and B can exhibit critical phenomena.

It is thought to be more common in hydrothermal systems to have
intersection of fusion curves and critical curves as shown in Fig. 12(b).
Two intersections at P and Q must occur because in mixtures of compositions approaching either pure component the lines A+L+V or B+L+V
must both lie below the critical curve. Unlike the situation in Fig. 12(a),
at the two intersections critical phenomena are shown by solutions in
contact with the solid phase (i.e. by saturated solutions) while between P
and Q critical phenomena can arise only in supersaturated (i.e. metastable) solutions. Whereas to the left of P and the right of Q one may have
solid in equilibrium with both liquid and vapour (the vapour–liquid
equilibria are shown by the tie-lines), between P and Q one can have only
fluid solution and solid. This can influence choice of experimental conditions. At saturation equilibrium and constant temperature the amounts of
water and solid can be varied within limits without altering the vapour
pressure or the composition of either the liquid or its vapour. One
changes only the relative amounts of solid, saturated solution and vapour.
On the other hand, between the critical end points P and Q one can have
only two phases, which are fluid solution and solid. In this divariant
system if the amount of water present is changed so will the pressure and
density of the fluid also change and therefore the concentrations of

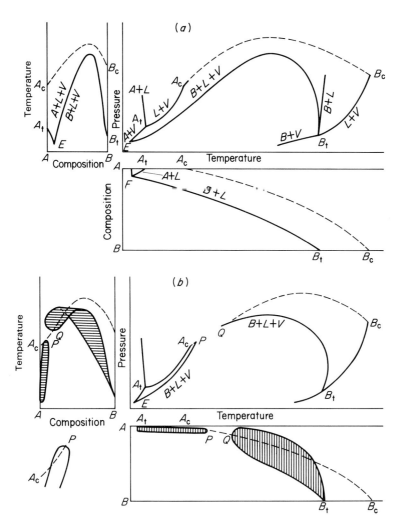

Fig. 12. (a) Projections on temperature–composition and temperature–pressure planes of a solid model expressing the phase equilibrium relations in a binary system of two components A and B. The critical or plait curve (the dashed lines) is not intersected by the solubility curve. A_t, B_t and A_c, B_c denote respectively the triple and critical points for the two components. E denotes the binary eutectic and L and V denote liquid and vapour. (b) In this case the critical curve is intersected by the solubility curve at points P and Q. The tie-lines connect co-existing vapour and liquid compositions.[95]

dissolved solids. More possibilities of obtaining different minerals from the same reactants arise, but exact reproduction of the conditions will be more difficult.

The occurrence of situations in which critical and fusion curves intersect is thought to be helped by two conditions: the volatility of the two components should be very different; and the solubility of the less volatile component in the more volatile component should be small near the critical point of the more volatile one. These conditions for pure water and a silicate or silica (e.g. water and quartz) are met and so the behaviour shown in Fig. 12(b) should be common. The second critical end-point, Q, may, however, involve very high pressures.

When a solute is present the critical temperature of the solvent is raised in a manner analogous to the elevation of the boiling point. The elevation is proportional to the mole fraction of the solute. Thus at the critical temperature of water if the solubility in water is low the saturated solution at P will show critical phenomena at a temperature and pressure near those of pure water.

As already noted, pressures developed over pure water or aqueous solutions in an autoclave may depend upon the degree of filling. The greater the initial fill the lower the temperature at which the liquid, by expanding, fills the autoclave so that the pressure rises thereafter to very large values. In autoclaves with >32% initial fill the liquid–gas meniscus moves up with rising temperature and the autoclave becomes fully occupied by liquid below the critical temperature (~374°C for pure water). If the autoclave is less than 32% filled the liquid–gas meniscus moves downward with increasing temperature and all liquid gasifies below 374°C. Although this gas may have density and solvent properties not unlike those of conventional liquids, almost all hydrothermal crystallizations have involved water densities above 0.32 g cm^{-3} and the most successful have involved densities above 0.65 g cm^{-3}.[94]

7. Alkalinity of Mineralizing Solutions in Relation to Mineral Types

The observations of §6 regarding solubility of SiO_2 in water at pH ~ 7 as compared with its solubility in alkaline media, and the similar observations for other amphoteric oxides (Fig. 10) have shown that even at super-critical temperatures water alone is not a good enough solvent for ready crystallization of most refractory compounds. Accordingly, reagents are added which complex the solute strongly enough to dissolve more of it, but without making the complex the stable solid phase. For this

TABLE 9

Likely relations between pH, temperature and occurrence of some hydrothermally formed minerals[96]

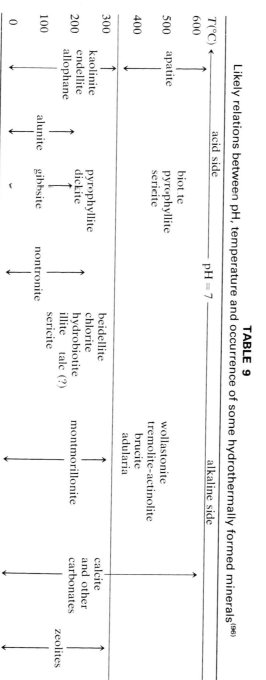

purpose OH$^-$ is an extremely valuable mineralizer. However, this does not mean that minerals cannot be formed under neutral or acid conditions. The presence or absence of alkali can indeed direct the crystallizations towards specific mineral types. Stringham[96] attempted to relate the occurrence of some minerals with the temperature and pH of mineralizing solutions. His conclusions are summarized in Table 9. Although the positions of the minerals both as regards temperature and pH cannot be very accurately defined from the field evidence, the sequence of positions can be considered qualitatively correct. Among aluminosilicates conditions near neutral pH lead only to phyllosilicates, in which Al has a strong preference for 6-co-ordination with oxygen and Si remains in tetrahedral co-ordination. The more acid the medium, at lower temperatures particularly, the fewer the exchangeable cations in the layer silicate. On the mildly alkaline side the layer silicates still persist up to around 350°C, but as pH increases beyond that yielding montmorillonite layer silicates are no longer in evidence, being replaced by tectosilicates such as zeolites in which both Al and Si are in tetrahedral co-ordination with oxygen. It can therefore be expected that, as mentioned in §1, zeolite formation will require both water and excess alkalinity and that temperatures mostly below 350°C will be involved.

8. Examples of Hydrothermal Syntheses and Transformations

Hydrothermal reactions have been studied with various ends in view. They serve to indicate the relative stabilities of minerals in mineral assemblages; to delineate crystallization fields and persistence of minerals under conditions which may or may not represent thermodynamic equilibrium among the species formed; to prepare a given mineral by different pathways; to explore the chemical transformations a given mineral may undergo; to prepare novel mineral-type compounds; and to grow large crystals of materials valuable for their optical, piezo-electric, electric, magnetic, fluorescent or thermal properties. Success has been achieved in all the above directions. Instances of synthesis of novel materials will be given later in connection with zeolites, but several examples of hydrothermal chemistry will be given here to illustrate its richness.

8.1. *Clay Minerals and Artificial Weathering*

The weathering processes indicated in Table 2, and the temperature and pH dependences recorded in Table 9 are on the whole supported by the conditions of synthesis of various clay minerals. Despite the importance of these minerals in soil chemistry and mineralogy, in the ceramic industry, as drilling muds and as catalysts and sorbents, systematic investigations of their syntheses are comparatively recent in the history of hydrothermal reactions. They often crystallize poorly to give very small particles of ill-defined shape so that identification is difficult. Eventually X-ray analysis became a sufficiently refined diagnostic method in the clay mineral area, although ambiguities can still remain, for example with beidellites and montmorillonites. Table 10 records a number of earlier syntheses with the authors' identifications of their products. The acidity of the mineralizing solutions leading to kaolinites is in line with the fact that this mineral has virtually no intracrystalline exchangeable cations. Norton[105,106,109] developed a procedure involving less acid conditions for accelerated weathering in which carbonic acid solutions under high partial pressures of CO_2 (often 500 or 800 p.s.i.) were calculated through a bed of mineral reactant, which was raised above the level of the mother liquor. Temperatures between 250 and 350°C were normally employed and the reactants were felspars such as orthoclase, albite, anorthite; felspathoids like leucite, pollucite and nepheline; zeolites (analcime and scolecite); and natural volcanic glass. Certain of the layer silicates formed were free of exchangeable cations (dickite, pyrophyllite and kaolinite) but others belonged to the smectite or mica-like groups and therefore had a considerable content of exchangeable cations (beidellite, montmorillonite and sericite).

Alternatively to accelerated weathering of alkali-rich minerals by acid solutions, clay minerals can readily be synthesized from aluminosilicate gels and glasses. Numerous syntheses have been made in this way post-dating the examples given in Table 10. In absence of strong alkali such as $(Na, K)OH$, the following hydrous layer silicates have been synthesized:[114]

Kaolinite, $Al_2O_3.2SiO_2.2H_2O$
Hydralsite, $2Al_2O_3.2SiO_2.H_2O$
Pyrophyllite, $Al_2O_3.4SiO_2.H_2O$
Talc $3MgO.4SiO_2.H_2O$
Montmorillonite, variable Mg-Al-Si
Clinochlore $5MgO.Al_2O_3.3SiO_2.4H_2O$
Serpentine $5MgO.Al_2O_3.3SiO_2.4H_2O$.

When a strong alkali is added in regulated amounts to the aluminosilicate gels, layer silicates with exchangeable cations between the siliceous layers appear. In one study[115] at temperatures above 200°C beidellites were reported with exchange capacities ranging between $N/2$ to over $2N$ where N corresponds with the "normal" anhydrous composition

TABLE 10

Earlier hydrothermal syntheses yielding clay minerals

Reactant mixtures	Conditions	Reported products	Ref.
Hydrated gels Al_2O_3, $nSiO_2$ from Na_2SiO_3 and $AlCl_3$ ($n = 3$–6)	5 days at 250°C 5 days at 300°C	Kaolinite	(97, 98, 99)
Potash felspar + 5% HF	1 day at 225°C	Kaolinite or dickite	(100)
Potash felspar + N or N/2 HCl	250 to 115 h at 330° and 320°C	Kaolinite	(101, 102)
Anorthite + N/2 HCl	60 h at 340°C	Kaolinite	(101, 102)
Leucite + N/2 HCl	60 h at 320°C	Kaolinite	(101, 102)
Co-precipitated Al_2O_3, SiO_2 gels + aq	7 and 10 days at 310°C; 11 days at 250°C	Kaolinite	(103)
$Al_2O_3 + SiO_2 + $ aq	10 days at 310°C	Kaolinite	(103)
$Al_2O_3 + SiO_2 + $ aq ($Al_2O_3 : SiO_2 = 1:1$ to $10:1$)	1 day at 300 to 350°C	Kaolinite	(104)
1.5 g potash felspar + 80 cm³ N/2 HCl	115 h at 320°C;	Kaolinite	(11)
	60 h at 400°C	Kaolinite, pyrophyllite	(11)
2 g plagioclase felspar + 80 cm³ N/2 HCl	60 h at 320°C	Kaolinite	(11)
Petalite or spodumene + circulating CO_2 aq	250–325°C	Kaolinite	(105)
Analcime + circulating CO_2 aq	300 and 500°C	Kaolinite, dickite and quartz	(106)
Felspars + HCl aq	300–525°C	Kaolinite below 350°C (with some pyrophyllite)	(107)
$KAlO_2 + SiO_2$ (from SiF_4) + aq (range of compositions prepared)	450°C	Pyrophyllite (hieratite, K-felspar, kaliophilite and opal with some mixtures)	(8)
$Al_2O_3 + SiO_2 + $ aq ($Al_2O_3 : SiO_2 = 1:2$ to $1:10$)	7 days at 400°C	Pyrophyllite	(104, 108)
1·5 g K-felspar + 80 cm³ N/2 HCl	60 h at 500°C, 90 h at 530°C	Pyrophyllite and possible boehmite	(11)
2 g plagioclase + 80 cm³ N/2 HCl	70 h at 470°C	Pyrophyllite and possible kaolinite	(11)
	90 h at 540°C	Pyrophyllite and boehmite	(11)

Starting material	Conditions	Product	References
Anorthite and circulating CO_2 aq	24 days at 275°C	Pyrophyllite	(109)
Felspars + HCl aq	300–525°C	Pyrophyllite, ν. kaolinite up to 350°C	(107)
Gels Al_2O_3.nSiO_2 ($n = 1, 2$ and 4) + aq	7 to 11 days at 345–265°C	Dickite (+boehmite from aluminous gels)	(103)
Analcime + circulating CO_2 aq	300°C	Dickite, kaolinite and quartz	(106)
Natural volcanic glass + circulating CO_2 aq	300°C	Dickite and sericite	(106)
Gels Al_2O_3, SiO_2 + aq	12 days at 350°C, 10 days at 390°C	Beidellite	(103)
Al_2O_3(gel) + SiO_2(glass) + aq	12 days at 350°C	Beidellite	(103)
γ-Al_2O_3 + SiO_2(glass or gel) + aq	12 and 18 days at 350°C, 10 days at 390°C	Beidellite	(103)
Boehmite + SiO_2(glass or gel) + aq	10 days at 390°C	Beidellite	(103)
Kaolinite + aq	14 days at 390°C	Beidellite	(103)
Albite + circulating CO_2 aq	250 and 350°C	Beidellite (up to 20% conversion)	(106, 105)
Pollucite + circulating 0·8 N HCl aq	300°C	Beidellite or montmorillonite (70% conversion)	(106)
Bayerite or hydrargyllite + amorphous SiO_2 (in molar ratio 1:4) + N/10 NaOH	1 day at 300°C	Montmorillonite	(104), (108), (110)
Al_2O_3 gel + SiO_2 (ratio 1:4) w. systematically varied alkali (Na_2O. K_2O. CaO or MgO) + aq	16 to 24 h at 300°C	Montmorillonite (sometimes with kaolinite)	(111, 112)
Petalite + circulating CO_2 aq	300°C	Montmorillonite and quartz	(106)
Gel 1 to 6 Na_2O + Al_2O_3, 4SiO_2 + aq	14 days at 400°C	Montmorillonite	(113)
Orthoclase + circulating CO_2 aq	280–320°C	Sericite	(109)
Leucite + circulating CO_2 aq	300°C	Sericite	(105, 106)
Nepheline + circulating CO_2 aq	250–325°C	Sericite	(105, 106)
Scolecite + circulating CO_2 aq	300°C	Sericite	(105, 106)
Natural volcanic glass + circulating CO_2 aq	300°C	Sericite and dickite	(106)

$Na_{0.33}Al_2[Al_{0.33}Si_{3.67}O_{10}(OH)_2]$, sodium being the exchangeable cation. With increasing temperature only the N compound remained as a single phase up to 425°C. Saponites with exchange capacities N and $2N$ were also prepared below ~550°C. The best concentrations of alkali for forming smectites differed among K, Na, Ca and Mg hydroxides and had the greatest range for $Mg(OH)_2$ (the least basic). In accordance with Table 9, if the alkali content was made too large tectosilicates began to appear. For example, at the low-soda end of the system $Na_2O–Al_2O_3–SiO_2–H_2O$, in addition to the layer silicates sodium montmorillonite and paragonite mica, the tectosilicates albite, nepheline, nepheline hydrate I, and such zeolites as analcime, natrolite, mordenite, hydroxy cancrinite and hydroxy sodalite[116,117] were made. Thus the requirements for synthesis both of kandites and smectites are adequately established. The patent literature also provides syntheses of layer silicates, such as hectorite,[118] $Na_{0.66}[Mg_{5.34}Si_8Li_{0.66}O_{20}(OH)_4]mH_2O$, and related smectites.[119]

Considerable work has been performed in producing clay minerals in which isomorphous replacements are effected by elements not usually found in naturally occurring clay minerals. Boron substitution proved possible to give saponites of composition $Na_x[Mg(B_xSi_{(4-x)}O_{10}(OH)_2]-mH_2O$ where $x = 0.33$ and 0.66, as well as boron phlogopite, $K[Mg_3(BSi_3)O_{10}(OH)_2]$ and boron muscovite, $K[Al_2(BSi_3)O_{10}(OH)_2]$, at temperatures of 350, 710 and 500°C respectively, under high pressures.[120] In other investigations[121,122] nickel-bearing antigorite $(Ni_3(OH)_4Si_2O_5)$, pyrophyllite $(Ni_3Si_4O_{10}(OH)_2)$ and montmorillonite $(Ni_3Si_4O_{10}(OH)_2)mH_2O$ have been reported, as well as cobalt-bearing antigorite, poorly crystallized montmorillonite[123] and manganese-bearing saponite.[124] The conditions for these syntheses were usually mild (about 100°C) with pH ranges from ~5.5 to 9.

In other unusual compositions inorganic cations were replaced by the organic cations, tetraethyl-, tetrapropyl- and tetrabutylammonium.[125] Smectites were prepared hydrothermally from compositions quaternary base $–Al_2O_3–SiO_2–H_2O$ at 210 to 300°C. In addition the organo-smectites were made with Ga_2O_3 in place of Al_2O_3. In the same way at about 100°C hectorite-type smectites were made from compositions quaternary base $–LiF–MgO–SiO_2–H_2O$ in which the bases were $N(CH_3)_4OH$ and $N(C_2H_5)_4OH$. The SiO_2 was also replaced by GeO_2. Evidently smectites can undergo many of the isomorphous substitutions which have been observed with micas.[126]

8.2. Hydrothermal Micas

The main attempts to produce large mica crystals have grown it from melts,[127,128] but small crystals have often been reported under milder conditions in hydrothermal reactions. As already noted[116] paragonite co-crystallized with other phases from gels of oxide composition $Na_2O.Al_2O_3.nSiO_2$ aq. The temperature range was 360–450°C and $1 \leqslant n \leqslant 7$. The best yields of paragonite were at the highest temperatures and smaller values of n and the pH values of the mother liquors after the crystallizations were in the range 7·0–8·3. In a second exploration homogeneous glasses varying in oxide composition from $Na_2O.Al_2O_3.SiO_2$ to $Na_2O.Al_2O_3.6SiO_2$ were crystallized hydrothermally. Various minerals were formed, and paragonite and nepheline hydrate I ($NaAlSiO_4 0·5H_2O$) were two of these which co-crystallized between about 260 and 350°C under pressures from 12 to 45 thousand p.s.i. from glass compositions $NaAlSiO_4$ and water.[129]

Fluor-hydroxyl phlogopite micas were synthesized when gel mixtures having compositions $KMg_3AlSi_3O_{10}(OH)_2$, or these compositions with one OH replaced by F, were treated hydrothermally at temperatures between 500 and 800°C under water pressures from 600 to 1200 bar.[130] The F content in the fluor-hydroxyl forms was found to depend on the pH of the co-existing liquid according to the approximate relation

$$pH = 7·79 - 2·21x$$

and the lattice constant C_0 could be related to x by

$$C_0(F_x) = C_0(OH) - 0·098x$$

In these relations x denotes the number of hydroxyls replaced by F in the above phlogopite composition.[131]

Hydrothermal growth of potassium micas was investigated by Warshaw,[132] whose work will serve as an example of such studies. The system $K_2O–MgO–Al_2O_3–SiO_2$ was investigated at temperatures above 250°C and at pressures above 10 000 p.s.i. Single layer monoclinic (1M) and three-layer trigonal (3T) polytypes of muscovite mica were made; for instance the former from gel of muscovite composition at 495°C and 14 000 p.s.i. for 7 days; the latter at 450°C and 13 500 p.s.i. for 14 days, followed by 8 days at 555°C. Illite-type micas were also prepared in this work, which could be micas with some random interstratification of smectite layers. They tend to be poorer in K and richer in H_2O than muscovites. A muscovite type mica was also observed in an investigation of reactions between kaolin or synthetic faujasite and various phosphates.[133] The mica co-crystallized with kalsilite ($KAlSiO_4$) from compositions 0·4–0·5 g kaolin, 7 g K_3PO_4 and 7 ml H_2O heated to 200°C for two days. The general behaviour was a co-crystallization of aluminophosphates and aluminosilicates. The aluminophosphates could sometimes be separated from the silicates by treating the products with appropriately dilute acid.

An extensive investigation of hydrothermal synthesis of muscovite micas was made by Yoder and Eugster[134] who also summarized examples of earlier syntheses. They worked in the system $K_2O–Al_2O_3–SiO_2–H_2O$ between 200 and 900°C, and at pressures from 2 to 30 thousand p.s.i. In addition to the micas, felspars such as sanidine and microcline were formed, also the felspathoids leucite ($KAlSi_2O_6$), kaliophilite ($KAlSiO_4$), kalsilite ($KAlSiO_4$) and other compounds, especially corundum (Al_2O_3) and occasionally sillimanite (Al_2SiO_5) and mullite (Al_4SiO_8). The muscovites obtained were randomly-stacked one-layer monoclinic (1Md), one-layer monoclinic (1M) and two-layer monoclinic (2M) polytypes. The stability ranges of these polytypes could not be established because transformations were too sluggish, but the transformation 1Md → 1M → 2M was effected. The upper stability limit for mica formation appeared to be near a curve passing through points 625°C, 5000 p.s.i. water pressure; 665°C, 15 000 p.s.i.; and 715°C, 30 000 p.s.i.

Like the pyrolytically formed micas, those of hydrothermal origin show ability to undergo various isomorphous replacements. Earlier reference

TABLE 11
Crystallizations in presence of NH_4OH or NH_4F[136]

T (°C)	Time (days)	Gel compositions		
		$3NH_4OH.Al_2O_3.2SiO_2$	**$3NH_4F.Al_2O_3.2SiO_2$**	**$3NH_4OH.Ga_2O_3.2SiO_2$**
250	5	Mod.–good NH_4K		Good NH_4K
	10	Good NH_4-K		v. good NH_4-K
	20	Good NH_4-K		v. good NH_4-K
300	2	Mod. NH_4-K		Good NH_4-K
	5	Mod. NH_4-K		
	11	Good NH_4-K		
350	2	Good NH_4-K		v. good NH_4-K
	5	Good NH_4-K	Good NH_4-K	v. good NH_4-K
	10			v. good NH_4-K
450	2	Good NH_4-K		Mod. NH_4-K
	5	Good NH_4-K	Good NH_4-K	Good NH_4-K
		$3NH_4OH.Ga_2O_3.2GeO_2$	**$3NH_4OH.Al_2O_3.2GeO_2$**	**$3NH_4OH.O.33NaOH.Al_2O_3.2SiO_2$**
300	2	Poor NH_4-K	v. good NH_4-F	Mod. (NH_4, Na)-K
	5	Poor–mod. NH_4-K	v. good NH_4-F	
	11	Mod. NH_4-K		
350	2	Mod. NH_4-K	v. good NH_4-F	Good (NH_4, Na)-K
	5	Good NH_4-K		v. good (NH_4, Na)-K
		$3NH_4OH.Ga_2O_3.2GeO_2$	**$3NH_4OH.Al_2O_3.2GeO_2$**	**$3NH_4OH.O.33NaOH.Al_2O_3.SiO_2$**
450	2	Mod. NH_4-K	Good NH_4-F and NH_4-J	Good (NH_4, Na)-K
	5 ⎫ 10 ⎭	v. good NH_4-K	Good NH_4-J and mod. NH_4-F	v. good (NH_4, Na)-K
	10	v. good NH_4-K	v. good NH_4-J	

was made (§8.1, and Ref. (120)) to boron phlogopite and muscovite. As further instances one may refer to work of Klingsberg and Roy[135] who prepared a Ga-phlogopite mica, and of Barrer and Dicks[136] who crystallized aqueous gel compositions $3NH_4OH.(Al, Ga)_2O_3.2(Si, Ge)O_2$. Good yields of micas were obtained with NH_4 replacing K or Na. Gallosilicate and gallogermanate ammonium micas were made under synthesis conditions summarized in Table 11. The reference letters K, F and J in the table denote respectively mica, an unidentified hexagonal phase and an andalusite-type phase. The unit cell dimensions and ammonium contents of the micas were:

	$NH_4Al_2[(Al, Si_3)O_{10}(OH)_2]$	$NH_4Ga_2[(Ga, Si_3)O_{10}(OH)_2]$	$NH_4Ga_2[(Ga, Ge_3)O_{10}(OH)_2]$
a (Å)	5·18	5·22	5·42
b (Å)	8·96	9·18	9·33
c (Å)	10·49	10·67	10·96
β	101·4°	101·3°	101·4°
% NH_4 (obs.)	4·6	3·4	2·7
% NH_4 (calc.)	4·77	3·56	2·8

The cell dimensions clearly reflect the respective chemical compositions. Another interesting substitution in the mica structure was observed when a zinc mica was formed from an aqueous gel–water mixture of composition $K_2O:ZnO:SiO_2:H_2O = 0·4:1·0:1·8:16$.[137] The mixture was crystallized for 4 days at 220°C. The X-ray powder pattern resembled that of a synthetic magnesium phlogopite[138] and the unit cell dimensions were:

$$a(Å) = 5·338 \pm 0·002$$
$$b(Å) = 9·249 \pm 0·003$$
$$c(Å) = 10·035 \pm 0·005$$
$$\beta = 99°55' \pm 2'$$

Analytical evidence was obtained that some zinc may be in the tetrahedral layers substituting for silicon. As a further example, Klingsberg and Roy[135] prepared a nickel phlogopite as the 3T polytype with Ni in the octahedral sites.

8.3. *Crystalline Forms of Silica*

Quartz, cristobalite and tridymite are in terms of their chemical composition the simplest tectosilicates. Quartz crystals have frequently appeared in hydrothermal systems, the earliest example perhaps being its reported synthesis in 1845[139] from silicic acid gel and water. Other hydrothermal syntheses from a variety of reactants and over a range of conditions appeared at intervals throughout the 19th and into the 20th century.

Quartz was often only one of many species formed and the objectives were broadly exploratory. In this early work the success of Chrustschoff[140,141] deserves mention. He heated 10% silica sol in glass tubes at 250–300°C each day for six months and reported crystals up to 8 mm in length.

Several studies have been made in which the progress of crystallization of amorphous silica has been followed. In one of these[142] the silica, in the form of powdered silicic acid, was heated with water in the range 330–440°C, at pressures of 15 000–59 000 p.s.i. and for times of 6–840 hours. The path followed was:

almost amorphous silica → cristobalite → keatite → quartz.

Pressure had a strong effect on the time for conversion to quartz, and also on the amount of silica dissolved at 400°C from a quartz crystal in one hour (Fig. 13). Siffert and Wey[143] extended these results, using silica gel derived from $Si(OC_2H_5)_4$ and with additions of Li, Na or K hydroxide in small amounts. At 300°C and with a 3% KOH solution the evolution of the crystalline phases is shown in Fig. 14. The occurrence of cristobalite and keatite before the appearance of quartz was confirmed. The three alkalis produced the cristobalite stage with varying effectiveness, the best mineralizer for cristobalite being KOH. The temperature was 300°C and the yields of cristobalite were measured after 20 hours.

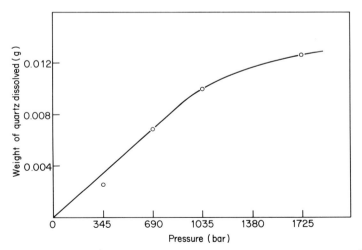

Fig. 13. Effect of pressure on the rate of solution of quartz at 400°C.[142] The ordinate is the amount dissolved in one hour.

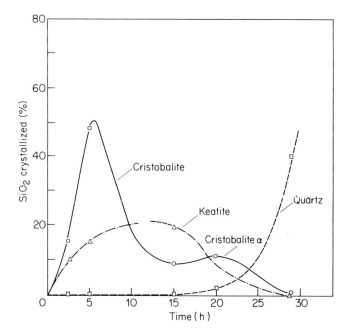

Fig. 14. Transformation of amorphous SiO_2 to quartz, cristobalite and keatite as a function of time. Temperature 300°C, in presence of 3% KOH aq.[143]

The results of Heydemann[144] differ in part from those above. Crystallization of amorphous silica was examined in an aqueous alkaline environment between 100 and 250°C. The alkalis were again Li, Na and K hydroxides, or the carbonates of these elements and of Rb and Cs. The crystallization sequence:

$$\text{amorphous silica} \rightarrow SiO_2\text{-}X \rightarrow \text{cristobalite} \rightarrow \text{quartz}$$

was reported. The SiO_2-X was not keatite or melanophlogite[145] but gave sharp X-ray powder lines. It crystallized as thin plates and its identity with known species could not be established. The above three studies appear to exemplify well for crystalline silicas the Ostwald rule of successive transformations. Metastable polymorphs appear before the final stable polymorph, in this case quartz.

Recently two porous crystalline silicas have been made as the Al-free end members of the silica-rich zeolites ZSM-5[146] and ZSM-11.[147] They have been termed silicalites 1[148] and 2[149] respectively. Silicalite 1 was grown from colloidal silica autoclaved at 100–200°C with tetrapropyl-ammonium hydroxide (TPAOH) as mineralizer. The silicalite precursor crystals so obtained have typical compositions such as $2(TPAOH).48SiO_2$ and to obtain the crystalline silicalite 1 the organic material is removed by heating in air at 500–600°C. Silicalite 2 was prepared by autoclaving silicic acid for 3 days at 170°C with tetrabutylammonium hydroxide as mineralizer. The precursor crystals to silicalite 2 were oval, 2–3 μm long and 1–1·5 μm in diameter. As prepared they contained, like the precursor to silicalite 1, some of the organic base which was again removed by heating in air. If suitable mineralizing template ions can be found, there seems no reason why other porous silicas could not be formed. The structures of both silicalites are known since the frameworks are those of zeolites ZSM-5 (orthorhombic) and ZSM-11 (tetragonal) respectively. Lattice constants of particular forms of the zeolites and of their silica end members are given below:

	a	b	c
Silicalite 1	20·07	19·86	13·36
ZSM-5[150]	20·1	19·9	13·4
Silicalite 2	20·04	20·04	13·38
ZSM-11[151]	20·12	20·12	13·44

The ZSM-5 topology produces a combination of intersecting linear and zig-zag channels access to which is controlled by 10-ring windows (i.e. windows of ten linked (Al, Si)O_4 tetrahedra). In the ZSM-11 topology the intersecting channel system comprises only linear channels, again with entry controlled by 10-ring windows (Chapter 6, Fig. 4).

The silicalites, traversed by 6 Å wide channels, are the most open forms of crystalline silica so far produced by direct crystallization and will sorb aromatic molecules such as p-xylene as well as n-alkanes and n-alcohols. Another porous crystalline silica, which occurs rarely in Nature and has not yet been synthesized is melanophlogite.[145] This is cubic with the structure of gas hydrate of type 1.[152] At high pressure and temperature, cristobalite is also sufficiently porous to accommodate significant amounts of helium, but quartz appears to be too dense to take up even helium.[153]

8.4. Felspars and Felspathoids

Felspars (e.g. albite, orthoclase, sanidine, anorthite) and felspathoids (e.g. nepheline, kalsilite, kaliophilite, leucite and pollucite) form important tectosilicate groups characterized by comparatively dense frameworks open enough to contain cations but not water, in contrast with the zeolites. The alkali content is high and they can be synthesized hydrothermally (Table 3). Because of their anhydrous nature they can also be made from melts and by other high-temperature methods. Under hydrothermal conditions their synthesis requires similar alkaline reaction mixtures to those needed for zeolite synthesis (cf. Chapter 5, and see Table 9). However their crystallization fields extend to higher temperatures than those of zeolites, in accordance with the general tendency that anhydrous or less hydrated species appear increasingly from hydrothermal reaction the higher the temperature. The ratios of the numbers of framework oxygens to numbers of zeolitic water molecules for several tectosilicates are:

Faujasite	$O:H_2O = 1.7$
Chabazite	$O:H_2O = 2.0$
Heulandite	$O:H_2O = 3.0$
Mordenite	$O:H_2O = 3.4$
Analcime	$O:H_2O = 6.0$
Felspars and some felspathoids	$O:H_2O = \infty$

Thus the most hydrous compounds, faujasite and chabazite, should have upper temperature limits to their crystallization fields which are not as high as those limits for mordenite or analcime, and analcime in turn should have the limit below those of felspars or felspathoids. This expectation is supported by the examples of syntheses of the four anhydrous tectosilicates given in Table 12, where the temperatures are often well above those usually involved in zeolite formation (Table 9). The table shows that hydrothermal syntheses of felspars have been recorded since 1879[154] and indeed continue to be studied up to the present time.

The variety of the reaction mixtures from each of which the felspar or felspathoid can be formed indicate that, provided alkaline conditions are maintained and temperature and pressure are in the correct regions, reaction is reproducible in nucleating and crystallizing the same compound. However, there are often competing species and in addition varying the time of reaction may result, at a given temperature, in the phase initially formed being replaced later by one or more additional and presumably more stable ones. For example the leucite produced by Yoder and Eugster[134] in four-hour treatments in the range 800–1000°C was wholly replaced by sanidine in more prolonged treatments. Although often formed at rather low temperatures, the synthetic felspars, when examined, have been found to be the high-temperature disordered forms.[169] A possible exception could be the triclinic microcline-type felspar made by Euler and Hellner.[162] This synthesis was of larger crystals made in a temperature gradient over the lengthy period of 45 days. In more nearly isothermal conditions many nuclei form, yielding only small crystals which have reached their limiting size rather rapidly.

TABLE 12

Examples of hydrothermal syntheses of some felspars and felspathoids

Date	Reactant mixtures	Conditions	Reported products	Refs.
1879	K_2SiO_3, little KOH, gel from adding $AlCl_3$ to K_2SiO_3+aq	Heated below dull red heat.	Orthoclase (+quartz)	(154, 155)
1881	SiO_2 sol, dialysed $Al(OH)_3$ sol+aq	300°C	Orthoclase (+quartz)	(156)
1887				
1902	SiO_2 (5 g), $KAlO_2$ (15 g)+aq	520°C	Orthoclase (+quartz)	(157)
1906	4 g of mixture of KOH, SiO_2, $Al(OH)_3$ in proportions of $KAlSi_3O_8$+aq	500°C	Orthoclase (+quartz)	(158)
1931	1 g leucite ($KAlSi_2O_6$), 1 g SiO_2+aq	395°C	Orthoclase	(24)
1936	Montmorillonite, $KHCO_3$+aq	245°, 272° and 300°C	Orthoclase	(25)
1950	Leucite with sat. Na_2CO_3+K_2CO_3	16 h at 195–200°C	Potash felspar	(159)
1956	Gels K_2O, Al_2O_3, $nSiO_2$+aq, with and without extra KOH. $3 \leqslant n \leqslant 12$	300° to 450°C. Best felspar at 400–450°C	Orthoclase or sanidine and other phases	(160)
1960	Obsidian glass, 0·25 M K_2CO_3	200–300°C	Orthoclase and quartz	(161)
1961	Rock 68% alkali felspar, 25% chlorite+aq	45 days in temp. gradient, 600°C hot end, 440°C cooler end	Triclinic K- felspar (with biotite mica)	(162)
1965	Kaolin with KH_2PO_4 aq	3 to 7 days 250–350°C	Potash felspar (with aluminophosphates)	(133)
1883	NaOH, SiO_2, Al_2O_3 in albite ratio with extra Na_2SiO_3+aq.	432, 505 and 517°C	Albite	(163)
1896	Muscovite+NaOH with SiO_2:Na_2O = 2:1, 3:1 or 4:1+aq	500°C	Albite (with analcime or quartz)	(164)
1902	SiO_2(5 g), $NaAlO_2$(10 g)+aq	520°C	Albite	(157)

Year	Starting materials	Conditions	Product	Ref.
1911	Na_2O (0.19 g), Al_2O_3 (0.31 g and 0.22 g), SiO_2 (0.7 g) + aq	450°	Albite	(8)
1931	1 g analcime, 1 g SiO_2 + aq	395°C for 3 days	Albite	(24)
1931	1 g nepheline, 2 g SiO_2 + aq	380°C for 7 days	Albite	(24)
1952	Gel Na_2O, Al_2O_3, $nSiO_2$ ($4 \leqslant n \leqslant 10$) + aq	330–450°C	Albite and other phases	(116)
1961	Glasses Na_2O, Al_2O_3, $nSiO_2$ ($3 \leqslant n \leqslant 6$) + aq	600–700°C	Albite and nepheline	(129)
1965	Gel Na_2O, Al_2O_3, $4SiO_2$ + Na_3PO_4 + aq	440°C for 4 and 11 days	Albite (and aluminophosphates)	(133)
1890	Mica, calcined SiO_2, KOH + aq	500°C	Leucite (with orthoclase and nepheline)	(165)
1918	K_3AlO_3 + K_2SiO_3 mixtures + aq or $Al(OH)_3$ + KOH + K_2SiO_3 + aq	350–440°C	Leucite (with orthoclase, kaliophilite and other phases).	(166)
1931	Kaliophilite + SiO_2 + aq	3 days at 395°C; more SiO_2 and 5 days at 395°C	Leucite	(24)
1955	Orthorhombic $KAlSiO_4$ + kaolinite + aq	4 h at 800–1000°C	Leucite (with sanidine and corundum)	(134)
1956	Gel K_2O, Al_2O_3, $nSiO_2$ + KOH aq ($2 \leqslant n \leqslant 5$)	250–400°C	Leucite, kalsilite	(160)
1890	Muscovite + NaOH aq	500°C	Nepheline (with kaliophilite)	(165)
1899	Muscovite + NaOH aq	36 h at 510°C	Nepheline	(167)
1902	2.5 g Analcime + 25 g (SiO_2, $NaAlO_2$) + aq	430°C for 110 h	Nepheline	(158)
1918	10 g labradorite, 1 g NaOH, 0.3 g K_2CO_3 + aq	330°C	Nepheline (with other phases)	(166)
1929	Paragonite + excess NaOH + aq	200, 300 and 400°C	Nepheline	(168)
1952	Gel Na_2O, Al_2O_3, $nSiO_2$ ($1 \leqslant n \leqslant 5$)	360–450°C	Nepheline	(116)
1961	Glass $NaAlSiO_4$ + aq	420–700°C	Nepheline	(129)

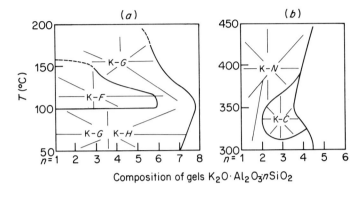

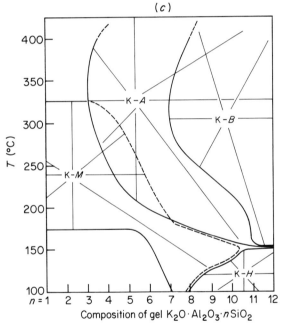

Fig. 15. Crystallization fields of some phases grown from hydrous gels $K_2O.Al_2O_3.nSiO_2 + KOH + aq.$[160]

K-*A* = potassium felspar	K-*G* = chabazite type zeolite
K-*B* = α-quartz	K-*H* = bayerite
K-*C* = leucite	K-*M* = phillipsite type zeolite
K-*F* = edingtonite type zeolite	K-*N* = kalsilite

Evidently these conditions of growth favour Al, Si disorder on the tetrahedral framework sites, and accord with Goldsmith's simplexity argument.[169]

Although zeolite, felspar and felspathoid crystallization fields can overlap, as noted earlier those of felspars and felspathoids persist to much higher temperatures. This is seen further in Fig. 15[160] which shows for hydrous gels of oxide composition $K_2O.Al_2O_3.nSiO_2 + KOH + aq$ the crystallization fields observed for a number of felspar, felspathoid and zeolite phases. Table 12 also records a number of preparations involving conversion of one mineral into another. Such transformations occur in remarkable variety by hydrothermal reaction, and conditions can often be found in which these pathways can be reversed. For example, in Table 12, transformations of montmorillonite and kaolinite to potash felspars are recorded, whereas Table 10 gives instances of conversion of felspars to clay minerals. Transformations of minerals like kaolinite to zeolites and of zeolites to other zeolites or non-zeolites will be considered subsequently (Chapter 5).

By pyrolysis and by hydrothermal reaction Goldsmith[170] showed that Ga could replace Al and Ge could replace Si in felspars and felspathoids. His work extends that already reported (§§8.1 and 8.2) in which these replacements were found in smectites and micas. It will be seen later that

TABLE 13
Isomorphous substitutions of Al and Si in felspars[170]

	Melting point (°C)	Melting process
Albites		
$NaAlSi_3O_8$	1118 ⎫	
$NaGaSi_3O_8$	1015 ⎬	
$NaAlGe_3O_8$	1067	All congruent
$NaGaGe_3O_8$	952 ⎭	
Orthoclases		
$KAlSi_3O_8$	1170	Incongruent, to leucite + liquid
$KAlGe_3O_8$	1122	
$KGaGe_3O_8$	~1000	
$KGaSi_3O_8$	~1000	Incongruent, to Ga-leucite + liquid
Anorthites		
$CaAl_2Si_2O_8$	1553	
$CaGa_2Si_2O_6$	1323	
$CaAl_2Ge_2O_6$	>1400	
$CaGa_2Ge_2O_6$	1321	

they also occur in zeolites (Chapter 6), so that such framework replacements are widespread. The compounds obtained by Goldsmith are summarized in Table 13. Two modifications of $KAlGe_3O_8$ and $KGaGe_3O_8$ were obtained. It was also reported that fully substituted (Ga, Ge)-, Ga- or Ge-anorthites were metastable.

The syntheses described in this section illustrate how some important non-zeolite mineral types have been crystallized. When taken in comparison with zeolite syntheses considered more fully in later chapters a broader perspective of hydrothermal reactions is obtained.

9. Big Crystals

Giant crystals occur naturally in zoned pegmatites, the biggest being near the zone centre. It is likely that the zones were formed by fractional crystallization of a magma with incomplete reaction between successive crops of crystals and pegmatite fluid rich in hyperfusible components. The crystals appear to have developed successively from the outer parts of the pegmatite body inward towards its centre, as have the lithologic zones that contain them.[171] Giant crystals of dimensions in feet or tens of feet and weights in tens, hundreds or even thousands of pounds have been found of beryl, soda and potash felspars, quartz, spodumene and biotite, muscovite and phlogopite micas. To a lesser extent giant crystals occur of allanite, amblygonite, apatite, columbite-tantalite, fluorite, hornblende, hypersthene, monazite, petalite, calcic plagioclase, topaz, tourmaline, triphylite-lithiophylite, triplite and other species.[171]

In the laboratory one cannot expect to synthesize giant crystals, but nevertheless, crystals of considerable dimensions have been made. In hydrothermal synthesis the biggest crystals have been produced by using a temperature gradient in the autoclave, with one or more seed crystals at the cold end and smaller crystals of the same kind as feed material at the hot end (Fig. 16). An alkaline aqueous medium serves to dissolve the nutrient and to convey it to the colder end where it is deposited upon the seeds. The hotter and cooler ends may be partially separated by a baffle. It is a requisite, with the seed at the cooler end, that solubility should decrease with falling temperature so that deposition can occur on the seed. The rate of growth must be regulated by adjusting the temperature drop between top and bottom because too rapid deposition can result in defects in the growing crystals. The autoclaves are of steel and are protected from corrosion by the alkaline liquors by means of a silver or platinum lining.

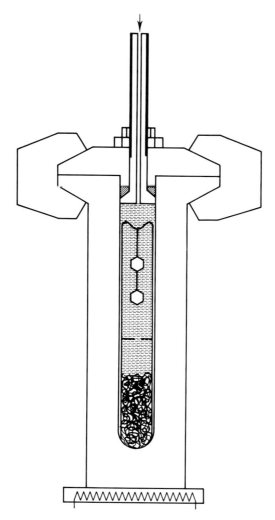

Fig. 16. Autoclave for hydrothermal growth on a seed crystal, using small crystals of the same kind as nutrient and a temperature gradient.[172]

The synthesis of optical grade piezo-electric quartz crystals illustrates the use of the above method, as the following examples show:

(i) A quartz sphere as seed was grown into a faced crystal[173] using $3 \text{ M Na}_2\text{CO}_3$ solution as solvent, with temperatures of 397 and 376°C at the bottom and the top of the autoclave.

(ii) Quartz seeds were cut parallel to 0001. Aqueous Na_2CO_3 was again the solvent, the hot part of the autoclave was at 400°C and the cool end was at 360°C. The silver-lined autoclave was of 5 cm internal diameter and 25 cm long. To obtain good quality quartz, deposition rates were initially restricted to 0·35 mm per day on each 0001 face, but subsequently rates approaching 1 mm per day were successful. A 15 g crystal grown from a 2 g seed was of a quality equal to that of the best natural quartz.[174]

(iii) Large single crystals of piezo-electric quality and weighing more than a pound were grown in 70 days[175] in a silver-lined autoclave 4 in. in diameter and 4 feet long. The end containing the quartz nutrient was separated from the main chamber by a baffle and was held at 400°C. The top of the main chamber where the seed was located was at about 350°C. Caustic soda and sodium carbonate solutions were used as solvents. The caustic soda solution proved to be the best medium for obtaining optical quality crystals.

Corundum (Al_2O_3) has been grown on seed crystals in essentially the same way as described above, in an autoclave with the seed at about 500°C and the base of the autoclave at about 580°C.[176] Deposition on the seed was achieved with potassium, rubidium and caesium carbonate solutions as solvents. The % gain in weight with potassium carbonate is given below:

Concentration	Duration (days)	% wt gain of seed
1·0 N	14	46
2·0 N	14	115
2·0 N	28	157
3·0 N	14	195

Doping the crystals with chromium by adding $K_2Cr_2O_7$ to the K_2CO_3 aq yielded deep pink ruby. The dichromate slowed the rate of growth of ruby but did not prevent it. Sapphire, and also ruby blood red in colour and of ~1 cm diameter, have been made,[177] as has emerald,[1,178] again using aqueous alkaline media as solvent and a temperature gradient between seed and nutrient crystallites. In the same way Euler and Hellner[162] made adularian felspar crystallites as plates 0·2 mm across. Even under more nearly isothermal conditions spontaneous nucleation can sometimes give crystals of appreciable dimensions. These include analcime icosatetrahedra about 0·5 mm in diameter, nosean hydrate as 1·5 mm blocks nepheline hydrate plates 2 mm across and 1 cm long rods of sodium silicate.[179] Ingerson and Tuttle[180] heated co-precipitated zinc and manganese hydroxides to 500–600°C in an autoclave with water and silica under pressures of 1000 to 4000 p.s.i. Single crystals of fluorescent willemite (Zn_2SiO_4), activated by manganese, were grown as rods up to 3 cm in length. There seems no reason why, if desired, many refractory species could not be grown as single crystals of significant dimensions by the hydrothermal method.

Rates of crystallization in a temperature gradient and under hydrothermal conditions have for non-zeolites been examined most fully in the case of quartz, although the formal relations should apply qualitatively to hydrothermal growth of other refractory species. For deposition of quartz from nutrient crystallites of quartz onto a seed crystal, growth rates have been represented by the expression[94]

$$R_{hkl} = k_{hkl}\alpha\,\Delta s$$

R_{hkl} is the rate of growth in a particular crystallographic direction, k_{hkl} a rate constant for that direction, Δs is the degree of supersaturation and α is a dimensionless conversion constant. R depends *inter alia* on the nature of the solvent and its concentration or pressure in the same way as Δs. Thus over modest temperature intervals each is proportional to ΔT, the difference in temperature between the zones of the autoclave where the nutrient material is dissolved and where it is deposited on the seed crystal.

Activation energies typically varied between 42 and 84 kJ mol^{-1} and are too high for control by pure diffusion of dissolved nutrient in the aqueous phase. The rate of deposition was independent of the percentage open area of the baffle (Fig. 16) so that convective transport was too fast to be rate-controlling. In general the lowest rates of deposition occurred on faces of lowest reticular atom density. These observations strongly suggest a rate-limiting step or sequence of steps which is chemical in nature and which occurs at the crystal–fluid interfaces. These observations apply of course to growth on a pre-formed crystal seed. They do not give information about spontaneous nucleation of crystallization in nearly isothermal systems (Chapter 4).

10. Concluding Remark

The purpose of this chapter has been to indicate some principles and the generality and power of hydrothermal reactions, in the context of typical silicate groups, hydrous and anhydrous. Zeolite formation, considered in more detail in subsequent chapters, can only be effected hydrothermally, and at the low-temperature end it will then be seen that hydrothermal synthesis of zeolites merges with ordinary solution chemistry. Except in the temperature and pressure ranges involved hydrothermal chemistry is not thermodynamically different from the ordinary chemistry of aqueous solutions. Hydrothermal conditions can, however, provide the acceleration of a reaction which, although thermodynamically possible at temperatures below 100°C, would require a geological time-scale to occur. This allows one to explore the chemistry of refractory substances in aqueous conditions and has opened for exploration an area of great variety.

References

1. E.A.D. White, *Quart. Rev. Chem. Soc.* (1961) **15**, 1.
2. E.g., G.W. Morey, *J. Amer. Ceram. Soc.* (1937) **20**, 285. Also Ref. (10).
3. F.A. Mumpton and W.C. Ormsby, *in* "Natural Zeolites, Occurrence, Properties, Use" (Ed. L.B. Sand and F.A. Mumpton), p. 113. Pergamon Press, Oxford, 1978.
4. R.L. Hay, *in* "Natural Zeolites, Occurrence, Properties, Use" (Ed. L.B. Sand and F.A. Mumpton). p. 135. Pergamon Press, Oxford, 1978.
5. R.C. Surdam and R.A. Sheppard, *in* "Natural Zeolites, Occurrence, Properties, Use" (Ed. L.B. Sand and F.A. Mumpton), p. 145. Pergamon Press, Oxford, 1978.
6. A. Iijima, *in* "Natural Zeolites, Occurrence, Properties, Use" (Ed. L.B. Sand and F.A. Mumpton), p. 175. Pergamon Press, Oxford, 1978.
7. M. Kastner and S.A. Stonecipher, *in* "Natural Zeolites, Occurrence, Properties, Use" (Ed. L.B. Sand and F.A. Mumpton), p. 199. Pergamon Press, Oxford, 1978.
8. E.g., E. Baur, *Z. Anorg. Chem.* (1911) **72**, 119.
9. A. Lacroix "Montagne Pelée et ses Eruptions". Macon Cie, Paris, 1902.
10. R.M. Barrer, *Faraday Soc. Discussion No.* 5 (1949) 326.
11. R. Schwarz and G. Trageser, *Naturwiss* (1935) **23**, 512.
12. W. Noll, *Centr. Miner.* (1934) 80; *Miner. Petr. Mitt.* (1934) **45**, 175; *Miner. Petr. Mitt.* (1936) **48**, 210; *Neues Jahrb. Miner. Geol. Beil. Bd.* (1935) **70**, 65; *Chem. Erde* (1936) **10**, 129.
13. R.H. Ewell and H. Insley, *J. Res. Nat. Bur. Stds* (1935) **15**, 173.
14. F.H. Norton, *Amer. Mineral* (1937) **22**, 1; *Amer. Mineral* (1939) **24**, 1; *Amer. Mineral.* (1941) **26**, 1.
15. C.J. v. Nieuwenberg and H. Pieters, *Rec. Trav. Chim. Pays-Bas* (1929) **48**, 27.
16. F.G. Straub, *Ind. Eng. Chem.* (1936) **28**, 113.
17. R.M. Barrer, *J. Chem. Soc.* (1948) 2158.
18. E. Flint, W. Clarke, E.S. Newman, L. Shartsis, D. Bishop and L.S. Wells, *J. Res. Nat. Bur. Stds* (1945) **36**, 63.
19. C.E. Imhoff and L.A. Burkhardt, *Ind. Eng. Chem.* (1943) **35**, 873.
20. G. van Praagh, *Geol. Mag.* (1947) **84**, 98.
21. A.C. Walker, *J. Amer. Ceram. Soc.* (1953) **36**, 250.
22. J. Wyart, *Bull. Soc. Fr. Minér.* (1943) **66**, 479.
23. R. Weil, *Compt. Rend* (1925) **181**, 423.
24. C.J. v. Nieuwenberg and H.B. Blumendahl, *Rec. Trav. Chim. Pays-Bas* (1931) **50**, 989.

25. J.W. Gruner, *Amer. Mineral.* (1936) **21,** 511.
26. J. Wyart and S. Scavnicar, *Bull. Soc. Fr. Miner. Cristallogr.* (1957) **80,** 395.
27. A. van Valkenberg and C.E. Weir, *Bull. Geol. Soc. Amer.* (1957) **68,** 1808.
28. H.S. Yoder, *Amer. J. Sci.* (1952) (Bowen Volume Pt 2) 569.
29. C. Frondel and R.L. Collette, *Amer. Mineral.* (1957) **42,** 759.
30. M. Michel-Levy, *Compt. Rend.* (1949) **228,** 1814.
31. C. Frondel and R.L. Collette, *Amer. Mineral.* (1957) **42,** 754.
32. R.A. Laudise and A.A. Ballman, *135th Mtg Amer. Chem. Soc., Boston,* April, 1959.
33. E. Hayek, W. Bohler, J. Lechleitner and H. Petter, *Z. Anorg. Allg. Chem.* (1958) **295,** 241.
34. T. Noda and S. Sugiyama, *J. Chem. Soc. Japan* (1943) **46,** 931, 1082.
35. J. Moroziewicz, *Tsch. Min. Mitt.* (1899) **18,** 20.
36. H.J. Emeleus and J.S. Anderson, "Modern Aspects of Inorganic Chemistry", pp. 210 *et seq.* Routledge, London, 1938.
37. G.W. Morey, *J. Amer. Ceram. Soc.* (1943) **17,** 145.
38. C. Doelter "*Mineralchemië.* Steinkopf (1912), Vol. 1, p. 601.
39. G.A. Rankin and H.E. Merwin, *Amer. J. Sci.* (1918) **45,** 301.
40. D.C. Stockbarger, *Faraday Soc. Discussion No. 5* (1949) 294.
41. R.G. Rudness and R.W. Kebler, *J. Amer. Ceram. Soc.* (1960) **43,** 17.
42. J.W. Nielsen, *J. Appl. Phys.* (1958) **29,** 390.
43. J.W. Nielsen and E.F. Dearborn, *Phys. Chem. Solids* (1958) **5,** 202.
44. G. Montel and G. Chaudron, *Compt. Rend.* (1951) **233,** 318.
45. A. Michel-Levy and J. Wyart, *Compt. Rend.* (1938) **206,** 261; (1939) **208,** 1030, 1594; (1940) **210,** 733; (1941) **212,** 89; *Bull. Soc. Fr. Minér.* (1947) **70,** 164, 168.
46. N. W. Taylor, *J. Amer. Ceram. Soc.* (1934) **17,** 155.
47. W. Jander and J. Petri, *Z. Elektrochem.* (1938) **44,** 747.
48. E.C. Robertson, F.J. Birch and J.F. MacDonald, *Amer. J. Sci.* (1957) **255,** 115.
49. H.W. Leverenz, "Luminescence of Solids", Wiley and Sons, Chichester, 1949.
50. J.W. Greig, H.E. Merwin and E.S. Shepherd, *Amer. J. Sci.* (1933) **25,** 61.
51. D.C. Reynolds and S.J. Czyzak, *Phys. Rev.* (1950) **79,** 543.
52. W.W. Piper, *J. Chem. Phys.* (1952) **20,** 1343.
53. R.M. Barrer and J.F. Cole, *J. Chem. Soc. A* (1970) 1516.
54. R.M. Barrer, J.F. Cole and H. Sticher, *J. Chem. Soc. A* (1968) 2475.
55. R.M. Barrer, *Transactions of 7th Internat. Ceramic Congress, London,* 1960, p. 379.
56. G. Peyronel, *Z. Krist.* (1936) **95,** 274.
57. W.T. Holser, *J. Phys. Chem.* (1954) **58,** 316.
58. G.M. Anderson, *Geochim. Cosmochim. Acta* (1964) **28,** 713.
59. A.J. Majumdar and R. Roy, *Geochim. Cosmochim. Acta* (1956) **10,** 311.
60. R.M. Barrer, *J. Phys. Chem. Solids* (1960) **16,** 84.
61. J.H. van der Waals and J.C. Platteuw, *Advances in Chem. Phys.* (1959) **2,** 1.
62. R.M. Barrer and G. Bratt, *J. Phys. Chem. Solids* (1959) **12,** 130.
63. R.M. Barrer and P.J. Reucroft, *Proc. Roy. Soc.* (1960) **258A,** 431, 449.
64. R.M. Barrer and A.J. Walker, *Trans. Faraday Soc.* (1964) **60,** 171.
65. C.W. Burnham, Bull. *Soc. Fr. Minéral. Cristallogr.* (1974) **97,** 223.
66. C.W. Burnham and N.F. Davis, *Amer. J. Sci.* (1971) **270,** 54.

67. C.W. Burnham, *Geol. Soc. Amer., Special Paper* (1963) **76,** 26.
68. E.G. Lebedev and N.I. Khitarov, *Geokhim.* (1964) No. 3, 195.
69. G.P. Orlova, *Internat. Geol. Rev.* (1964) **6,** 254.
70. A.J. Moulson and J.P. Roberts, *Trans. Faraday Soc.* (1961) **57,** 1208.
71. D.L. Hamilton, C.W. Burnham and E. Osborn, *J. Petrol.* (1964) **5,** 21.
72. C.W. Burnham and R.H. Jahns, *Amer. J. Sci.* (1962) **260,** 721.
73. S. Brunauer, "The Adsorption of Gases and Vapours", p. 150. Oxford University Press, 1944.
74. J.R. O'Neil and H.P. Taylor Jr, *Amer. Mineral.* (1967) **52,** 1414.
75. R.M. Barrer, *in* "Molecular Sieves", p. 39. Society for Chemical Industry, London, 1968.
75a. T. Moeller and R.O'Connor, "Ions in Aqueous Systems", Ch. 5. McGraw-Hill, New York, 1972.
76. A.J. Ellis and W.S. Fyfe, *Rev. Pure App. Chem.* (1957) Vol. 7. 261.
77. A.J. Ellis, unpublished, Ref. (76).
78. J. Verhoogen, *Econ. Geol.* (1938) **33,** 34, 775.
79. M.L. Gavrish and I.S. Galinker, *Dok. Akad. Nauk USSR* (1955) **102,** 89.
80. G.C. Kennedy, *Econ. Geol.* (1950) **45,** 629.
81. A.J. Ellis, unpublished, Ref. (76).
82. G.W. Morey, F.R.D. Boyd, J.L. England and W.T. Chen, *J. Amer. Chem. Soc.* (1955) **77,** 5003.
83. M.H. Lietzke, *J. Amer. Chem. Soc.* (1957) **79,** 267.
84. F.C. Kracek, G.W. Morey and H.E. Merwin, *Amer. J. Sci.* (1938) **35A,** 143.
85. A. Seidell "Solubilities of Inorganic and Metal Organic Compounds" Van Nostrand, New York, 1940.
86. N.B. Keevil, *J. Amer. Chem. Soc.* (1942) **64,** 841.
87. J.A. van Lier, P.L. de Bruyn and J.Th.G. Overbeek, *J. Phys. Chem.* (1960) **64,** 1675.
88. G.W. Morey, R.O. Fournier and J.J. Rowe, *Geochim. Cosmochim. Acta* (1962) **26,** 1029.
89. R.O. Fournier and J.J. Rowe, *Amer. Mineral.* (1962) **47,** 897.
90. G.M. Anderson and C.W. Burnham, *Amer. J. Sci.* (1965) **263,** 494.
91. T.H. Elmer and M.E. Nordberg, *J. Amer. Ceram. Soc.* (1958) **41,** 517.
92. I. Friedman, *J. Amer. Chem. Soc.* (1948) **70,** 2649.
93. I. Friedman, *Amer. Mineral.* (1949) **34,** 583.
94. R.A. Laudise and E.D. Kolb, *Endeavour* (1969) **28,** 114.
95. G.W. Morey and E. Ingerson, *Econ. Geol.* (1937) **32,** 607.
96. B. Stringham, *Econ. Geol.* (1952) **47,** 661.
97. W. Noll, *Naturwiss.* (1932) **20,** 366.
98. W. Noll, *Centr. Miner.* (1934) 80.
99. W. Noll, *Mineral. Petr. Mitt.* (1934) **45,** 175.
100. A.E. Badger and A. Ally, *J. Geol.* (1932) **40,** 745.
101. R. Schwartz, *Naturwiss.* (1933) **21,** 252.
102. R. Schwartz and G. Trageser, *Z. Anorg. Chem.* (1933) **215,** 190.
103. R.H. Ewell and H. Insley, *J. Res. Nat. Bur. Stds* (1935) **15,** 173.
104. W. Noll, *Neues Jb. Min. Geol., Beil. Bd.* (1935) **70,** 65.
105. F.H. Norton, *Amer. Mineral.* (1939) **24,** 1.
106. F.H. Norton, *Amer. Mineral.* (1941) **26,** 1.
107. J.W. Gruner, *Econ. Geol.* (1944) **39,** 578.
108. W. Noll, *Fort. Mineral. Krist. Petr.* (1935) **19,** 46.

109. F.H. Norton, *Amer. Miner.* (1937) **22,** 1.
110. W. Noll, *Naturwiss.* (1935) **23,** 197.
111. W. Noll, *Chemie der Erde* (1936) **10,** 129.
112. W. Noll, *Mineral. Petr. Mitt.* (1936) **48,** 210.
113. R. Roy and L.B. Sand, *Amer. Mineral.* (1956) **41,** 505.
114. D.M. Roy and R. Roy, *Amer. Mineral.* (1955) **40,** 147.
115. M. Koizumi and R. Roy, *Amer. Mineral.* (1959) **44,** 788.
116. R.M. Barrer and E.A.D. White, *J. Chem. Soc.* (1952) 1561.
117. L.B. Sand, R. Roy and E.F. Osborn, *Econ. Geol.* (1957) **52,** 169.
118. Pfizer Inc., B.P. 1,321, 338, (1973).
119. D.A. Hickson to Chevron Research Co., U.S.P. 3,887,454, (1975).
120. V. Stubican and R. Roy, *Amer. Mineral.* (1962) **47,** 1166.
121. P. Franzen and J.J.B. van Eyk van Voorthuysen, *Transactions of the Int. Cong. of Soil Sci., Amsterdam,* 1950, Vol. III, p. 34.
122. J.J.B. van Eyk van Voorthuysen, *Rec. Trav. Chim. Pays-Bas* (1951) **70,** 793.
123. S. Caillère, S. Hénin and J. Esquevin, *Clay Minerals Bull.* (1958) **3,** 232.
124. S. Caillère, S. Hénin and J. Esquevin, *Bull. Gr. Fr. des Argiles* (1959) **11,** 53.
125. R.M. Barrer and L.W.R. Dicks, *J. Chem. Soc. A* (1967) 1523.
126. E.g. T. Noda, *J. Amer. Ceram. Soc.* (1955) **38,** 147.
127. *U.S. Bureau of Mines Reports* 5283 (1956) and 5337 (1957).
128. N. Daimon, Y. Ito and M. Hirao, *Mem. Facult. Eng., Nagoya Univ.* (1960) **12,** 136.
129. P. Saha, *Amer. Mineral.* (1961) **46,** 859.
130. T. Noda and N. Yamanishi, *Kogyo Kagaku Zasshi* (1964) **66,** 289.
131. T. Noda and M. Ushio, *Kogyo Kagaku Zasshi* (1964) **67,** 292.
132. C.M. Warshaw, "Clays and Clay Minerals—Proceedings of the 7th National Conference", p. 303. Pergamon Press, Oxford, 1960.
133. R.M. Barrer and D.J. Marshall, *J. Chem. Soc.* (1965) 6621.
134. H.S. Yoder and H.P. Eugster, *Geochim. Cosmochim. Acta* (1955) **8,** 225.
135. C. Klingsberg and R. Roy, *Amer. Mineral.* (1957) **42,** 629.
136. R.M. Barrer and L.W.R. Dicks, *J. Chem. Soc. A* (1966) 1379.
137. R.M. Barrer and W. Sieber, *J. Chem. Soc. Chem. Comm.* (1977) 905.
138. F. Seifert and W. Schreyer, *Amer. Mineral.* (1965) **50,** 1114.
139. Schafheutl, *Munchner gelehrte Auzeigen* (1845) p. 557; Ref. in P. Niggli and G. Morey, *Z. Anorg. Chem.* (1913) **83,** 369.
140. K. von Chrustschoff, *Bull. Soc. Mineral. Fr.* (1887) **10,** 31.
141. K. von Chrustschoff, *Neues Jb. Mineral.* (1887) **1,** 205.
142. R.M. Carr and W.S. Fyfe, *Amer. Mineral.* (1958) **43,** 908.
143. B. Siffert and R. Wey, *Silicates Industriels* (1967) (12) 415.
144. A. Heydemann, *Beit. Mineral. Petrog.* (1964) **10,** 242.
145. B.J. Skinner and D.E. Appleman, *Amer. Mineral.* (1963) **48,** 854.
146. Mobil Oil Corp. Neth. Pat. 7,014,807; (1971).
147. R.J. Argauer and G.R. Landolt, U.S.P. 3,702,886; (1972).
148. E.M. Flanigen, J.M. Bennett, R.W. Grose, J.P. Cohen, R.L. Patton, R.M. Kirchner and J.V. Smith, *Nature* (1978) **271,** 512.
149. D.M. Bibby, N.B. Milestone and L.P. Aldridge, *Nature* (1979) **280,** 664.
150. G.T. Kototailo, S.L. Lawton, D.H. Olson and W.M. Meier, *Nature* (1978) **272,** 437.

151. G.T. Kokotailo, P. Chu, S.L. Lawton and W.M. Meier, *Nature* (1978) **275,** 119.
152. B. Kamb, *Science* (1965) **148,** 232.
153. R.M. Barrer and D.E.W. Vaughan, *Trans. Faraday Soc.* (1967) **63,** 2275.
154. Ch. Friedel and E. Sarasin, *Bull. Soc. Minéral. Fr.* (1879) **2,** 113.
155. Ch. Friedel and E. Sarasin, *Bull. Soc. Minéral. Fr.* (1881) **4,** 171.
156. K.v. Chrustschoff, *Compt. Rend.* (1887) **104,** 602.
157. E. Baur, *Zeit. Phys. Chem.* (1902) **42,** 567.
158. C. Doelter, *Tsch. Mineral. Petrog. Mitt.* (1906) **25,** 79.
159. R.M. Barrer and L. Hinds, *Nature* (1950) **166,** 562.
160. R.M. Barrer and J.W. Baynham, *J. Chem. Soc.* (1956) 2882.
161. R. Kiriyama and H. Iwasaki, *J. Geol. Soc. Japan* (1960) **66,** 242.
162. v.R. Euler and E. Hellner, *Zeit. Krist.* (1961) **115,** 28.
163. Ch. Friedel and E. Sarasin, *Compt. Rend.* (1883) **97,** 290.
164. G. Friedel, *Bull. Soc. Minéral. Fr.* (1896) **19,** 5.
165. Ch. and G. Friedel, *Bull. Soc. Minéral. Fr.* (1890) **13,** 129.
166. J. Konigsberger and W.J. Muller, *Zeit. Anorg. Chem.* (1918) **104,** 1.
167. G. Friedel, *Bull. Soc. Minéral. Fr.* (1899) **22,** 20.
168. E. Gruner, *Z. Anorg. Chem.* (1929) **182,** 319.
169. J.R. Goldsmith, *J. Geol.* (1953) **61,** 439.
170. J.R. Goldsmith, *J. Geol.* (1950) **58,** 518.
171. R.H. Jahns, *Amer. Mineral.* (1953) **38,** 563.
172. E.A.D. White, *Hilger Journal* (1965) **9,** 3.
173. D.R. Hale and C.S. Hurlbut, *Amer. Mineral.* (1949) **34,** 596.
174. C.S. Brown, R.C. Kell, L.A. Thomas, N. Wooster and W.A. Wooster, *Nature* (1951) **167,** 940.
175. A.C. Walker, *J. Amer. Ceram. Soc.* (1953) **36,** 250.
176. J. Butcher and E.A.D. White, *Min. Mag.* (1964) **33,** 974.
177. R.A. Laudise and A.A. Ballman, *J. Amer. Chem. Soc.* (1958) **80,** 2655.
178. E.M. Flanigen, D.W. Breck, N.R. Mumbach and A.M. Taylor, *Amer. Mineral.* (1967) **52,** 744.
179. R.M. Barrer, *Trans. British Ceram. Soc.* (1957) **56,** 155.
180. E. Ingerson and O.F. Tuttle, *Amer. J. Sci.* (1947) **245,** 313.

Chapter 3

Reactants in Zeolite Synthesis and the Pre-nucleation Stage

1. Introduction

Thermodynamic variables in the synthesis of minerals in general and of zeolites in particular are the temperature, pressure, and overall chemical composition of the reactant mixtures. These variables do not necessarily determine the products obtained in hydrothermal reactions because the reactant mixtures may be heterogeneous and because nucleation appears to be kinetically rather than thermodynamically determined and controlled. When competing species appear the tendency is for the least stable phase to crystallize first and then to be replaced by a more stable form, and so on, until a final, most stable product results (Ostwald's rule of successive transformations). This, however, is still only a partial description because a number of non-thermodynamic factors may decisively influence the kinetic stage of nucleation and so the compounds which form. As will be described in Chapter 4 these include the treatment of reactants prior to crystallization, their chemical and physical nature and the influence of mineralizers, templating cations and additives (e.g. salts and dyestuffs). There is a pre-nucleation stage, sensitive to such influences, in which germ nuclei of various kinds may form and disappear without becoming viable for crystal growth. These fluctuations may eventually result in one or more kinds of germ nucleus reaching a critical size after which they grow spontaneously (Chapter 4, §2).

Because of the above situation a variety of sources of cations, aluminium and silicon has been used to prepare reaction mixtures, which differ greatly in physical and chemical nature. As much information as possible is desirable about the constitution of reactants and the nature of the dissolved species in a hydrothermal magma. This in turn leads to the examination of solutions of individual components in the eventual hope of understanding some of the molecular events which govern crystallization. Such information, as in much of solution chemistry, is difficult to obtain, but information bearing on this aspect is given in what follows.

2. Reactants

Among the sources of cations, Al_2O_3 and SiO_2 are those given in Table 1. Zeolite syntheses have frequently involved soluble silicates and soluble aluminates. For this reason the nature of such solutions will be briefly considered. Silica sols and silica gels have been another common source of silica and their behaviour when treated with alkali is relevant, as is the nature of the amorphous aluminosilicate gels usually formed on mixing soluble silicates or colloidal silica and soluble aluminates. When the source of alumina is $Al(OH)_3$ the hydroxide is usually dissolved in alkali to yield the soluble aluminate.

2.1. Aluminate Solutions

Two studies of aluminate solutions[1,2] have summarized earlier results. Measurements of electrical conductance, hydrogen electrode data, elec-

TABLE 1
Some sources of cations, aluminium and silicon in zeolite crystallization

Charge compensating cations	Aluminium	Silicon
Alkali metal hydroxides	Metal aluminates	Silicates and silicate
Alkaline earth hydroxides	$Al(OH)_3$, Al_2O_3, $AlO.OH$	hydrates
and oxides	Al alkoxides	Water glass
Other oxides and	Al salts	Silica sols
hydroxides	Glasses	Silica gels
Salts (fluorides,	Sediments	Silica and other
halides, carbonates,	Minerals, especially	synthetic glasses
phosphates, sulphates,	clay minerals,	Silicon esters
etc.)	felspathoids, felspars	Tuffs and volcanic glasses
Organic bases and	and other zeolites	Minerals, including clay
ammonium hydroxide,		minerals, felspathoids,
especially quaternary		felspars and other
bases		zeolites
Silicates and aluminates		Basalts and mineral
Mixtures of two or more		mixtures
of the above.		Sediments
		Combinations of two or
		More of the above

trometric titrations, osmotic and cryoscopic properties, viscosity, and optical and dialysis behaviour have led on balance to the view that in alkaline solutions the important anion is $Al(OH)_4^-$, but that as concentrations of Na_2O in sodium aluminate solutions are increased above 25% this ion tends to dehydrate to give anions such as AlO_2^-. Earlier interpretations[3,4] in terms of polymeric or colloidal structures in these alkaline media were ruled out on grounds of lack of direct experimental evidence. Instead sodium aluminate was considered to behave as a practically fully dissociated salt of Na^+ and $Al(OH)_4^-$ or a mixture of $Al(OH)_4^-$ and AlO_2^- according to the alkali concentration. The infra-red and Raman spectra of sodium aluminate dissolved in H_2O and D_2O and the ^{23}Na and ^{27}Al n.m.r. spectra of 0·5–6 M sodium aluminate were also interpreted as being consistent with $Al(OH)_4^-$ as the dominant anion in the more dilute range.[2] Around and above 1·5 M, however, new infra-red bands appeared and the ^{27}Al resonance was considerably broadened, but without a significant change in chemical shift. These new effects were considered in terms of two possible dehydration steps:

$$Al(OH)_4^- \rightarrow AlO(OH)_2^- + H_2O$$
$$2Al(OH)_4^- \rightarrow [(OH)_3Al-O-Al(OH)_3]^{2-} + H_2O$$

Although the evidence was not conclusive the second of the above dehydration processes was considered to be more probable. It was pointed out that in crystals of potassium aluminate[5] a dimeric anion occurs with an Al–O–Al angle of 132°.

In sufficiently acid solutions the aluminium exists as hydrated Al^{3+}. If, although still maintaining the solutions on the acid side, one increases the pH polymeric species containing Al appear. The degree of aggregation depends on the ratio $m = OH^-/Al^{3+}$, although, especially at higher m, equilibrium is only slowly established.[6] From solutions with $m = 2·5$ addition of Na_2SeO_4 precipitates cubic crystals with $a = 18·0$ Å. X-ray structural determinations[7] showed the presence of polyvalent cations in each of which a central AlO_4 tetrahedron is surrounded by twelve AlO_6 octahedra sharing edges. The composition of the selenate is

$$Na[Al_{13}O_4(OH)_{24}(H_2O)_{12}](SeO_4)_4.13H_2O$$

If the partial neutralization is performed with NH_4OH and is followed by addition of concentrated $(NH_4)_2SO_4$ solution tetragonal crystals are precipitated having $a = b = 19·8$ Å and $c = 19·6$ Å, with a composition

$$(NH_4)_7[Al_{13}O_4(OH)_{24}(H_2O)_{12}](SO_4)_7.xH_2O$$

In each of these two basic salts one sees that the hydrated cation

$[Al_{13}O_4(OH)_{24}(H_2O)_{12}]^{7+}$ occurs. The Raman spectra of partially hydrolysed $AlCl_3$ solutions and of the above crystals supports the view that this cation exists in aluminium chlorhydrol solutions. It should have a diameter of about 9 Å, and indeed when the interlayer cations of smectites are exchanged with aluminium chlorhydrol solutions the interlayer spacings expand to approximately this value.[8] It has further been stated that, as the pH of chlorhydrol solutions increases from 2 to 6, the aluminous cation changes by progressive hydrolysis according to the formula

$$[Al_{13}O_4(OH)_{24+y}(H_2O)_{12-y}]^{+(7-y)}$$

where $y \geqslant 0$. These complex cations may co-exist with simpler species such as Al^{3+}, $Al(OH)^{2+}$ or $Al(OH)_2^+$. Huang and Keller[8a] considered various ionic quotients involving such ions. They estimated that unit values of concentration ratios of certain pairs of ions occurred at the following pH values:

$$Al(OH)^{2+}/Al^{3+} = 1 \quad \text{at pH } 4{\cdot}89$$
$$Al(OH)_2^+/Al^{3+} = 1 \quad \text{at pH } 4{\cdot}28$$
$$Al(OH)_4^-/Al^{3+} = 1 \quad \text{at pH } 5{\cdot}87_5$$

At other pH values the concentration ratios read as in Table 2.

One may summarize the various observations as follows. In strongly acid solutions the octahedral cation $Al(H_2O)_6^{3+}$ is dominant. At pH 1–4 it

TABLE 2

Estimated concentration ratios of ion pairs as functions of pH[8a]

pH	$\dfrac{Al(OH)_4^-}{Al^{3+}}$	$\dfrac{Al(OH)^{2+}}{Al^{3+}}$	$\dfrac{Al(OH)_2^+}{Al^{3+}}$
1	$10^{-19{\cdot}5}$	$10^{-3{\cdot}89}$	$10^{-6{\cdot}56}$
2	$10^{-15{\cdot}5}$	$10^{-2{\cdot}89}$	$10^{-4{\cdot}56}$
3	$10^{-11{\cdot}5}$	$10^{-1{\cdot}89}$	$10^{-2{\cdot}56}$
4	$10^{-7{\cdot}5}$	$10^{-0{\cdot}89}$	$10^{-0{\cdot}56}$
5	$10^{-3{\cdot}5}$	$10^{0{\cdot}11}$	$10^{1{\cdot}44}$
6	$10^{0{\cdot}5}$	$10^{1{\cdot}11}$	$10^{3{\cdot}44}$
7	$10^{4{\cdot}5}$	$10^{2{\cdot}11}$	$10^{5{\cdot}44}$
8	$10^{8{\cdot}5}$	$10^{3{\cdot}11}$	$10^{7{\cdot}44}$
9	$10^{12{\cdot}5}$	$10^{4{\cdot}11}$	$10^{9{\cdot}44}$
10	$10^{16{\cdot}5}$	$10^{5{\cdot}11}$	$10^{11{\cdot}44}$
11	$10^{20{\cdot}5}$	$10^{6{\cdot}11}$	$10^{13{\cdot}44}$
12	$10^{24{\cdot}5}$	$10^{7{\cdot}11}$	$10^{15{\cdot}44}$
13	$10^{28{\cdot}5}$	$10^{8{\cdot}11}$	$10^{17{\cdot}44}$
14	$10^{32{\cdot}5}$	$10^{9{\cdot}11}$	$10^{19{\cdot}44}$

is more important than hydroxy ions such as $Al(OH)^{2+}$, $Al(OH)_2^+$ or $Al(OH)_4^-$. In the pH range 2–6 polymeric cations like those in chlorhydrol may also be present. However, the acid side is not relevant for zeolite formation. From pH 6 upwards Al^{3+} becomes wholly insignificant, polymeric ions are no longer important and $Al(OH)_4^-$ or very simple dehydration products such as AlO_2^- are dominant. The important $Al(OH)_4^-$ is tetrahedral in configuration and so should favour tectosilicate formation with silicates, in which only the tetrahedral TO_4 groups (T = Al or Si) occur, polymerized into 3-dimensional crystalline networks.

2.2. Silicates and Silicate Solutions

On the alkaline side it has been seen that aluminate solutions are relatively simple with $Al(OH)_4^-$ as the dominant anionic species. With silicates this is certainly not the case and various polymeric species are known to occur. Accordingly one may consider in some detail methods of investigating the silicate anions in solid silicates and in solutions of silicates.

Structural studies have shown a variety of island silicate anions in solid silicates, such as those in Table 3. The examples given refer to compounds of relatively simple chemical composition. Some of the anions of column 3 occur in natural minerals of greater chemical complexity. The dimeric ion $Si_2O_7^{6-}$ is present in hemimorphite, $Zn_4Si_2O_7(OH)_2.H_2O$ and in thortveitite, $Sc_2Si_2O_7$. The cyclic trimer (3-ring anion) occurs in benitoite, $BaTiSi_3O_9$ and wadeite, $K_2ZrSi_3O_9$. Cyclic tetramer (4-ring anion) is found in axinite, $Ca_2(Fe^{II}, Mn)Al_2(OH)BO_3Si_4O_{12}$ and cyclic hexamer (6-ring anion) in beryl, $Be_3Al_2Si_6O_{18}$, cordierite, $Al(MgFe^{II})_2$-$(Si_5Al)O_{18}$ and tourmaline, $(Na, Ca)(Li, Mg, Al)_3(Al, Fe, Mn)_6(OH)_4$-$(BO_3)_3Si_6O_{18}$. Two 6-rings linked to form a hexagonal prism occur in osumilite, $(K, Na, Ca)(MgFe^{II})_2[(Si_{1-x}Al_x)_{12}O_{30}]^{-12(1-x)}$, and in milarite $(K, Ca_2, Be_2, Al)Si_{12}O_{30}.\frac{1}{2}H_2O$.

When silicates are dissolved either in water or in acid, according to their solubility characteristics, one may expect, at least transiently, the silicate anion or silicic acid corresponding with that in the parent compound. Therefore there appears to be scope for investigating silicates like those in Table 3 as materials providing specific anionic units in zeolite synthesis. This aspect has been little explored. The stability of anions or silicic acids under the necessary alkaline conditions is important and leads to considerations of their rates of polymerization and depolymerization, and ways of finding what kinds of anions may co-exist. Some methods of studying these aspects are illustrated below.

TABLE 3
Some island anions in silicates

Number	Compound	Anion	Ref.
1	β-Ca_2SiO_4	Monomeric	
2	Li_4SiO_4	Monomeric	
3	MgK_2SiO_4	Monomeric	
4	$Ca_2SiO_{4\alpha}$-hydrate	Monomeric	(9)
5	$Na_2H_2SiO_4$	Monomeric	(10)
6	γ-Ca_2SiO_4	Monomeric	(11)
7	$Na_2H_2SiO_4.8H_2O$	Monomeric	(12)
8	Ca_3SiO_5	Monomeric	(13)
9	$Ca_2Na_2Si_2O_7$	Dimeric	
10	$Na_6Si_2O_7$	Dimeric	
11	$Ba_3Si_3O_9$	Cyclic-trimeric	(14, 15)
12	$Na_4Cd_2Si_3O_{10}$	Cyclic-trimeric	(16)
13	$Ca_3Si_3O_9$	Cyclic-trimeric	(17)
14	$Si_4O_3Cl_{10}$	Linear tetrameric	(18)
15	$K_4H_4Si_4O_{12}$	Cyclic-tetrameric	(19)
16	$Cu_6Si_6O_{18}.6H_2O$	Cyclic-hexameric	(20)
17	$[Cu(en)_2]Si_8O_{20}{}^a$	Double 4-ring	(21)
18	Na_2SiO_3	Chain polymeric	(22)
19	Li_2SiO_3	Chain polymeric	(23)

^{a} "en" denotes ethylene diamine.

(i) *The molybdate method* Monomeric silicic acid reacts readily with first-order kinetics with acid molybdate solution to give 12:1 molyb-dosilicic acid. With polysilicic acids the reaction velocity depends on the degree of polymerization because this influences the rate at which monomer becomes available by hydrolysis.[24,25,26] The rate of formation of the 12:1 acid is also different for cyclic and linear polymers having the same number of Si atoms, indicating different rates of hydrolysis to monomer. Rates were measured for the silicic acids derived from compounds 1, 2, 3 and 4 of Table 3 (monomeric acid); from 9 and 10 (dimeric acid); from 15 (cyclic tetramer); and from 18 and 19 (chain polymers). At 25°C first-order rate constants for forming the 12:1 heteropoly acid starting from monomeric, dimeric and cyclic tetrameric silicic acids were 1·7, 0·90 and 0·67 min^{-1} respectively. Semi-log plots of the percentages of unreacted silica are shown as functions of time in Fig. 1[26] for specific monomeric, dimeric, linear-polymeric and layer silicates respectively. The layer silicates reacted extremely slowly (slow production of monomeric silicic acid). The kinetics give indications of the polymer present, but in solutions where various polymeric species are simultaneously present the

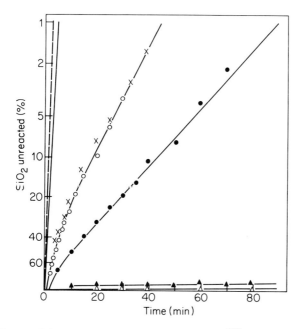

Fig. 1. Rates of formation of silicomolybdic acid[26] from monosilicate (-----); disilicate (——); polysilicate Na_2SiO_3 ($\times$, $\bigcirc$); polysilicate Li_2SiO_3 ($\bullet$); and phyllosilicate $K_2Si_2O_5$ ($\blacktriangle$) and $Na_2Si_2O_5$ ($\triangle$).

argument cannot be more than qualitative. In all cases the $12:1$ acid $H_8[Si(Mo_2O_7)_6]$ together with water of crystallization is the final product.

(ii) *Paper chromatography* Dilute solutions of silicic acid hydrolyse or polymerize only slowly at pH 2·3 and at 0°C. Accordingly, polymer mixtures have been analysed by paper chromatography using as examples of eluants 800 mg of trichloracetic acid in 100 cm^3 of dioxane and 9 cm^3 of water; or 160 mg of trichloracetic acid, 100 cm^3 of dioxane and 3 cm^3 of water.[27] The test substances were compounds 6, 7 and 8 of Table 3 (monomer), compound 11 (cyclic trimer), compound 15 (cyclic tetramer) and compound 16 (cyclic hexamer). The results were characteristic of the different species.

(iii) *Trimethylsilylation* Lentz[28] introduced a chemical method for identifying the anions present in water or acid-soluble silicates. The hydroxyl groups of the silicic acid liberated from the parent silicate can react with trimethylsilyl chloride or the corresponding ether to give the trimethyl ester. Provided ester formation is rapid compared with polymerization and depolymerization of the silicic acid the ester formed will serve to characterize the anion in the original silicate. Silicates were dissolved in hydrochloric acid in the presence of hexamethyldisiloxane, $(CH_3)_3Si–O–Si(CH_3)_3$, which served both to produce the esters and to dissolve these. Lentz also added isopropanol which likewise served as a solvent and was considered to retard side reactions. The esters were adequately stable to hydrolysis and heat and soluble in various organic liquids. Those of lower molecular weight silicic acids could be separated and analysed by gas–liquid partition chromatography at temperatures up to 300°C.

Gotz and Masson[29] modified the method by using trimethylsilyl chloride to form the esters and hexamethyldisiloxane to serve only as solvent. The addition of water was necessary especially when anhydrous minerals were trimethylsilylated, in order to generate the hydrochloric acid needed to form the silicic acid corresponding with the anion in the parent silicate. The trimethylsilyl alcohol liberated is itself able to form esters with the silicic acid. The amounts of water added required careful regulation to minimize side reactions.

Tamas et al.[30] investigated trimethylsilylation in a range of solvents again using $(CH_3)_3SiCl$ and $(CH_3)_3Si–O–Si(CH_3)_3$ as the esterifying agents. Among all the solvents used dimethylformamide gave the best results in that it largely suppressed side reactions and was a good solvent for by-products such as the $CaCl_2$ and $ZnCl_2$ liberated by hydrochloric acid from β-Ca_2SiO_4 and hemimorphite $(Zn_4(OH)_2Si_2O_7.H_2O)$. However, not all silicates reacted satisfactorily with the reagent mixture in dry dimethylformamide, for example metakaolinite and natrolite. On the other hand, silicates containing SiO_4^{4-} and $Si_2O_7^{6-}$ anions yielded the trimethylsilyl esters in absence of water and with very little formation of products other than the ester based on the anion in the parent silicate.

The importance of side reactions in the Lentz method is illustrated by the work of Calhoun et al.,[31] who used the Lentz procedure with $Ag_{10}Si_4O_{13}$, which contains the linear tetrameric anion $Si_4O_{13}^{10-}$.[32] The main product was the trimethylsilyl ester of the tetrameric cyclic anion $Si_4O_{12}^{8-}$. But when the Lentz procedure as modified by Tamas et al. was subsequently employed the ester of the linear tetrameric anion was obtained virtually free of by-product esters.[33] Likewise H-Pb_2SiO_4 which contains the tetrameric cyclic anion[34] yielded only the trimethyl-silyl ester of this anion.[33] The foregoing results suggest that the Lentz method or modifications of it can be of considerable interest in investigating silicate solutions, but that side reactions are often likely to be significant.

(iv) ^{29}Si *nuclear magnetic resonance spectroscopy* High-resolution ^{29}Si n.m.r. spectroscopy has been very successful in elucidating the structures of silicic acids or silicate anions present in solution. The ^{29}Si atoms give chemical shifts which differ according to the position of this atom in the structure. Thus the shifts fall into subdivisions with well separated ranges[35,36] for monosilicates (Q^0), disilicates and chain end groups (Q^1), middle groups in chains (Q^2), chain-branching positions (Q^3) and three-dimensional cross-linked framework positions (Q^4). The high resolution obtained with liquids and solutions has been extended to crystalline solids using high-speed magic-angle sample spinning with high-power proton decoupling and, wherever possible, polarization transfer.[36a] The resolution achieved was 1 p.p.m. The total range of ^{29}Si chemical shifts is from about -60 to -120 p.p.m. with Si(CH$_3$)$_4$ as reference material. Although ionization and cation influence are reflected in the shifts the isotropic ^{29}Si chemical shifts in solids and solutions are in general alike. The analytical sub-divisions are illustrated in Fig. 2 for a series of reference solids and solutions. Although within each Q-division (Q^0 to Q^4) there is variability among the solids and solutions examined, in no case was there overlap between the shifts for one division and those of another. Provided this remains true for an unlimited number of reference solids and solutions, high-resolution n.m.r. will be an important probe for the solution chemistry of silicates. For solid silicates it will be a useful subsidiary to X-ray structure determinations.

From the viewpoint of zeolite synthesis, dissolved aluminosilicate precursors to nucleation are of particular interest. Here results obtained with solid aluminosilicates may serve as a guide because it has been found that there is a strong paramagnetic influence of 4-co-ordinated Al substituting for Si in the second co-ordination sphere of Si. Thus in aluminosilicates there may be different extents of substitution of Si by Al giving rise to non-equivalent Q^4 units of different degree of Al substitution with the

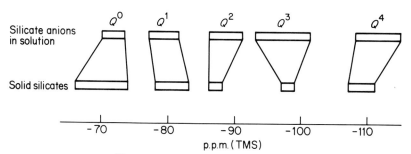

Fig. 2. Ranges of ^{29}Si chemical shifts of different structural units of silicate anions in solutions and in solid silicates.[36a]

TABLE 4

^{29}Si isotropic Q^4 chemical shifts in tectosilicates (p.p.m. difference from Si(CH$_3$)$_4$) based on Table 3 of Chapter 6

Second co-ordination sphere of Si^a	Maximum ranges in Q^4	Mean of values	No. of values in mean	Intervals in means $Q^4(n\text{Al}) - Q^4((n-1)\text{Al})$
4Al	−80·1 to −86·5	−83·7	17	
3Al, 1Si	−86·4 to −94·0	−89·4	18	5·7
2Al, 2Si	−92·5 to −99·4	−95·4	19	6·0
1Al, 3Si	−96·7 to −105·3	−100·7	25	5·3
4Si	−103·1 to −110	−106·6	17	6·5

^{a} The first coordination sphere consists of four O atoms.

different chemical shifts illustrated in Table 4. If in solution there were dissolved aluminosilicate precursors giving the Q^4 (nAl) sub-divisions where $n = 1$ to 4 with the range of values in Table 4, these ranges could now overlap with Q^1, Q^2 and Q^3 in Fig. 2 for purely silica species, because substitution of Si by Al in the second co-ordination sphere progressively lowers the ^{29}Si chemical shift, as seen in Table 4. However, evidence of aluminosilicate species would be available if, for example, tetramethylammonium aluminate solution were added to tetramethylammonium silicate solution, which gives tetramethylammonium aluminosilicate solution (not gel). The high-resolution ^{29}Si chemical shifts of such a solution compared with those of the original tetramethylammonium silicate solution should show whether aluminosilicate species were present.

We may consider examples of the use of high-resolution ^{29}Si FT n.m.r. to investigate silicate solutions. Figure 3 shows the spectrum of a solution of sodium metasilicate.[35] The peak M was the only one remaining when the ratio SiO$_2$/Na$_2$O = r was reduced to 0·16 (high relative excess of alkali) but was almost absent from solutions with $r > 3$. It was therefore ascribed to a monomeric species (Q^0). Observations on the variation of peak areas with r indicated that peaks E and D arose from terminal Si (Q^1) and that D most probably arose from dimer. The cluster of peaks C was attributed to chain silicons (Q^2). B was ascribed to positions of chain branching (Q^3) while the broad peak Q found when $r = 3.8$ was attributed to positions of double-chain branching (Q^4).

The ^{29}Si n.m.r. method was extended by a further investigation of silicate solutions with various Si/Na or Si/K ratios, confirming the five silicon environments of the previous paragraph,[37] which occurred with characteristic non-overlapping ranges in the chemical shifts (cf. Fig. 2).

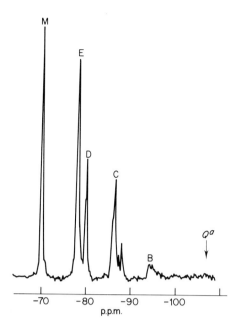

Fig. 3. ^{29}Si Fourier transform n.m.r. spectrum at 19·9 MHz of 1·35 mol dm^{-3} Na$_2$SiO$_3$ (SiO$_2$/Na$_2$O = 1) at 26°C (66430 transients, $\pi/2$ pulse, repetition rate 0·8 s). a denotes the position of resonance observed for solutions with a high ratio SiO$_2$/Na$_2$O, equal for example to 3·8, but not observed for SiO$_2$/Na$_2$O = 1.$^{(35)}$

Within a given subdivision Q^n ($n = 0$ to 4) there was finer-scale signal-splitting arising from the influence of differing neighbouring groups. It was therefore concluded that there is an equilibrium among many condensed anionic silicate species. The equilibrium was considered to shift towards anions with a high content of branching and cross-linking groups as alkali/SiO$_2$ decreased. Potassium silicate solutions also gave ^{29}Si FT n.m.r. spectra indicating various anionic species$^{(37)}$ (Fig. 4). The spectrum of a freshly prepared solution containing 8 M KOH, 3 M SiO$_2$ and 0·02 M Cr^{III} in H$_2$O + D$_2$O was explained by the presence of five species at the following concentrations:

monomer	1·45 M
dimer	0·22 M
cyclic trimer	0·16 M
substituted cyclic trimer	0·08 M
linear trimer	0·07 M
linear tetramer	0·03 M

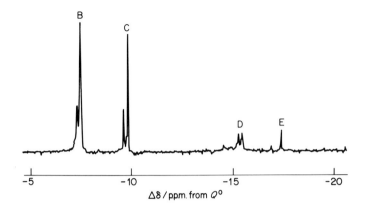

Fig. 4. ^{29}Si n.m.r. spectrum of a freshly prepared silicate solution:[37] 8 mol dm^{-3} KOH; 3 mol dm^{-3} SiO$_2$; and 0·02 mol dm^{-3} Cr^{3+}; solution in H$_2$O+D$_2$O at 30°C. 5466 90° pulses, $T_{ac} = 4$ s and $T_d = 2$ s, i.e. total accumulation time 9·1 h.

That the cations present may influence the kind of anionic species in solution appears to be further shown by the study of dissolved quaternary ammonium silicates. Quaternary ammonium and other organic bases can very successfully replace all or part of inorganic bases in zeolite syntheses, as discovered by Barrer and Denny[38] and subsequently extensively explored (Chapter 4, §5.4 and Chapter 5, Tables 2, 9 and 11). The anionic constitution of 1·8 M tetramethylammonium (TMA) silicate solutions having TMA/Si = 1 was investigated using paper chromatography; the molybdate method; trimethylsilylation and mass spectroscopy; and X-ray structure determination of crystals of TMA-silicate grown from the solution. The crystals, of composition (TMA)$_8$Si$_8$O$_{20}$.69H$_2$O contained only the anion Si$_8$O$_{20}^{8-}$ which was the double 4-ring found in the framework of Linde zeolite of type A. In freshly prepared solutions from the TMA-silicate crystals, therefore, this should be the anion initially present. With increasing dilution other anions were detected[39,40] which were considered to be monomer, dimer, cyclic tetramer and cyclic hexamer. Subsequently, from solutions with TMA/Si > 1 a new silicate crystallized together with that having double 4-rings.[41] It contained the anion Si$_7$O$_{19}^{10-}$, which could have any one of a number of configurations, none of which could be decisively chosen.

Solutions of tetraethylammonium (TEA) silicates have likewise been studied, using paper chromatography, ^{29}Si n.m.r. and trimethylsilylation to identify the anions present.[41a] Concentrated TEA silicates with TEA/Si from 28 to 1 were reported to contain **mainly** double 3-ring anions with small amounts of mono-, di-, tri-, tetra-, cyclotri-, cyclotetra-, double 4-ring and other polycyclic anions. From the solutions a crystalline product $[TEA]_6Si_6O_{15}.57H_2O$ was obtained. Concentrated TEA silicate solutions having TEA/Si from $0·8$ to $0·6$ contained mainly double 3-, 4-, 5- and possibly 6-ring anions, but now the crystals obtained were always a TEA silicate containing double 4-ring anions.

The systems $N(n-C_4H_9)_4OH-SiO_2-H_2O$ and $N(i-C_5H_{11})_4OH-SiO_2-H_2O$ have also been examined.[42] From the solutions a tetra-alkylammonium silicate crystallized containing the anion $Si_{10}O_{25}^{10-}$. Trimethylsilylation of the solution of the new silicate yielded the ester $((CH_3)_3Si)_{10}Si_{10}O_{25}$. Paper and thin-layer chromatography of the trimethyl-silylation products indicated that other species were present either in the original solutions or formed during the esterification. The structure pro-posed for the anion $Si_{10}O_{25}^{10-}$ was a double 5-ring (pentagonal prism). A particular interest in these unusual silicate anions resides in the possibility of using them to introduce precursor species of known structure into reaction mixtures during zeolite synthesis, and so to influence nucleation and the crystals formed. Evidence of this has been obtained for zeolite A.[43] One may readily model the anionic framework of this zeolite from double 4-rings (cubic units) and monomeric anions of aluminate and silicate. The silicate $(TMA)_8Si_8O_{20}.69H_2O$ provides such cubic units. It is therefore of interest that when this silicate was added to aqueous very alkaline hydrogels made from sodium aluminate and sodium silicate solutions rich in the monomer ions, improved syntheses of silica-rich zeolite A were achieved, free from co-crystallizing species (Chapter 4, Table 7). To obtain the best results the $(TMA)_8Si_8O_{20}.69H_2O$ had to be the last component added. This would be expected to preserve the $Si_8O_{20}^{8-}$ units in the reaction mixture as long as possible.

The examples given have illustrated how, by varying the choice and concentration of soluble silicates and alkali one may vary the silicate anions initially present. The examples also suggest that condensation and hydrolysis reactions occur especially in alkaline media so that preserving anions present in the parent solid silicate may not be easy. In this connection Stade and Wieker's[44] study of the kinetics of degradation of aqueous polysilic acid is relevant. They used the molybdate method previously described. Activation energies were obtained and the hyd-rolysis was found to be catalysed by nucleophiles. Maximum stability of polysilic acid was found at $0°C$ and pH $3·3$, the half-life then being 53 days. These observations refer to concentrations up to about 10^{-3} M, where the end product is monosilicic acid. Above this concentration layer silicates were reported to form by condensation, with cyclosilicates as intermediates.

Thus, to retain as long as possible the anions in freshly made solutions of soluble silicates the temperature should be as low as possible. The aluminate may then be added quickly with stirring to yield a gel which may incorporate and so preserve in the network some of the silicate units from the parent solution. However, evidence has accumulated (Chapter 4) that nucleation takes place from dissolved precursor species, so that preservation of the original silicate units in gel may not prevent subsequent reactions involving the supernatant solution, or solution of gel to give new dissolved species. The alkaline gels and supernatant solutions are often aged at or near room temperature before raising the temperature to that at which crystallization takes place and the longer the interval involved the less certain it becomes how much "memory" the systems retain of the original species. There is a strong interaction between aluminates and silicates which can generate, in addition to the gel, dissolved aluminosilicate anions (§2.5).

Silicate solutions which are very viscous can be made in which the ratios SiO_2/M_2O (M = K, Na or Li) are as high as 25. In the range 4–25 these solutions have been termed polysilicates. The solutions are an equilibrium mixture of cations and of negatively charged particles of colloidal silica so small thay they can be regarded also as polyanions. There is a borderline between polysilicate solutions and silica sols stabilized with alkali. The distinction can be drawn at a particle size of silica below which the enhanced solubility of silica as $Si(OH)_4$ results in spontaneous growth of the particles unless more alkali is added.[45]

Complementing the four methods already described for investigating silicate solutions are other methods including hydrogen electrode studies[46,47] and equilibrium ultra-centrifugation.[48] Measurements with hydrogen electrodes in carbonate-free sodium silicate solutions in 0·5 M and 3·0 M $NaClO_4$ at 25 and 50°C and in 0·5 M NaCl at 25°C, in each case over a range of silicate concentrations, were interpreted as indicating both monomeric and polymeric ions. The monomeric species were considered to be $Si(OH)_4$, $SiO(OH)_3^-$ and $SiO_2(OH)_2^{2-}$ and the polymerized species to have two or four Si atoms. Aveston[48] submitted dissolved silicate compositions $Na_mH_{4-m}SiO_4$ in NaCl solutions with m between 0·64 and 1·04 to equilibrium centrifugation. He reported weight average degrees of polymerization which increased rapidly as m decreased or concentration increased. The results were compatible with an extended series of polymers having $m = 0·5$ in addition to monomeric and tetrameric species as proposed in Refs 46 and 47.

Comparison of the behaviour of aluminate and silicate solutions on the alkaline side shows, from the examples given, that silicate solutions can contain many kinds of anion, whereas aluminate solutions contain primarily monomeric ions Al(OH)$_4^-$ or simple dehydration products such as AlO$_2^-$.

2.3. *Silica Sols with Aluminate Solutions*

Instead of soluble silicates, including water glass, silica sols of high silica content (e.g. 30% by weight of SiO$_2$) have found wide use as the source of silica. When mixed with aluminate solutions a gel is as a rule quickly formed which must initially be more heterogeneous than gels made from well stirred mixtures of solutions of soluble silicates and aluminates. However, in the alkaline conditions of zeolite synthesis the colloidal silica can react further with alkali to bring silica into solution and to help to homogenize the hydrogel during the ripening phase. A definitive study of the reaction between silica sol and alkali was made by Polak and Stobiecka.[49] Addition of appropriate concentrations of sodium hydroxide solution to 30% silica sol first gave a dense precipitate of gel in which the silica then slowly depolymerized with the alkali to build up the concentration of dissolved silica. These silica gels treated with the alkali for five minutes, three hours and one day, respectively, were thoroughly mixed with sodium aluminate solutions to give the same overall composition for each system, the oxide formula being 2·4Na$_2$O, Al$_2$O$_3$, 8·0SiO$_2$, 96H$_2$O.

Each preparation was then aged for one day, after which time the chemical compositions (SiO$_2$ and Al$_2$O$_3$) of supernatant liquid and of solid respectively were analysed. These compositions are shown in Figs 5 and 6. Prior to any crystallization to zeolite each composition depended on the time over which the parent precipitated silica gel had been aged with the sodium hydroxide solution before adding the aluminate. After crystallizing to zeolite, however, the composition of the crystals was independent of the length of the above preliminary treatments with caustic soda.

In a second series of experiments (P1 to P9) the standard oxide composition of the previous paragraph was made from aluminate and that silica gel which had been treated with NaOH aq for five minutes only. The resultant aluminosilicate hydrogels were then aged over periods of four hours to one year and the compositional changes of liquid and solid were again monitored as functions of time. In the liquid phase for the first two days there was little dissolved silica (about 0·2%), but between two and seven days this amount rose sharply to about 8% and thereafter changed little, being 10·4% after 360 days. Over the first 7 days of ageing the dissolved Al$_2$O$_3$ dropped sharply from 2% to a very low value. In the solid the Al$_2$O$_3$ content rose in the same time from 6% to 20% and then

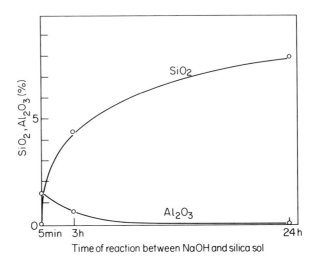

Fig. 5. The chemical compositions of the liquid phase before crystallization. Parent gel oxide composition in all experiments was $2\cdot4Na_2O$,
Al_2O_3, $8SiO_2$, $96H_2O$. The silica sols used to make the parent mixture
were treated for 5 min, 3 h and 24 h respectively with NaOH aq. before
the aluminosilicate composition was found (the abcissa in the figure).
The compositions of the supernatant liquids w.r.t. SiO_2 and Al_2O_3 after
ripening for 24 h are the ordinate.[49]

stayed nearly constant while the silica in the same period dropped from
about 89% to 63% and thereafter remained nearly constant. Thus, over
seven days the compositions of both solid and liquid were evolving,
yielding a liquid much enriched in dissolved silica but with little dissolved
alumina and hence relatively free of aluminosilicate species; and a hydrogel much enriched in alumina. When the ageing study was made using
the silica hydrogel treated for 24 hours with caustic soda to make the
standard oxide mix with sodium aluminate (experiment SK3), the limiting
compositions of liquid and solid phases were reached after only about one
day's ageing of the resultant aluminosilicate hydrogels.

2.4. *Aluminosilicate Gel and Metakaolinite*

The gels formed by mixing soluble silicates, including water glass, or silica
sols with soluble aluminates have no long-range order, as shown by the
absence of X-ray diffraction. Nevertheless, the co-ordination number of
the Al, which is four in tectosilicates and four and six in clay minerals, is

of considerable interest. Phosphorescence spectroscopy[50] and X-ray scattering[51,52] have been used to investigate this aspect. The phosphorescence spectra were examined for trace amounts of Fe^{III} or Cr^{III}. Fe^{III} was considered to substitute in traces for Al^{III} in the 4-co-ordination state.[50] In one instance $Al(OH)_3$ was dissolved in NaOH solution and added to amorphous silica doped with 0·02% of Fe^{III}. After refluxing the mixture for 90 hours the gel was filtered and dried at 120°C. It was still amorphous to X-rays. Other gels were prepared in various ways in each case with additions of trace amounts of Fe^{III} or Cr^{III}.

The conclusion from the spectroscopic and X-ray studies of such gels was that they comprise aluminosilicate networks in which Al is in tetrahedral co-ordination with oxygen only under sufficiently alkaline conditions. In neutral or acidic environment tetrahedral Al was not found. The alkaline solution required to give a tetrahedral network could be either NaOH or NH_4OH, so that the base need not be stronger than NH_4OH.[50] It was not, however, possible to produce a tetrahedral

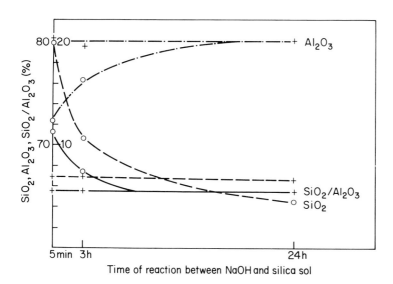

Time of reaction between NaOH and silica sol

Fig. 6. The chemical compositions of the solid phase corresponding with those of the liquid phase shown in Fig. 5. The figure also shows the compositions of the zeolite crystals finally formed, and indicates that these compositions did not depend upon the time of treatment of the silica sols with NaOH aq before making up the parent gel compositions.[49] – – –, SiO_2; —·—, Al_2O_3; ——, SiO_2/Al_2O_3. ○, before crystallization; +, after crystallization.

network when the ammonium aluminosilicate gel was heated sufficiently to eliminate the NH_4^+ (as NH_3) so as to replace this ion by protons. Instead the phosphorescence spectra suggested that the parent framework had been transformed into another structure. It was considered that the tetrahedral framework is not stable to heating unless charge balance is achieved with alkali metal or like cations which are themselves stable to heat.

It was further observed that the Fe^{III} phosphorescence signal from metakaolinite was identical with that from alumina–silica cracking catalysts heated to 400°C or above. Metakaolinite is a dehydroxylated product obtained from kaolinite heated to about 500°C. The loss of hydroxyl water does not involve a sharp transition, but is a function of the temperature and duration of heating. Metakaolinite has lost the c-axis periodicity of kaolinite and is essentially amorphous. The hexagonal layer network, $[Si_2O_5^{2-}]_n$ is probably retained in a rather distorted configuration but the octahedral co-ordination of the aluminous layers is altered strongly through water loss, sometimes with partial separation of amorphous alumina. Whatever the detailed consitution of metakaolinite may be, it is an excellent raw material for zeolite synthesis,[53,54] whereas the parent kaolinite with the Al in 6-co-ordination is much less reactive (Chapter 5).

2.5. Supernatant Solutions

It has been seen that silica is freely soluble in aqueous alkaline solutions (Chapter 2, Table 8) and that the solutions may contain a variety of anionic species (§§2.2 and 2.3). From the viewpoint of zeolite synthesis, however, the chemical nature of the supernatant solutions in contact with the aluminosilicate gels can be of even greater interest. It was suggested in 1966[55] that the precursors which nucleate zeolite crystals and maintain their growth may be polyhedral or polygonal aluminosilicate anions, because the development of such open and complex anionic frameworks as those in zeolites seemed unlikely if the building blocks were simply dissolved AlO_4^{5-} and SiO_4^{4-} tetrahedra. Aluminosilicate precursors in solution could form by the action of aqueous alkali upon the co-existent gel.

It is of course not necessary to have gel. Thus with tetra-alkylammonium hydroxides clear solutions can be prepared containing both parent silicate and aluminate in concentrations up to 2 mol dm^{-3}.[56] There is no precipitation of a gel phase like those found when NaOH or KOH replace the quaternary organic bases. With quaternary bases potentiometric measurements[57] indicated that at the moment of mixing aluminate and silicate solutions reaction occurs between them so that the term "aluminosilicate solution"[58] may be justified. Alkali metal aluminosilicate gels often separate into a supernatant liquid and a gel solid, as studied by Polak and Stobiecka.[49] For aged sodium aluminosilicate gels they reported a co-existent liquid very low in alumina but rich in silica (§2.3). This raises the question whether in this instance Al does or does not form aluminosilicate precursors, and if it does what is the Al/Si ratio in such species.

Ueda and Koizumi[59] obtained a partial answer to the above questions. They found that sodium aluminosilicate gels, initially formed on mixing the components, dissolved within 5 minutes at 100°C to give seemingly limpid solutions provided the Al/Si ratio was sufficiently low. They worked with compositions of oxide formulae $10Na_2O$, $(0\cdot05-0\cdot15)Al_2O_3$, $(6-16)SiO_2$, $(168-592)H_2O$ and investigated compositional boundaries between initial gels which wholly dissolved and those which did not (Fig. 7). When the composition contained no alumina, no amorphous phase formed; but as the ratio Al_2O_3/Na_2O was increased, the regions in which

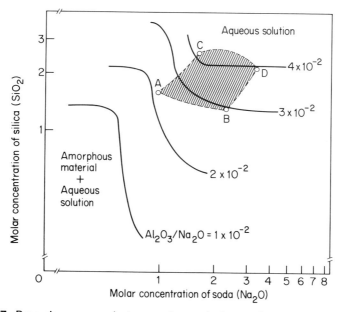

Fig. 7. Boundary curves between clear solution and regions where gel and solution coexisted for various initial molar concentrations of SiO_2 and Na_2O at constant ratios Al_2O_3/Na_2O. The gels were prepared from 10 M NaOH, sodium aluminate and silica sol. The diagram is based on 250 experimental values.[59]

gel co-existed with solution enlarged in the direction of higher concentrations of silica and caustic soda. The area ABDC in Fig. 7 indicates the composition range investigated for zeolite formation when $Al_2O_3/Na_2O \leqslant 1 \cdot 5 \times 10^{-2}$. According to the figure this area is well within the range of limpid solutions free of gel. The experiments thus show that sodium "aluminosilicate" solutions can exist in very alkaline media but that the solubility ranges are very much less than those observed with tetra-alkylammonium "aluminosilicates".

From the clear sodium "aluminosilicate" solutions at 100°C three zeolite species appeared in order of increasing concentration of caustic soda: gismondine-type Na-P; analcime; and sodalite hydrate. It thus appears that certain zeolites such as analcime may be able to nucleate homogeneously and grow without intervention of gel, although in this and analogous experiments giving zeolite Na-X[59a] it has not been shown that the solutions do not scatter a Tyndall cone of light. Homogeneous nucleation does not preclude heterogeneous nucleation on or from gel if conditions are such as to maintain co-existent gel and solution. In another investigation[60] aluminosilicate compositions much richer in alumina than those of Ueda and Koizumi were studied in which $SiO_2/Al_2O_3 \sim 4$ and the bases were Li_2O, Na_2O or K_2O. The mixtures were made with sufficient water and concentrations of alkali to give initially clear liquids, which were then heated at 80°C. These liquids were metastable with respect to gel and zeolites. After an induction time disc-shaped flakes of gel formed in suspension which, when the base was NaOH, give some electron beam diffractions characteristic of smectite layer silicates. Next, zeolite crystals began to form attached to the flakes. Lacunae in the flakes then developed around the zeolite crystals suggesting consumption of the material of the flakes to allow zeolite growth. Transport across lacunae to crystals must presumably have been that of dissolved material. With LiOH as base the zeolite was Li-*ABW*; with NaOH, Na-P and sodalite hydrate appeared; with KOH the product was the chabazite type zeolite K-G. The experiments can be interpreted to mean heterogeneous nucleation of zeolite in or against gel with crystal growth fed by dissolved chemical species. Alternatively, however, zeolite nuclei formed homogeneously could have adhered to the flakes which provide a ready local supply of nutrient solution. Additional strong evidence has been obtained that crystal growth of faujasites Na-X and Na-Y involves reaction between dissolved species and the crystal surface.[60a] If seeds of Na-X were suspended in the solution before gelling, the gel formed around the seeds and then very little crystal growth occurred. If the seeds were added after gelling, excellent yields of Na-Y resulted.

The above and other similar investigations,[60b,c] while supporting the view that crystals grow by deposition of dissolved reactants, so far give no idea of the precursor species in solution. However Ueda and Koizumi carried their investigation further with a detailed study of analcime formation at 100°C. They found that the ratios SiO_2/Al_2O_3 in the zeolite were governed by the *initial* ratios of soda, alumina and silica in the reaction mixtures and that these ratios remained constant, independent of the depletion of nutrient with the progress of crystallization. Since the ratios SiO_2/Al_2O_3 of the crystals fell within the range 3·24 to 4·39, the crystals were very much richer in Al_2O_3 than the solution, and the only reasonable explanation would seem to be that dissolved aluminosilicate

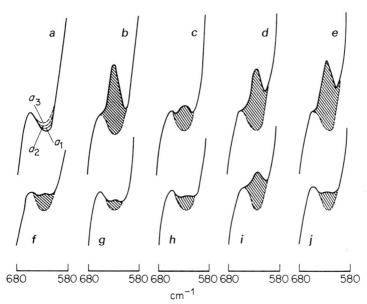

680   580 680   580 680   580 680   580 680   580

cm^{-1}

Fig. 8. Raman spectra of sodium silicate, sodium aluminate and their mixtures. The shaded areas are due to the $Al(OH)_4^-$ ion.[63]

(a) [NaOH] = 0·5 mmol/l. a_1[Si] = 5 mmol/l. a_2[Si] = 10 mmol/l.
                                                            a_3[Si] = 20 mmol/l.
(b)   [Al] = 20 mmol/l.   [NaOH] = 50 mmol/l.
(c)   [Al] = 20 mmol/l.   [Si] = 20 mmol/l.   [NaOH] = 100 mmol/l.
(d)   [Al] = 20 mmol/l.   [Si] = 20 mmol/l.   [NaOH] = 500 mmol/l.
(e)   [Al] = 20 mmol/l.   [Si] = 20 mmol/l.   [NaOH] = 1000 mmol/l.
(f)   [Al] = 5 mmol/l.   [NaOH] = 12·5 mmol/l.
(g)   [Al] = 5 mmol/l.   [Si] = 5 mmol/l.   [NaOH] = 25 mmol/l.
(h)   [Al] = 5 mmol/l.   [Si] = 5 mmol/l.   [NaOH] = 100 mmol/l.
(i)   [Al] = 10 mmol/l.   [NaOH] = 25 mmol/l.
(j)   [Al] = 10 mmol/l.   [Si] = 10 mmol/l.   [NaOH] = 50 mmol/l.

TABLE 5

Complexing of Al(OH)$_4^-$ with silicate anions[63]

Total dissolved Si (mmol dm^{-3})	Total dissolved Al (mmol dm^{-3})	Initial NaOH (mmol dm^{-3})	% Al(OH)$_4^-$ complexed as aluminosilicate anion
5	5	25	50 ± 10
5	5	100	15 ± 10
10	10	50	60 ± 10
20	20	100	75 ± 10
20	20	500	20 ± 10
20	20	1000	10 ± 10

species had formed initially, the compositions of which were fixed by that of the overall reaction mixture but were richer in Al than the reaction mixture; and that these aluminosilicate species were the chemical nutrients for the growing crystals.

Physico-chemical methods have also served to demonstrate the existence of aluminosilicate species in the liquid phase present after mixing soluble silicate and aluminate. These include comparing pH curves obtained during neutralization of silicate or aluminate solutions with those of their mixtures;[61] the changes in electrical conductivity when dissolved silicate and aluminate are mixed;[62] and laser-Raman spectroscopy of these solutions and their mixtures.[63] Results obtained by the latter method are shown in Fig. 8. The spectra were recorded in the 580–680 cm^{-1} range. The polarized band at 625 cm^{-1} (shaded in the figure) is believed to arise from the anion Al(OH)$_4^-$. The figure shows that addition of silicate to the aluminate solution suppresses this band, an effect attributed to formation of aluminosilicate species. From Fig. 8 an estimate was made of the amount of Al(OH)$_4^-$ which had been incorporated into aluminosilicate anions (Table 5). The indication from Table 5 is that, for Si/Al = 1 and for a given ratio NaOH/Al, the greater the concentration of dissolved Al the larger the percentage of Al(OH)$_4^-$ which is complexed; and that for a given concentration of dissolved Al, as NaOH/Al increases so the percentage of Al complexed decreases. The concentrations of dissolved silicon and aluminium were less than 0·02 mol dm^{-3} because above this limit polymerization could result in formation of gel.[64] If similar investigations were made using the much more concentrated tetra-alkylammonium "aluminosilicate" solutions,[56] it should be possible to extend this and other methods of study, such as ^{27}Al and ^{29}Si n.m.r. employed successfully with aluminate and silicate solutions (§§2.1 and 2.2), to give further information about aluminosilicate anions.

3. Concluding Remarks

In order to understand zeolite nucleation and growth at the molecular level as much information as possible is needed about precursor states and species. Therefore, the preceding account has been given of the chemical nature of solutions of aluminates and silicates, of aluminosilicate gels and of the liquids co-existing with such gels. As the work described makes clear, such information is not easy to obtain, but encouraging progress is being made. If dissolved precursor species include polymeric ions like those already observed in crystalline silicates (cf. Table 3) and in solution (§2.2), but with Al replacing Si on some of the tetrahedral positions, then as noted previously polymerization to give more complex species is not difficult to visualize. This is illustrated in Fig. 9,[55] in which

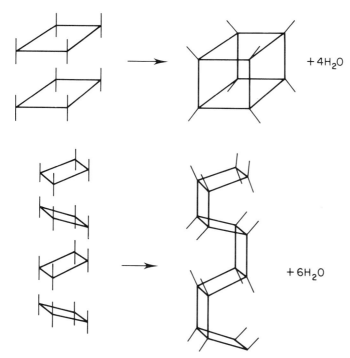

$+ 4H_2O$

$+ 6H_2O$

Fig. 9. The condensation of 4-ring anions to give (a) cubic anions, and (b) part of the double "crankshaft" chain found in felspars and in phillipsite-harmotome zeolites.[55]

there is shown the notional condensation of two 4-ring anions to give the cubic unit found in zeolite A; and also the condensation of several 4-ring anions to give a section of the double-crankshaft chain found in the felspars and the phillipsite-harmotome zeolites. An n-ring means as previously a ring formed by joining n tetrahedra of AlO_4 and SiO_4 by sharing apical oxygens. Assuming the Al–O–Al avoidance rule of Lowenstein,[65] Table 6[55] gives the number of possible compositions for

TABLE 6
Some possible aluminosilicate anions[55]

Anion type	Compositions	Maximum charge	Total configurations
3-ring[a]	$3Si(1)$[b]	-6	2
	$1Al, 2Si(1)$	-7	
4-ring	$4Si(1)$	-8	3
	$1Al, 3Si(1)$	-9	
	$2Al, 2Si(1)$	-10	
5-ring	$5Si(1)$	-10	3
	$1Al, 4Si(1)$	-11	
	$2Al, Si(1)$	-12	
6-ring	$6Si(1)$	-12	4
	$1Al, 5Si(1)$	-13	
	$2Al, 4Si(1)$	-14	
	$3Al, 3Si(1)$	-15	
Cube (double 4-ring)	$8Si(1)$	-8	6
	$1Al, 7Si(1)$	-9	
	$2Al, 6Si(2)$	-10	
	$3Al, 5Si(1)$	-11	
	$4Al, 4Si(1)$	-12	
Hexagonal prism (double 6-ring)	$12Si$	-12	25
	$1Al, 11Si(1)$	-13	
	$2Al, 10Si(7)$[c]	-14	
	$3Al, 9Si(8)$[d]	-15	
	$4Al, 8Si(6)$[c]	-16	
	$5Al, 7Si(1)$	-17	
	$6Al, 6Si(1)$	-18	

[a] An n-ring is composed of n tetrahedra (SiO_4 or AlO_4) linked by sharing apical oxygens.
[b] Figures in brackets are numbers of distinguishable configurations.
[c] Two pairs of these configurations are non-superposable mirror images.
[d] Three pairs of these configurations are mirror images.

several of the simpler precursor anions. It is also seen that optical enantiomorphs can appear among the chemical configurations of the hexagonal prism. The validity of the Al–O–Al avoidance rule of Lowenstein has been challenged, albeit incorrectly (Chapter 6, §4). For example, in the cubic unit in zeolite A it was thought[66,67] that each such unit has two Al–O–Al bonds. If the rule is disregarded, more compositions and chemical configurations than those in Table 6 would be possible.

In the compounds in Table 1, the silicon occurs in 4-co-ordination with respect to oxygen. However, silicon could be introduced in 6-co-ordination. Complexes with this co-ordination are reported when o-dihydroxyphenols of which the simplest is catechol, $(OH)_2C_6H_4$, react in mildly alkaline solution with silica gel, silicates and even quartz and felspars.[68,69,70] The reaction with crystalline tectosilicates in particular is slow; the alkalinity is supplied by ammonium, pyridinium or alkylammonium bases; and air should be excluded during reaction. The need for absence of air from the reaction mixture with silicon present as the complex could be a disadvantage in zeolite synthesis, and the stability of the complex in more strongly alkaline media may be low. The complex has been formulated as

$$\left[Si\left(\begin{matrix}O\\O\end{matrix}\right)_3\right]^{2-} 2RNH_3^+$$

and this formula has been verified for crystals in which RNH_3^+ was replaced by pyridinium ion.[71]

The fluorsilicates represent a stable group in which Si is in octahedral co-ordination with fluorine yielding ions SiF_6^{2-}. Many of the heavy metal fluorsilicates are readily soluble in water although this is less so for Na, K, Rb, Cs and Ba fluorsilicates. These latter, in the alkaline solutions appropriate for zeolite synthesis, may react with some of the excess alkali to give, for example with NaOH, sodium fluoride and sodium silicate. In some areas of synthetic mineralogy fluorides are regarded as good mineralizers,[72] but such a role has been little investigated in zeolite synthesis, nor has the effect of introducing the silicon as a fluorsilicate. Indeed fluorsilicates may be stable enough to co-exist with tectosilicates. For example, an early claim by Baur[73] to have made faujasite hydrothermally from silica and potassium aluminate appears instead to have been a preparation of hieratite, K_2SiF_6,[74] which like faujasite is cubic. Baur often made his silica gels by the reaction between water and SiF_4 and the silica then apparently retained some HF or H_2SiF_6, Fluoride ion (as NH_4F, KF or NaF) was included in quantity in the reaction mixture in a synthesis[75] of crystals of the porous silica, silicalite 1, related structurally to zeolite ZSM-5. The presence of F^- at a pH of 11 or less allowed the preparation of 2–15 μm crystals even at 100°C. The growth was accelerated as compared with the rate in absence of F^- so that a positive mineralizing role of this ion seems to have been demonstrated in this instance.

Finally, it is noted that penta-co-ordinated Si occurs in compounds such as I, II and III below:

Structures have been determined for crystals of I when R is C_6H_5[76] and m-$C_6H_4NO_2$;[77] for II when R = CH_3;[78] and for III when R = C_6H_5.[79] No attempt to use penta-co-ordinated Si as the source of silicon in the preparation of zeolites has so far been made.

References

1. J.R. Glastonbury, *Chemistry and Industry* (1961) 121.
2. R.J. Moolenaar, J.C. Evans and L.D. McKeever, *J. Phys. Chem.* (1970) **74,** 3629.
3. von E. Herrmann, *Z. Anorg. Allg. Chemie* (1953) **274,** 81.
4. E. Calvert, H. Thibon, A. Maillard and P. Boivinet, *Bull. Soc. Chim. Fr.* (1950) **17,** 1308.
5. G. Johansson, *Acta Chem. Scand.* (1966) **20,** 505.
6. D.N. Waters and M.S. Henty, *J. Chem. Soc. Dalton* (1977) 243.
7. G. Johansson, *Acta Chem. Scand.* (1960) **14,** 771.
8. D.E.W. Vaughan and R.J. Lussier, *in* "5th Int. Zeolite Symposium, Naples", (Ed. L.V.C. Rees), p. 94. Heyden, London, 1980.
9. L. Heller, *Acta Cryst.* (1952) **5,** 724.
10. cf L.G. Sillen and N. Ingri, *Acta Chem. Scand.* (1959) **13,** 758.
11. D.K. Smith, A. Majumdar and F. Ordway, *Acta Cryst.* (1965) **18,** 787.
12. P.B. Jamieson and L.S. Dent Glasser, *Acta Cryst.* (1961) **14,** 1298.
13. J.W. Jeffrey, *Acta Cryst.* (1952) **5,** 26.
14. W. Hilmer, *Naturwiss.* (1958) **45,** 238.
15. F. Liebau, *Neues Jahrb. Mineral. Abh.* (1960) **94,** 1209.
16. M.A. Simonov, Y.K. Egorov-Tismendo and N.V. Belov, *Dokl. Akad. Nauk. SSSR* (1968) **179,** 1329.
17. F.J. Trojer, *Z. Krist.* (1969) **130,** 185.
18. W.C. Schumb and A.J. Stevens, *J. Amer. Chem. Soc.* (1950) **72,** 3178.
19. W. Hilmer, *Acta Cryst.* (1964) **17,** 1063.

20. von H.G. Heide, K. Boll-Dornsberger, E. Thilo and E.M. Thilo, *Acta Cryst.* (1955) **8,** 425.

21. N.A. Toropov, Y.I. Smolin, Y.F. Shepelev and I.K. Butikova, *Krist. SSSR* (1972) **17,** 15.

22. A. Grund and H. Pizy, *Acta Cryst.* (1952) **5,** 837.

23. H. Seemann, *Acta Cryst.* (1956) **9,** 251.

24. G.B. Alexander, *J. Amer. Chem. Soc.* (1953) **75,** 5655.

25. T.L. O'Connor, *J. Phys. Chem.* (1961) **65,** 1.

26. E. Thilo, W. Wieker and H. Stade, *Z. Anorg. Allg. Chem.* (1965) **340,** 261.

27. W. Wieker and D. Hoebbel, *Z. Anorg. Allg. Chem.* (1969) **366,** 139.

28. C.W. Lentz, *Inorg. Chem.* (1964) **3,** 374.

29. J. Gotz and C.R. Masson, *J. Chem. Soc.* A (1970) 2683; (1971) 686.

30. F.D. Tamas, A.K. Sarkar and D.M. Roy, *in* "Hydraulic Cement Pastes: their Structure and Properties", p. 55. Cement and Concrete Association, Slough, 1976.

31. H.P. Calhoun, C.R. Masson and M. Jansen, *J. Chem. Soc. Chem. Comm.* (1980) 576.

32. M. Jansen and H. L. Keller, *Angew. Chem. Int. Ed. Engl.* (1979) **18,** 464.

33. L.S.D. Glasser and E.E. Lachowski, *J. Chem. Soc. Chem. Comm.* (1980) 973.

34. J. Gotz, D. Hoebbel and W. Wieker, *Z. Anorg. Chem.* (1975) **416,** 163.

35. R.O. Gould, B.M. Lowe and N.A. MacGilp, *J. Chem. Soc. Chem. Comm.* (1974) 720.

36. von G. Engelhardt, D. Zeigan, H. Jancke, D. Hoebbel and W. Wieker, *Z. Anorg. Allg. Chem.* (1975) **418,** 17.

36a. E. Lippmaa, M. Magi, A. Samoson, G. Engelhardt and A.-R. Grimmer, *J. Amer. Chem. Soc.* (1980) **102,** 4889.

37. R.K. Harris and R.H. Newman, *J. Chem. Soc. Faraday II* (1977) **73,** 1153.

38. R.M. Barrer and P.J. Denny, *J. Chem. Soc.* (1961) 983.

39. D. Hoebbel and W. Wieker, *Z. Anorg. Allg. Chem.* (1971) **384,** 43.

40. W. Wieker "Neue Entwicklungen der anorganische Chemie." V.E.B. Verlag de Wissenschaften, Berlin, 1974.

41. D. Hoebbel and W. Wieker, *Z. Anorg. Allg. Chem.* (1974) **405,** 267.

41a. D. Hoebbel, G. Garzo, G. Engelhardt, R. Ebert, E. Lippmaa and M. Alla, *Z. Anorg. Allg. Chem.* (1980) **465,** 15.

42. D. Hoebbel, W. Wieker, P. Francke and A. Otto, *Z. Anorg. Allg. Chem.* (1975) **418,** 35.

43. H. Kacirek and H. Lechert, *in* "Molecular Sieves II" (Ed. J.R. Katzer), p. 244. American Chemical Society Symposium Series No. 40, 1977.

44. H. Stade and W. Wieker, *Z. Anorg. Allg. Chem.* (1971) **384,** 53.

45. R.K. Iler, "The Chemistry of Silica", Ch. 2. Wiley, New York, 1979.

46. G. Lagerstrom, *Acta Chem. Scand.* (1959) **13,** 722.

47. N. Ingri, *Acta Chem. Scand.* (1959) **13,** 758.

48. J. Aveston, *J. Chem. Soc.* (1965) 4445.

49. F. Polak and E. Stobiecka, *Bull. Acad. Polonaise des Sciences, Serie des Sciences Chim.* (1978) **26,** 899.

49a. A.J. Léonard, S. Suzuki, J.J. Fripiat and C. de Kimpe, *J. Phys. Chem.* (1964) **68,** 2608.

50. G. T. Pott, *6th Int. Cong. on Catalysis, London, 12–16th July,* 1976, Paper B.3.

51. A.J. Léonard, P. Ratnasamy, F.D. Declerck and J.J. Fripiat, *Faraday Discussion No. 52* (1972) p. 98.

52. P. Ratnasamy and A.J. Léonard, *Catalysis Rev.* (1972) **6,** 29.
53. R.M. Barrer and D.E. Mainwaring, *J. Chem. Soc. Dalton* (1972) 1254, 1259, 2534.
54. R.M. Barrer, R. Beaumont and C. Colella, *J. Chem. Soc. Dalton* (1974) 934.
55. R.M. Barrer, *Chemistry in Britain* (1966) 380.
56. W. Sieber and W.M. Meier, *Helv. Chim. Acta* (1974) **57,** 1533.
57. W. Sieber, unpublished.
58. R.M. Barrer and W. Sieber, *J. Chem. Soc. Dalton* (1977) 1020.
59. S. Ueda and M. Koizumi, *Amer. Mineral.* (1979) **64,** 23.
59a. E.F. Freund, *J. Cryst. Growth* (1976) **34,** 23.
60. R. Aiello, R.M. Barrer and I.S. Kerr, "Adv. Chem. Series", No. 101, p. 44. American Chemical Society, 1971.
60a. H. Kacirek and H. Lechert, *J. Phys. Chem.* (1975) **79,** 1589.
60b. C.L. Angell and W.H. Flank *in* "Molecular Sieves II" (Ed. J.R. Katzer), p. 194. American Chemical Society Symposium Series No. 40, 1977.
60c. A. Culfaz and L.B. Sand, *in* "Molecular Sieves" (Ed. W.M. Meier and J.B. Uytterhoven), p. 140. Advances in Chemistry Series No. 121. American Chemical Society, 1973.
61. J.L. Guth, P. Caullet and R. Wey, *Bull. Soc. Chim. Fr.* (1974) (9–10), 1758.
62. J.L. Guth, P. Caullet and R. Wey, *Bull. Soc. Chim. Fr.,* 1974 (11), 2363.
63. J.L. Guth, P. Caullet, P. Jacques and R. Wey, *Bull. Soc. Chim. Fr.* (1980) (3–4), 121.
64. J.L. Guth, P. Caullet and R. Wey, *Bull. Soc. Chim. Fr.* (1975) (11–12), 2375.
65. W. Lowenstein, *Amer. Mineral.* (1954) **39,** 92.
66. G. Engelhardt, D. Zeigan, E. Lippmaa and M. Magi, *Z. Anorg. Allg. Chem.* (1980) **468,** 35.
67. E.A. Lodge, L.A. Bursill and J.M. Thomas, *J. Chem. Soc. Chem. Comm.* (1980) 875.
68. H. Bartels and H. Erlenmeyer, *Helv. Chim. Acta* (1964) **47,** 7.
69. R. Bach and H. Sticher, *Experientia* (1966) **22,** 515.
70. R. Bach and H. Sticher, *Schweiz. Landwirtschaft. Forschung* (1963) **2,** 139.
71. J.J. Flynn and F.P. Boer, *J. Amer. Chem. Soc.* (1969) **91,** 5756.
72. D.P. Grigor'ev, *Mém. Soc. Russe Minéral.* (1939) **68,** 363.
73. E. Baur, *Z. Anorg. Chem.* (1911) **72,** 119.
74. M. Schlaepfer and P. Niggli, *Z. Anorg. Chem.* (1914) **87,** 52.
75. E.M. Flanigen and R.L. Patton, to Union Carbide, 1978, U.S.P. 4073865.
76. J.W. Turley and F.P. Boer, *J. Amer. Chem. Soc.* (1968) **90,** 4026.
77. J.W. Turley and F.P. Boer, *J. Amer. Chem. Soc.* (1969) **91,** 4129.
78. F.P. Boer and J.W. Turley, *J. Amer. Chem. Soc.* (1969) **91,** 4134.
79. F.P. Boer, J.W. Turley and J.J. Flynn, *J. Amer. Chem. Soc.* (1968) **90,** 5102.

Chapter 4

Nucleation, Crystal Growth and Reaction Variables

1. Introduction

The chemical precursors to zeolite formation (dissolved silicate, aluminate and aluminosilicate species, and gel) are more complex and varied than are those involved in precipitation of ionic compounds like $BaSO_4$ or AgCl. Nevertheless the stages of zeolite formation and salt precipitation are very similar: small aggregations of precursors give unstable germ nuclei; some of these embryos become large enough to be stable nuclei; and spontaneous deposition of more material on such nuclei results in larger crystallites. Zeolite formation occurs as a rule more slowly than precipitation of AgCl or $BaSO_4$ from their supersaturated solutions because the zeolite crystal is not wholly ionic and must grow by condensation polymerization in three dimensions with formation of bonds Si–O–T (T = Al or Si) which are partially covalent. Crystallization is sufficiently slow for the kinetics to be followed and the factors influencing the kinetics investigated.

2. Nucleation

General properties of nucleation, which should apply also to zeolitization, can be summarized as below:[1]

  (i) Rates of nucleation increase with extent of undercooling, i.e. with increasing metastability. However, viscosity also tends to increase, often rapidly, as temperature falls so that effects of the degree of undercooling and of viscosity oppose one another in influencing nucleation rates. These rates can then pass through a maximum as temperature falls as illustrated in Chapter 2 Fig. 7.

(ii) An incubation period is observed, particularly in condensed phases, during which nucleation cannot be detected. Even in seeded solutions metastable regions of supersaturation can occur within which nucleation is not detectable. In many phase studies involving solutions, well-defined composition boundaries are found beyond which nucleation occurs freely so that under appropriate conditions the rate of nucleation increases extremely rapidly with degree of supersaturation.

(iii) The extent of the incubation time can be changed significantly by very small changes in composition.

(iv) The onset of nucleation often depends on the previous history of the system.

The appearance of viable nuclei, i.e. nuclei on which crystal growth occurs spontaneously, is the result of fluctuations in the magma or solution. These fluctuations result in the appearance and disappearance of embryos. These may build up in size until some reach the critical size needed for spontaneous further growth. They are the result of chemical aggregations and dis-aggregations of the precursor species referred to in Chapter 3, and in any given aqueous magma it is possible that various chemically distinct germ nuclei co-exist and that more than one kind of nucleus can reach the critical size with resultant co-crystallization of more than one zeolite.

The reason why germ nuclei must reach a critical size before becoming viable for spontaneous further growth can be considered as follows. In the liquid phase or in the gel there is an interfacial free energy, Δg_σ, between the nucleus and the surrounding medium. There may also, when the nucleus develops in a rigid gel, glass or crystal matrix, be a strain free energy, Δg_s, which arises from misfit between the nucleus and the medium in which it forms. For an aqueous aluminosilicate gel Δg_s should be small, but it will be included to make the treatment as general as possible. Both Δg_σ and Δg_s are positive in sign so that they tend to render the nucleus unstable. In the absence of Δg_σ and Δg_s there would only be a negative free energy of formation, Δg, for the potentially viable nucleus composed of j aluminosilicate units with associated cations. The net free energy of formation, Δg_j of the nucleus is thus

$$\Delta g_j = \Delta g + \Delta g_\sigma + \Delta g_s \tag{1}$$

Δg should be proportional to $-j$, Δg_s to $+j$ and Δg_σ to the surface area of

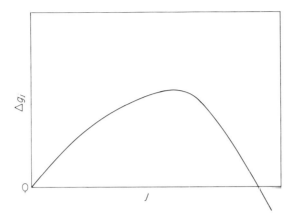

Fig. 1. Formal representation of the free energy of formation, Δg_j, of a germ nucleus composed of j structural units as a function of j.

the nucleus and so to $+j^{2/3}$. Thus

$$\Delta g_j = -Aj + Bj^{2/3} + Cj \qquad (2)$$

where A, B, and C are coefficients. For small values of j, Δg_j is thought to be positive in sign because $Bj^{2/3} + Cj > Aj$. Initially therefore Δg_j increases with j and the embryos are thermodynamically unstable. However $(A - C)j$, with $A > C$, increases more rapidly with j than $Bj^{2/3}$ so that Δg_j passes through a maximum and thereafter decreases as j increases (Fig. 1). Any nucleus reaching the maximum has an equal chance of adding or losing a further precursor unit, and a nucleus to the right of the maximum adds units with a decrease in free energy and loses units with an increase in free energy. Thus once a current of germ nuclei passes over the maximum the nuclei grow spontaneously. The maximum in Δg_j shown in Fig. 1 is reached when $d(\Delta g_j)/dj = 0$, and so when

$$j = \frac{8}{27} \frac{B^3}{(A - C)^3} \qquad (3)$$

The number of viable nuclei formed per unit time plotted against time would be expected initially, following the incubation period, to follow a rising curve. However, both nucleation and crystal growth consume precursor species, and hence compete for these. So as crystals grow it could be expected that a smaller and smaller proportion of precursors is available for fresh nucleation. Rates of nucleation should therefore pass through a maximum and then decline (see §4). Various expressions for rates of nucleation have been suggested to represent different situations.[2,3] They include the relations (where N is the number of nuclei at time t):

$$\frac{dN}{dt} = A \exp(-k_1 t) \qquad \text{(exponential decay)} \qquad (4)$$

$$\frac{dN}{dt} = k_1 \qquad \text{(constant rate)} \qquad (5)$$

$$N = \text{constant} \qquad \text{(instantaneous nucleation)} \qquad (6)$$

$$\frac{dN}{dt} = At^n \qquad \text{(power law)} \qquad (7)$$

$$\frac{dN}{dt} = A[\exp(Et) - 1] \qquad \text{(exponential increase)} \qquad (8)$$

It is seen that none of the above relations shows the maximum in the rate of nucleation suggested above, but an expression which has the form

$$\frac{dN}{dt} = At^n \exp(-t^p) \qquad (9)$$

has the required property; n and p are exponents. The part At^n indicates the power law nucleation rate, offset by the exponential term indicating the increasing competition for the precursor species by the crystals. Figure 2 shows the form taken by Eqn (9) for several values of A, n and p.

A number of attempts[4] have been made to calculate the current of germ nuclei across the maximum in the curve of Δg_j against j (Fig. 1). These vary according to the physical situation being considered. If the critical size corresponding to the maximum in Fig. 1 occurs when $j = k$, then the current dN/dt, in the steady state condition in which one imagines the critical nuclei to be removed from the system as fast as they are born, is given by an expression of the form[5]

$$\frac{dN}{dt} = B \exp(-E/RT) \exp(-N_0 \Delta g_k / RT) \qquad (10)$$

in which $\Delta g_k = \Delta g + \Delta g_\sigma + \Delta g_s$ (Eqn (1)) is the free energy of formation of a critical nucleus from k precursor species and E is an energy of activation which may be involved for example in transport and incorporation of precursors at the nucleation centre. N_0 is Avogadro's number and B is a coefficient. The theoretical evaluation of B is very uncertain and the postulated steady state production of nuclei would require to be represented by Eqn (5), which is most unlikely to be appropriate for the actual nucleation of zeolites (§4). However the expression does indicate

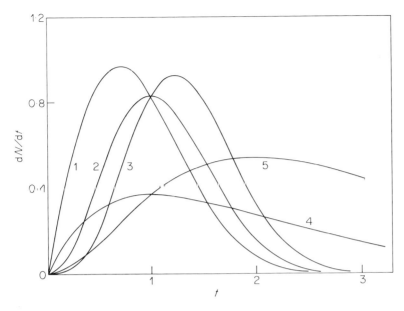

Fig. 2. Rates of nucleation, dN/dt, which pass through maxima.
Curve 1: $dN/dt = 4\pi^{-1/2}t\exp(-t^2)$ Curve 4: $dN/dt = t\exp(-t)$
Curve 2: $dN/dt = 4\pi^{-1/2}t^2\exp(-t^2)$ Curve 5: $dN/dt = t^2\exp(-t)$
Curve 3: $dN/dt = 4\pi^{-1/2}t^3\exp(-t^2)$

that the nucleation current can have a very large temperature coefficient since more than one exponential term can be involved.

3. Growth on Zeolite Seed Crystals

Provided conditions are right for the development of a given zeolite, growth by deposition upon seeds of the zeolite can be most effective. Crystal growth in general has the following characteristics:[1]

(i) The deposition rate on a seed or stable nucleus increases with the extent of supersaturation or undercooling, and is often increased by stirring. Large single crystals are not formed except at very small degrees of supersaturation or undercooling.

(ii) Under the same conditions different faces of a crystal often grow at differing rates. High-index faces grow more rapidly and so tend to disappear, as indicated formally in Fig. 3.[6] Crystal habit depends on this and may be influenced by impurities such as dyestuffs[7] which

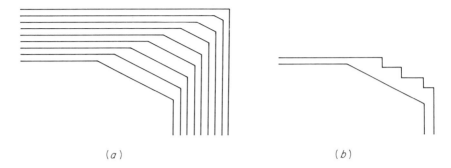

(a)   (b)

Fig. 3. Elimination of a high index surface (a) by straightforward more rapid growth of the high index than of the neighbouring lower index faces; and (b) by step formation.[6]

can be preferentially absorbed on certain faces and so modify the growth rates on those faces. Many interesting habit modifications occur among natural and synthetic zeolites. For example, in addition to forming as cubes, cubes with the corners bevelled off, or 24-hedra, analcime may be found in other forms, such as the 30- and 42-hedra of Fig. 4(a).[8] Likewise faujasite, which often forms as octahedra, can instead by using mixed NaOH and KOH in suitable proportions be grown in a tabular habit (Fig. 4(b)).[8a] Numerous illustrations of habit modifications of natural zeolites are given by Dana.[9]

(iii) As the overall rate of crystallization increases so the difference in growth rates of the various faces tends to diminish.

(iv) Flawed surfaces tend to grow more rapidly than corresponding surfaces without flaws.

(v) The maximum rate at which unflawed growth can be obtained on a particular surface decreases with increasing surface area. This behaviour suggests a practical limit to the size to which an unflawed single crystal could be grown in a reasonable time. This limitation can be partly removed over geological time. In Chapter 2 §9, when the formation of large crystals was considered, it was noted that exceptionally large crystals of certain silicates have been formed in pegmatites. Although not perfect, these include 12-foot "leaves" in a book of mica, a 19-ton crystal of beryl and a 40-ton crystal of spodumene. While large zeolite crystals are found in cavities in basalt, these are measured at most in inches rather than feet. Synthetic zeolites, which do not have the benefit of geological time for their formation, are usually crystalline powders in the size range $0.1\ \mu m$ to $15\ \mu m$. They are formed from mixtures representing a

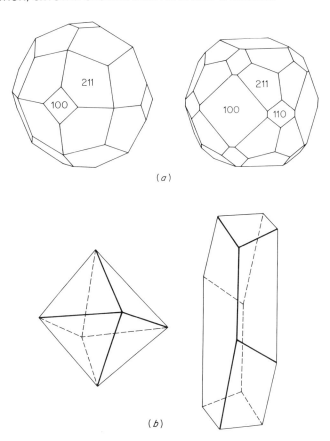

Fig. 4. (a) The 30-hedral and 42-hedral crystal habits in synthetic analcimes.[8] (b) Octahedral and tabular habits of faujasite.[8a]

high supersaturation and hence a very large number of nuclei develop.

As mentioned in §2, the consumption by growing crystals of precursor species will limit the availability of such species for forming fresh nuclei. Moreover the surface area provided by fresh nuclei will be small compared with that provided by the relatively much larger seed crystals. Thus seeding provides favourable conditions for measuring linear growth rates. Such studies have been made for cubic crystals of the faujasites zeolites X and Y.[10,11] Constant linear growth rates were observed over a considerable period of time (Fig. 5[12]). Linear growth rates, $k = 0.5\Delta l/\Delta t$, are given in Table 1.[11] These rates, other things being equal, depended strongly

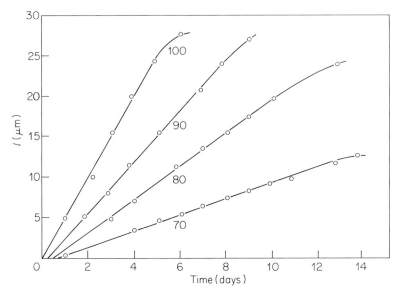

Fig. 5. Growth in crystal size of the largest crystals in the faujasite zeolite Na-X.[12] The crystals were grown from aluminosilicate gels. The temperatures on the curves are in °C.

upon the Si/Al ratio of the resultant crystals. They were also proportional to the concentration of dissolved silica, and when $k/[SiO_2]_{sol}$ was plotted against the Si/Al ratio the relatively smooth curve in Fig. 6[11] was obtained. The amount of dissolved silica is in turn strongly dependent upon the alkalinity (cf. Chapter 2, Table 8). The very slow linear growth rates of siliceous faujasites appear to place an upper limit to the Si/Al ratios obtainable in a reasonable time. The seed crystals used were of zeolite Na-X and were introduced in amounts needed to give the same initial mole fraction, x_0, of crystals in the magma. The seeds were selected to have the same mean dimension, $\bar{r}_0$, in all runs in Table 1. Figure 7[11] shows curves at different temperatures of the fraction of total crystallization as functions of time. No incubation period is observed and the rate of deposition on the seeds increases with the total surface available for deposition. The linear growth rate constants k were evaluated from these curves, starting with the assumptions below:

(i) The yield of the crystals increases at a rate proportional to the total free external surface $O(x)$ of the faujasite crystals in the mixtures. The linear growth rates on faujasite crystals of given Si/Al ratio are independent of crystal size.

(ii) The compositions of the several components forming the solid phase (faujasite, gel and any co-crystallizing species such as Na-P) do not differ appreciably. Then there are n_1 formula weights of faujasite $NaAlO_2.mSiO_2$ $(m>1)$; n_2 formula weights of $NaAlO_2.mSiO_2$ amorphous; and n_3 formula weights $NaAlO_2.mSiO_2$ of crystalline by-products, if any. The mole fraction x, of faujasite (including original seeds) at time t is then

$$x = n_1/\sum n_i \qquad (\sum n_i = n_1 + n_2 + n_3) \qquad (11)$$

(iii) The number Z of crystals of faujasite does not change during the deposition of zeolite and is equal to the number of seed crystals added. The crystals are idealized as spheres.

TABLE 1

Linear growth rates, k, for faujasites grown from different batch compositions at 88°C[11]

$\dfrac{SiO_2}{AlO_2^-}$	$\dfrac{(Na-Al)}{SiO_2}$	$\dfrac{H_2O}{AlO_2^-}$	$(SiO_2)_{sol}$ (M)	$10^3 k$ ($\mu m\, h^{-1}$)	$\dfrac{Si}{Al}$ in crystals
5·0	2·5	400	0·515	200	1·39
2·1	1·06	160	0·231	75	1·47
5·0	2·0	400	0·500	145	1·50
2·1	1·06	400	0·102	17	1·53
2·5	0·98	195	0·380	70	1·63
5·0	1·5	400	0·477	90	1·65
3·0	0·92	400	0·174	25	1·78
3·0	0·92	235	0·283	58	1·83
5·0	1·0	400	0·435	49	1·95
4·1	0·83	320	0·370	38	2·02
4·0	0·83	400	0·274	25	2·08
5·0	0·78	400	0·400	28	2·20
5·0	0·78	400	0·400	22	2·20
5·0	0·78	400	0·396	25	2·22
6·0	0·75	400	0·509	23	2·40
6·0	0·75	400	0·510	24	2·43
7·0	0·73	400	0·638	17	2·52
7·0	0·73	400	0·638	19	2·52
7·5	0·72	400	0·706	19	2·54
8·0	0·71	400	0·756	18	2·69
8·0	0·71	400	0·755	15	2·70
9·0	0·70	400	0·830	14	2·82
10·0	0·69	400	1·02	8	2·85
10·0	0·69	400	1·02	9	2·89
12·0	0·67	400	1·27	2·0	3·08
14·0	0·65	400	1·54	1·0	3·19
14·0	0·65	600	1·01	0·53	3·40

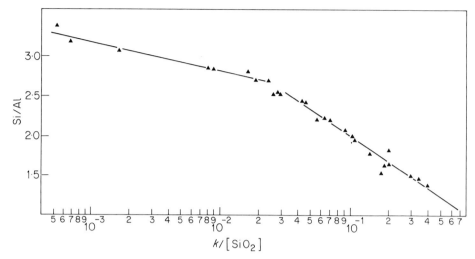

Fig. 6. The correlation between $k' = k/(SiO_2)_{sol}$ and the Si/Al ratios of faujasites. The temperature was 88°C.[11]

With these assumptions the expression[10]

$$\frac{dx}{dt} = 3kx^{2/3}x_0^{1/3}\bar{r}_0^{-1} \tag{12}$$

was obtained. When integrated and re-arranged this relation gives

$$x = x_0 + 3k\frac{x_0}{\bar{r}_0}t + 3k^2\frac{x_0}{\bar{r}_0^2}t^2 + k^3\frac{x_0}{\bar{r}_0^3}t^3 \tag{13}$$

and

$$k = \left[\left(\frac{x}{x_0}\right)^{1/3} - 1\right]\frac{\bar{r}_0}{t} \tag{14}$$

The rate constants for linear growth given in Table 1 are calculated from Eqn (14). The full lines in Fig. 7[11] are calculated from Eqn (13), which thus appears to give a consistently adequate representation of the experimental observations over a considerable range in x.

Inspection of Fig. 5,[12] where the linear growth rates were measured directly (as $0.5\Delta l/\Delta t$) for the largest crystals of dimension l, shows that towards the end of the crystallization, when one or more of the essential chemical nutrients is becoming exhausted, the constant linear rate of deposition of zeolite is no longer maintained. This of course is as expected, but the final decline in $k = 0.5\Delta l/\Delta t$ is not considered in deriving Eqn (13). Thus the equation is applicable only over the range in

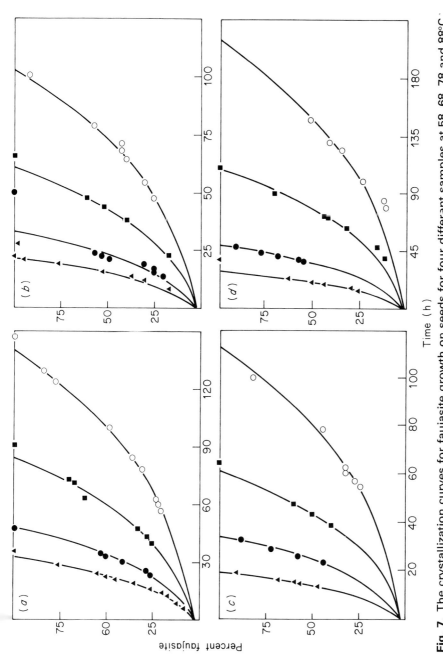

Fig. 7. The crystallization curves for faujasite growth on seeds for four different samples at 58, 68, 78 and 88°C. The full curves are calculated from Eqn (13).[11] (a) Sample of horizontal row 4 of Table 1. (b) Sample of row 7. (c) Sample of row 12. (d) Sample of row 19.

TABLE 2
Constant linear rates $k = 0.5\Delta l/\Delta t$
for growth of Na-X[12]

T (°C)	k (μm h^{-1})
70	0·0175
80	0·0375
90	0·0625
100	0·1071

x for which $0.5\Delta l/\Delta t$ is constant, and provided the basic assumptions (i), (ii) and (iii) above are reasonably accurate. In the measurements shown in Fig. 5 the constant linear growth rates were obtained for zeolite Na-X crystallizing from a reaction mixture of oxide composition $4.12Na_2O$, Al_2O_3, $3.9SiO_2$, $59H_2O$. The measurements were made on the largest Na-X crystals in the products at each of a series of times. Some rate constants are given in Table 2, a characteristic feature being the large temperature coefficient. The result for 90°C is the most appropriate one to relate to those for the most aluminous faujasites of Table 1. Because the compositions of the reaction mixtures are not identical, quantitative correspondence is not expected. The oxide compositions for the first two reaction mixtures of Table 1 were $6.75Na_2O.Al_2O_3.5SiO_2.200H_2O$ and $2.11Na_2O.Al_2O_3.2.1SiO_2.80H_2O$ with linear rates at 88°C of 0·20 and 0·075 μm h^{-1} as against 0·0625 μm h^{-1} at 90°C (Table 2) for the mixture of oxide composition $4.12Na_2O.Al_2O_3.3.9SiO_2.59H_2O$.

Seeding aqueous magmas is a known procedure[13,14] for eliminating or reducing the incubation time and so shortening the overall time required for the synthesis. Such a procedure can be most helpful where, as in synthesis of Na-Y faujasite, the incubation period starting from aqueous gels can be lengthy. The seeds, as Kacirek and Lechert[10,11] showed, can be of the aluminous and more easily prepared Na-X faujasite. One can even dispense with pre-formed crystals as the seeds; a portion of an amorphous mixture previously aged to contain germ or viable nuclei or a mother liquor from a previous crystallization can replace seeds and serve to shorten the induction period and hence the time for crystallization. The fresh crystalline material deposited on the surfaces of seeds can have a very different SiO_2/Al_2O_3 ratio from that of the seeds.[10,11] Although the unit cell dimensions of seed and crystalline deposit may differ somewhat, such a difference does not prevent growth. A further example was studied by Rollmann.[15] Synthesis of zeolite ZSM-5 was initiated from a magma containing aluminium in the usual way. The magma was then replaced by one free of aluminium and deposition on the parent ZSM-5 was continued of crystalline silica having the ZSM-5 structure.

4. Nucleation Combined with Crystal Growth

In absence of seed crystals nucleation must precede crystal growth. The curves showing yield of crystals against time are characteristically sigmoid in shape, as illustrated for mordenite in Fig. 8.[16] This behaviour has been observed also for zeolite A,[17] faujasites (as zeolites X and Y),[10,11,18] sodalite hydrate,[19] (Na-K)-phillipsites,[20] zeolite ZK-5,[21] the gismondine type zeolite Na-Pl[22] (also termed zeolite B), the mazzite type zeolite Ω,[23] zeolites K-F[23a] (edingtonite type), K-M[23a] (near-phillipsite) and Na-ferrierite.[23a] A method of analysing the nucleation and crystal growth parts of such S-shaped curves was developed by Zdhanov and Samulevich[12] in the case of zeolite Na-X. The method

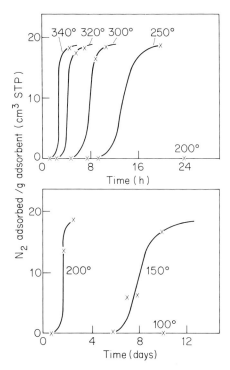

Fig. 8. The characteristic S-shaped crystallization curves illustrated for mordenite.[16] The strong effect of temperature is also shown; nitrogen sorption was used to estimate the yield of mordenite.

does not involve assuming an equation for nucleation kinetics and utilizes experimental measurements on the final crystal size distribution and linear growth rate measurements on the largest crystals to interpret the sigmoid curves. Figure 9(a) shows a histogram of the final size distribution. If n_i denotes the number of crystals in a sample in the ith mode (i.e. the group having "diameters" between d_i and $d_i + \Delta d_i$ and mean diameter $\bar{d}_i$), then the fraction of the total number N of crystals in this mode is $\alpha_i = n_i/N$. Curves 1 in Fig. 9(a) and (b) give the linear growth rates of the largest crystals. Each point on curve 1 of Fig. 9(a) was obtained by microscope measurements of the diameters of the 10 to 20 largest crystals observed in identical samples of reaction mixture reacted for different intervals of time from the start of heating.

As seen from curve 1, the period of constant linear growth rate for crystals of different size extended in the system studied (mixture of oxide composition $4 \cdot 12\text{Na}_2\text{O}.\text{Al}_2\text{O}_3.3 \cdot 9\text{SiO}_2.59\text{H}_2\text{O}$) for about 115 hours and thereafter declined asymptotically towards zero. It was assumed that not only during the period of constant linear growth rate but also during the final decay period crystals of all sizes still grew at the same but declining linear rate. With this assumption, curve 1 of Fig. 9(a) serves to determine the times at which crystals in that mode started to grow. This is the time when nuclei for crystals in that mode became viable. Figure 9(a) shows, for example, that crystals in the mode having $\bar{d} = 4 \cdot 5 \ \mu\text{m}$ nucleated at $t = 136$ h and crystals with $\bar{d} = 16 \cdot 5 \ \mu\text{m}$ nucleated at $t = 90$ h. From the times for nucleation of crystals of each mode the size distribution curve of Fig. 9(a) gives the nucleation rate curve 2 of Fig. 9(b). As predicted in §2 this rate passes through a maximum. It is greatest before any appreciable mass of crystals has formed and dies away sharply thereafter.

The ratio Z_t/Z_f of the mass Z_t of crystals at time t to the mass Z_f in the final product is also the ratio V_t/V_f of the corresponding volumes, and can be expressed as

$$\frac{Z_t}{Z_f} = \frac{V_t}{V_f} = \frac{\sum\limits_i \alpha_i [\bar{d}_i(t)]^3}{\sum\limits_i \alpha_i [\bar{d}_i]^3} \tag{15}$$

where $\bar{d}_i(t)$ is the mean diameter of the crystals of the ith mode found for time t from the linear growth rate curve. Thus one may evaluate Z_t/Z_f as a function of t with the result shown in curve 3 of Fig. 9(b). The crosses near curve 3 represent Z_t/Z_f determined in the usual way by X-ray powder photography. These points lie near the calculated sigmoid growth curve and so tend to support the assumption involved in the analysis that linear growth rates are **independent** of crystal size.

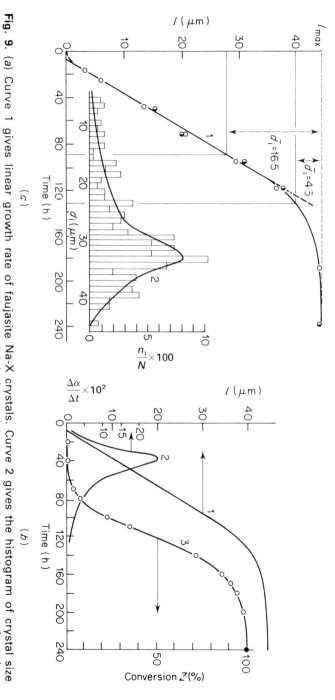

Fig. 9. (a) Curve 1 gives linear growth rate of faujasite Na-X crystals. Curve 2 gives the histogram of crystal size distribution in the final product. (b) Curve 1 is curve 1 of (a). Curve 2 is the curve of nucleation rate against time derived from curves 1 and 2 of (a) and curve 3 gives the yield of faujasite as a function of time derived from curves 1 and 2.[12]

The curves of crystal growth against time have also been represented by the model-based relation[24]

$$\frac{Z_t}{Z_f} = 1 - \exp\left(-k_1 t^n\right) \tag{16}$$

derived for random nucleation and a constant linear rate of crystal growth corrected for overlap as crystals impinge on each other. $n = 4$, $n > 4$ and $n < 4$ correspond with nucleation rates constant, increasing or decreasing with time respectively. Equation (16) can be considered in relation to a study by Ciric[17] of the synthesis of zeolite Na-A. The maximum slopes of curves of Z_t/Z_f against t are given in Table 3 for reaction mixtures having $SiO_2/Al_2O_3 = 1$. Maximum slopes occur at the inflexion points of the curves where $d^2(Z_t/Z_f)/dt^2 = 0$. If Eqn (16) is valid

$$\frac{d(Z_t/Z_f)}{dt} = k_1 n t^{(n-1)} \exp\left(-k_1 t^n\right) \tag{17}$$

and

$$\frac{d^2(Z_t/Z_f)}{dt^2} = k_1 n t^{n-2} \exp\left\{-k_1 t^n\left[(n-1) - k_1 n t^n\right]\right\} \tag{18}$$

TABLE 3
Crystallization   of   Na-A   at   100°C   from   mixtures   with $SiO_2/Al_2O_3 = 1$[17]

Average analyses of filtrate after crystallization (mol dm^{-3})			Maximum slope = $100 \times \dfrac{d}{dt}(Z_t/Z_f)$,
Na	Al$_2$O$_3$	SiO$_2$	(t in hours)
0·480	0·065	0·0060	7·7
0·440	0·065	0·0063	6·2
0·500	0·065	0·0072	7·7
0·870	0·068	0·0120	34·4
1·17	0·067	0·0156	63·5
1·17	0·070	0·0166	60
1·43	0·072	0·0217	86
1·54	0·072	0·0241	100
0·675	0·063	0·0090	17
1·26	0·068	0·0183	69
1·26	0·070	0·0192	77
1·30	0·070	0·0183	76
1·26	0·070	0·0183	69
1·30	0·070	0·0192	75
1·68	0·072	0·0270	120
1·19	0·067	0·0175	60·6
1·19	0·066	0·0173	63

At the inflexion point $t = t_i$, Eqn (18) thus gives

$$k_1 t_i^n = (n-1)/n \qquad (19)$$

and so at this point Eqns (16) and (19) give

$$\frac{Z_{t_i}}{Z_f} = 1 - \exp\left(-\frac{(n-1)}{n}\right) \qquad (20)$$

According to Eqn (20), Z_{t_i}/Z_f at the inflexion point takes the following values for several values of n:

n	2	3	4	5	6	∞
Z_{t_i}/Z_f	0·394	0·487	0·528	0·551	0·565	0·632

For zeolite A (Fig. 2 of Ref. (17); and line 5 of Table 3) the maximum value of $d(Z_t/Z_f)/dt$ was 0·635, for which $t = t_i$ was ~3·3 h, and Z_{t_i}/Z_f was about 0·8. This value is not possible for any value of n as the above figures show, so that a quantitative description of the curve of Fig. 2, Ref. 17, in terms of Eqn (16) is not possible. On the other hand, for Na-X curve 3 of Fig. 9(b) has its maximum slope at $Z_t/Z_f \sim 0·5$ and hence for $n \sim 4$. By the different procedure of curve fitting n was given as ~5.[12] Equation (16) cannot be a generally valid description of zeolite formation because it is not based on nucleation rates which increase to a maximum and then decrease.

Quantitative measurements on the kinetics of formation of zeolite A were further extended by Meise and Schwochow[3] who attempted to describe both the curves of yield of crystals against time and the final distribution of crystal sizes. Spherical particles were assumed for which the linear growth rate was taken to be constant, i.e.

$$r = kt \qquad (21)$$

where r is the radius of a given crystal. The rate of nucleation was assumed to be

$$\frac{dN}{dt} = A[\exp(Et) - 1] \qquad (8)$$

where A and E are coefficients. At reaction time t the normalized contribution $d(Z_t/Z_f)$ to the final yield Z_f arising only from nuclei formed at time τ in the interval $d\tau$ is

$$d(Z_t/Z_f) = d(V_t/V_f) = \frac{4\pi}{3V_f} k^3(t-\tau)^3 A[\exp(E\tau) - 1] d\tau \qquad (22)$$

In Eqn (22), $k(t-\tau)$ is the radius and dV_t the volume at time t of those nuclei formed at time τ (less than t) in the interval of $d\tau$. Reaction was

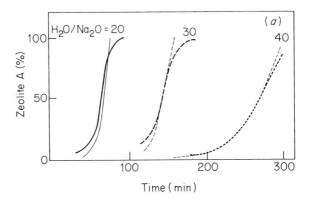

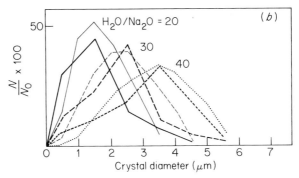

Fig. 10. (a) The influence of alkalinity upon curves of yield against time for zeolite A; and calculated curves for the best fit of Eqn (23). (b) Experimental size distribution curves for the final yields of A, and curves based on Eqn (24).[3] Bold curves are experimental; light curves are calculated.

assumed to be complete after a time t_e when $Z_t/Z_f = 1$. Integration of Eqn (22) over all values of τ between 0 and t gave

$$\frac{Z_t}{Z_f} = \frac{4\pi}{3V_f} \frac{Ak^3}{E} \left[\frac{6}{E^3} [\exp(Et) - 1] - \frac{6}{E^2} t - \frac{3t^2}{E} - t^3 - \frac{t^4}{4} \right] \quad (23)$$

while when $t = t_e$ the particle size distribution was taken to be

$$d(Z_t/Z_f) = \frac{4\pi}{3V_f} \frac{A}{k} r^3 \left\{ \exp\left[E\left(t_e - \frac{r}{k} \right) \right] - 1 \right\} dr \quad (24)$$

t_e is given by Eqn (23) on setting $Z_t/Z_f = 1$.

Equation (21) assumes a constant linear growth rate up to $Z_t/Z_f = 1$, although towards the end of crystallization Fig. 5 shows for zeolite X that there is a falling off in this linear rate, so that instead of a finite value of t_e the state when $Z_t/Z_f = 1$ is approached asymptotically. Also according to Fig. 9(b) nucleation rates for zeolite X pass through a maximum and then decline, whereas Eqn (22) predicts only an exponential increase in the rate. Accordingly t_e in Eqn (24) is an empirical coefficient and the equation (24) is expected to be valid only over the earlier stage of crystal formation. Likewise Eqn (24) will give a somewhat distorted size distribution. This is shown by Fig. 10,[3] the top half of which compares experimental curves of Z_t/Z_f vs t with those based on a programmed fit of Eqn (23). Similarly, the bottom half of the figure compares experimental and calculated size distributions. The qualitatively satisfactory agreements were obtained with the values of k, A and E given in Table 4. It would in view of Fig. 9(b) be of interest to employ Eqn (9) for the nucleation rate in place of Eqn (8) in an analysis of the above kind.

The sigmoid curves of yield against time, while typical of nucleation and growth of zeolites from gels in contact with supernatant liquid, are not the only form of kinetic curve that has been found. Thus, when quartz was the source of silica, the reaction between it, caustic soda and sodium aluminate to produce sodalite hydrate proceeded with no apparent induction time and at about the same rate as did the solution of the quartz.[25] The energy of activation of ~ 22 kcal mol^{-1} was close to 20 kcal mol^{-1} as reported for the solution kinetics of quartz.[26]

5. Crystallization Kinetics and Some Reaction Variables

Temperature, alkalinity (pH), composition of reaction mixtures, nature of reactants and pre-treatments can all affect crystallization kinetics and even the type of zeolite which forms. Some aspects of these influences are described in the following sections.

5.1. *Temperature*

The coefficients k, A and E in Table 4 and k in Table 2 all increase with temperature, indicating that linear rates of deposition (coefficient k) and rates of nucleation (coefficients A and E) both increase with rising temperature. The temperature coefficient of A is exceptionally large, but cannot be interpreted in terms only of an activation energy (cf. Eqn (10)). The decrease in induction time with temperature is illustrated in Fig. 8 for the synthesis of mordenite.[16] This behaviour is typical of that found with other zeolites such as Na-Y[27] or ZSM-5.[27a]

TABLE 4

k, A and E for zeolite Na-A as functions of alkalinity and temperature[3]

T (°C)	H_2O/Na_2O	k $(\mu m\,h^{-1})$	$A \times 10^6$	E
70	20	0·050	20	0·102
70	30	0·027	13·2	0·033
70	40	0·017	9	0·0115
60	20	0·03	1	0·062
70	20	0·05	20	0·102
80	20	0·06	200	0·132

The temperature coefficient of the linear growth rate, k, gives information regarding the growth mechanism. If the coefficient were to have the rather small value typical of the viscosity of the aqueous phase, corresponding with an activation energy E of ~3–4 kcal mol^{-1}, this would suggest control of k by transport of precursor species through a constant diffusion layer at the crystal surface. On seed crystals of Na-X k was found to be proportional to the concentration of dissolved silica, $(SiO_2)_{sol}$.[11] Plots of $\ln k' = \ln (k/(SiO_2)_{sol})$ against reciprocal temperature between 58 and 88°C gave the activation energies in Table 5. Likewise the values of k between 70 and 100°C in Table 2 give $E = 15\cdot4$ kcal mol^{-1}. The high values of E rule out diffusion control and suggest that one is measuring the rate of a chemical process in which appropriate precursor species are being generated or are undergoing condensation-polymerization at the crystal surfaces to extend the crystal lattice of faujasite. At least one essential precursor species exists in concentrations proportional to that of dissolved silica $(k' = k/(SiO_2)_{sol})$; and the rate of deposition is proportional to the total external area of the crystals. The crystals do not, therefore, grow by *in situ* direct conversion of gel. The larger the ratio Si/Al in the crystals formed the greater E appears to be (Table 5). The chemical process referred to above must

TABLE 5

E in kcal mol^{-1} for Na-Y, based on linear growth rates, k, on seed crystals of Na-X[11]

Batch composition	Si/Al in crystals grown	E
$NaAlO_2, 2\cdot1(Na_{1\cdot06}H_{2\cdot94}SiO_4), 400H_2O$	1·53	11·8
$NaAlO_2, 3\cdot0(Na_{0\cdot92}H_{3\cdot08}SiO_4), 400H_2O$	1·78	12·3
$NaAlO_2, 5\cdot0(Na_{0\cdot78}H_{3\cdot22}SiO_4), 400H_2O$	2·20	14·1
$NaAlO_2, 7\cdot5(Na_{0\cdot73}H_{3\cdot27}SiO_4), 400H_2O$	2·54	15·6

involve the formation of T–O–Si bonds (T = Al or Si):

$$\equiv\!\!\diagdown\!\!T\!-\!O^-Na^+ + HO\!-\!Si\!\diagdown\!\!\equiv \longrightarrow \equiv\!\!\diagdown\!\!T\!-\!O\!-\!Si\!\diagdown\!\!\equiv + NaOH$$

$$\equiv\!\!\diagdown\!\!T\!-\!OH + HO\!-\!Si\!\diagdown\!\!\equiv \longrightarrow \equiv\!\!\diagdown\!\!T\!-\!O\!-\!Si\!\diagdown\!\!\equiv + H_2O$$

and so it may be inferred that elimination of water or NaOH occurs with a lower E when T = Al than when T = Si. The above reactions between precursor molecules and the surface require a steric fit between, for example, the surface hydroxyls of the lattice at the crystal solution interface and corresponding hydroxyls of the impinging precursor molecules which successfully condense onto the growing surfaces. This requirement of fit provides a possible mechanism by which a zeolite structure, once nucleated, continues to grow in the correct topology. The surface reaction selects those molecules from solution which satisfy the requirement of fit. It would be difficult to visualize the continued development of a lattice as open as that of faujasite if, for example, the condensation-polymerization occurred from the showering down onto the surface of monomeric species like $Na^+Al(OH)_4^-$ or $(Na^+)_2[SiO_2(OH)_2]^{2-}$. Such a reaction would be more likely to yield compact tectosilicates such as felspars or quartz.

Based not upon linear growth rates but upon temperature coefficients of the rate of yield increase at 50% conversion Culfaz and Sand[28] reported apparent activation energies of 14–15 kcal mol^{-1} for mordenite and 19 kcal mol^{-1} for zeolite A. For formation of phillipsite Aiello et al.[29] estimated apparent activation energies in the range 13·5–15·4 kcal mol^{-1}, and Chao et al.[27a] for ZSM-5 estimated $E \sim 7$ and 11 kcal mol^{-1} for $SiO_2/Al_2O_3 = 70$ and for aluminium-free silica gel respectively. These energies are not, however, as soundly based as values determined from measurements of linear growth rates.

The induction times in zeolite formation decrease rapidly with increasing temperature (e.g. Fig. 8). In §4 it was seen that nucleation rates for faujasite formation passed through a maximum at an early stage of crystal deposition. The induction time must depend upon the kinetics of nucleation, but though qualitatively obvious from the sigmoid kinetic curves this sigmoid shape makes a quantitative definition difficult. One estimate of induction time was made by extrapolating the tangent representing the maximum rate of deposition of crystals to cut the time axis. However, this procedure is not based upon the rate coefficients of a quantitative formulation of nucleation kinetics so that the significance of the induction times is unclear. Equally so are apparent energies of activation of nucleation kinetics derived from such induction times. As examples of apparent energies and without prejudice to the unsolved problem of their meaning, the "activation energy" for "nucleation" of phillipsite was given[20] as 13·5 and 14·3 kcal mol^{-1} with and without stirring respectively; while for the Al-free end member of the ZSM-5 series (silicalite 1) this energy was given[27a] as 9·1 kcal mol^{-1}. Equation (8) expresses the rate of nucleation as exponentially dependent upon time and Table 4 gives the two rate constants involved in this simple model. Both coefficients increase with temperature; the pre-exponential constant A, in particular, shows a 200-fold rise between 60 and 80°C. However, Eqn (8) does not reproduce the curve of rate of nucleation of Fig. 9(b) with its well-defined maximum.

5.2. Alkalinity (pH)

Within the stability field of a given zeolite, increasing alkalinity at constant temperature influences the kinetics like increasing temperature at constant alkalinity. Thus Fig. 11[16] for mordenite can be compared with Fig. 9. At pH 13·3 the mordenite at first formed began to disproportionate into other species, but apart from this effect of extreme alkalinity the induction times decrease strongly and the maximum slopes tend to increase with pH. The strong dependence of the incubation times on OH$^-$ concentration was also observed when (Na, K)-phillipsites were crystallized at 100°C.[20] By appropriate additions of NaCl and KCl the total

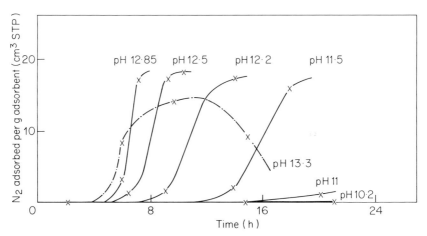

Fig. 11. Influence of pH upon curves of yield against time for mordenite synthesis at 300°C.[16]

$Na^+ + K^+$ in the mixture was maintained constant while the OH^- was varied. The incubation time was reported to vary inversely as $(OH^-)^2$ and the maximum rate of crystallization as $(OH^-)^{1.75}$. As a third example one may again consider Table 4 which shows for Linde zeolite Na-A that, in Eqn (8) $(dN/dt = A[\exp(Et-1)])$ for the nucleation rate dN/dt, both coefficients A and E increase strongly with increasing alkalinity. At 70°C and a given time t the nucleation rate based on Eqn (8) increases about 20-fold when the total Na_2O in the oxide composition of the reaction mixture is doubled. For comparison, dN/dt increases ~440-fold over the temperature interval 60–80°C.

For ratios $SiO_2/Al_2O_3 = 140$ and $H_2O/OH^- = 450$, 225 and 112 in the reaction mixtures at 94°C, Chao et al.[27a] found for ZSM-5 that the induction period was least for $H_2O/OH^- = 225$. Also, once crystal deposition had commenced, the rates of deposition were not very sensitive to these ratios. On the other hand, for a ratio H_2O/OH^- of 450, the induction times are not sensitive at 94°C to the ratio SiO_2/Al_2O_3 (between 70 and 140), but the rates of deposition increased as this ratio increased. Thus ZSM-5 zeolites appear to behave rather differently from mordenite, phillipsite or Na-A, at least in the effect of alkalinity.

In Chapter 2 §2 reasons were given why water in small and large amounts functions as a very effective mineralizing catalyst. The above three examples show that OH^- is also a powerful mineralizer. The OH^- ion is a good complexing agent which can through this property bring amphoteric oxides and hydroxides into solution, but which does not yield such stable complexes that further reaction involving them is prevented. Solution permits ready mixing of the complexes formed, examples of which are given in Chapter 2 §6. The decrease in nucleation time and enhanced rate of crystal growth with rising pH can be attributed at least in part to the much greater concentrations of dissolved species. This helps nuclei to develop more quickly from the more numerous encounters between precursor species in solution. Similarly the greater concentration of precursors allows faster surface nucleation and reaction at the crystal–solution interface. Table 8 in Chapter 2 shows how greatly the solubility of quartz increases with concentration of NaOH or Na_2CO_3. By contrast the solubility of quartz in pure water at 1 atm and 25°C has been estimated to be only 6 p.p.m.[30]

Other important effects of OH^- have been observed. Starting with silica-rich mixtures, the greater the concentration of this anion the more nearly the ratio SiO_2/Al_2O_3 approaches 2.[31,32] This is seen in Fig. 12[32] for some parent compositions $K_2O.Al_2O_3.nSiO_2$ + excess KOH + aq. The final products were the chabazite type zeolite K-G and the edingtonite type K-F. When SiO_2/Al_2O_3 is two the Al and Si should as far as possible alternate throughout the tetrahedral framework according to the Al–O–Al avoidance rule.[33] Thus, by increasing the OH⁻ concentration the tendency of Al and Si to be ordered on the tetrahedral sites is increased wherever the avoidance rule is valid.

Also, when the concentration of OH⁻ (as KOH) was kept constant, but the absolute amount was increased by increasing the volume of KOH solution added, the yields of crystals declined (Fig. 13[32]). The two solutions were 1 M and 6 M and respectively yielded K-G and K-F. Thus the solution retains unreacted but dissolved species in amounts increasing with the volume of KOH solution taken. The zeolites accordingly appear to show a solubility in their alkaline mother liquor. Measurements of the solubility of zeolites A and X in 0·02–0·5 M NaOH established for both

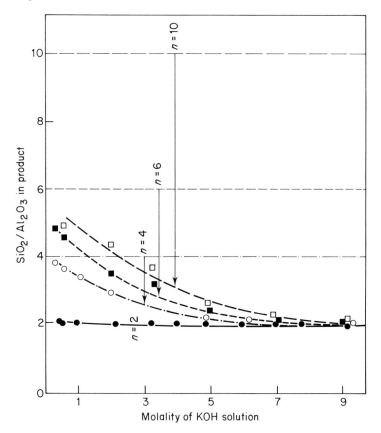

Fig. 12. Effect of concentration of KOH solution upon the ratio SiO_2/Al_2O_3 in the resultant crystals. The products are primarily chabazite type K-G and edingtonite type K-F.[32]

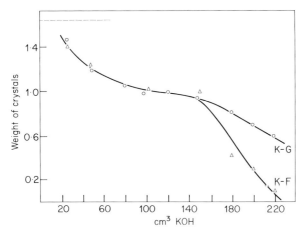

Fig. 13. Effect of the volume of KOH solution of constant concentration upon the yield of crystallization products from metakaolinite. ○, 1 M KOH, giving chabazite type K-G; △, 6 M KOH, giving edingtonite type K-F.[32]

zeolites an increase with rising temperature or pH.[34] As the ratio of liquid to solid was increased, the ratio Si/Al in the "equilibrium" solution approached more and more closely to that in the crystals. Side reactions such as selective extraction of Al were thus less and less important.

Finally, when zeolites form from aqueous, alkaline, silica-rich mixtures the zeolites usually have lower SiO_2/Al_2O_3 ratios than the parent mixtures (Fig. 12). On the other hand, if the reaction mixture is highly aluminous ($SiO_2/Al_2O_3 < 2$) the zeolite has SiO_2/Al_2O_3 equal to 2 or more and so is more siliceous than the parent mixture. Also (Chapter 3 §2.3) in a reaction mixture prior to crystallization and having $SiO_2/Al_2O_3 = 8$ this ratio for silica and alumina dissolved in the supernatant liquid was much greater than that of the gel, whereas when $SiO_2/Al_2O_3 = 1$ in the total mixture the supernatant liquid was much more aluminous than the gel (Table 3).

5.3. *Structure-Directing Role of Cations*

The kind of zeolite which crystallizes from aqueous alkaline magmas can be strongly influenced by the type of cation present. In aqueous solution cations can have structure-making or structure-breaking influences on water. Quaternary alkylammonium ions are regarded as structure-making in that the organic ion is believed to form water "icebergs" around itself. Small inorganic cations are energetically hydrated and so hold water molecules in charged clusters. In solutions containing dissolved silicate and aluminate situations can be envisaged in which the water is at least partially displaced by siliceous or alumino-silicate polymeric anions of various configurations, such as some in Chapter 3, Table 3. Precursors of this nature may explain the structure-directing role which the cations can exercise.

Table 6 indicates for a number of zeolites those cation environments in which they have been observed to form, taken mainly from results obtained in the author's laboratory.[35,36,37,38,39,40] The preferred cationic environments are given in the last column of the table. Among other factors such as alkalinity, temperature and overall composition, one may refer to the following conclusions:

(i) Sodic environments favoured the crystallization of sodalite and cancrinite hydrates, gismondine-types (Na-P), gmelinites, faujasites and zeolite A. Sometimes the combination of a small and a large cation served the same function as an ion of intermediate size. Thus zeolite A appeared in presence of $Li^+ + Cs^+ + TMA^+$ in place of Na^+.[41]

(ii) Mordenite formed in sodic, calcic and strontium-containing environments; the analcime topology was also obtained in a variety of cation environments, and so were edingtonite and phillipsite-type zeolites.

(iii) Chabazite and zeolite L were favoured by potassic and zeolites Li-ABW and Li-H by lithic environments. Various other zeolites (types related to thomsonite, epistilbite, heulandite, ferrierite, yugawaralite and also Ba-J and Ba-K) appeared in media containing alkaline earth metal cations, although not necessarily only with such cations.

(iv) Some zeolites were best grown in presence of two or more cations, examples being zeolite EAB (TMA-E), offretite and mazzite, all of which were favoured by $Na^+ + TMA^+$ mixtures. Also, in order to make the most silica-rich forms of zeolite A, the NaOH must be largely replaced by TMAOH.[42,43]

The results in Table 6 serve to illustrate the importance of cations in determining the reaction pathways and products. It is possible that the zeolites in Table 6 may result from still more varied cationic environments under the right conditions. A given zeolite framework may result from very distinct routes. An interesting example is the KFI (i.e. ZK-5) framework. It was first made in 1948[37] by recrystallizing synthetic analcime with excess $BaCl_2$ or $BaBr_2$ plus a little water (Species P and Q). Later it was made from aqueous aluminosilicate gels with

$$NaOH + \left[CH_3\overset{+}{N}\langle\hexagon\rangle\overset{+}{N}CH_3 \right] (OH^-)_2$$

TABLE 6

Zeolite syntheses in relation to cation environments[a]

Zeolite	Cations in reaction mixture	Preferred cations
Gismondine types	Na, (Na, NMe$_4$), (Na, Li), (Na, K), (Na, Ba), (Li, Cs, NMe$_4$)	Na
Gmelinite types	Na, Sr, (Ca, NMe$_4$), (Na, NMe$_4$)	Na
Faujasite types	Na, (Na, NMe$_4$), (Na, Li), (Na, Ba)	Na
Zeolite A types	Na, (Na, NMe$_4$), (Na, K), (Na, Ba), (Na, Ba, NMe$_4$) (Li, Cs, NMe$_4$)	Na
Zeolite (Na, NMe$_4$)-V	Na, (Na, NMe$_4$)	Na
Sodalite hydrates	Na, NMe$_3$H, NMe$_4$, (Na, NMe$_4$), (Na, K), (Na, Li), (Ca, NMe$_4$), (Li, Cs, NMe$_4$)	Na
Cancrinite hydrates	Na, (Na, Li), (Na, NMe$_4$), Sr, (Li, Cs, NMe$_4$)	Na
Zeolite EAB[b] (TMA-E)	(Na, NMe$_4$)	(Na, NMe$_4$)
Mazzite type (zeolite Ω)	(Na, NMe$_4$)	(Na, NMe$_4$)
Offretite types	(Na, K, NMe$_4$), (Na, Ba), (Na, Ba, NMe$_4$), (Li, Cs, NMe$_4$)	(Na, K, NMe$_4$)
Mordenites	Na, Ca, Sr	Na, alkaline earth ions
Analcimes and isotypes	Na, K, Rb, Cs, Tl, NH$_4$, Ca, Sr, (Na, K), (Na, Rb), (Na, Cs), (Na, Tl), (K, Rb), (Rb, Tl), (Li, Cs)	Various
Edingtonite types	K, Rb, Cs, (K, Na), (Na, Li), (K, Li), (Li, Cs, NMe$_4$), (Ba, Li), (Li, Cs)	K, Rb, Cs, Ba
Phillipsite types	K, Ba, Ca, NH$_4$, NMeH$_3$, NMe$_2$H$_2$, NMe$_3$H, NMe$_4$, (Na, K), (Na, NMe$_4$), (Ca, NMe$_4$), (Na, Ba)	K and others
Chabazite types	K, Sr, (K, Na), (K, Li), (K, Ba), (K, Na, NMe$_4$)	K
Zeolites L	(K, Na), K, Ba, (Ba, K), (Na, Ba)	K, Ba
Thomsonite type	Ca	Ca
Epistilbite type	Ca	Ca
Heulandite type	Sr	Sr
Ferrierite type	Sr	Sr
Yugawaralite type	Sr, (Ba, Li)	Sr, Ba
Zeolite Ba-J	Ba	Ba
Zeolite ZK-5 types	Ba, (Na, Ba), (K, Ba), (Li, Cs, NMe$_4$)	Ba
Zeolite Ba-K	Ba	Ba
Zeolite Li-ABW	Li, (Li, K), (Li, Na), (Li, Cs, NMe$_4$)	Li
Zeolite Li-H	Li	Li
Zeolite ZSM-2 type	(Li, Cs, NMe$_4$)	

[a] The examples in this table are taken mainly from the series Hydrothermal Chemistry of Silicates, Parts 1 to 22, and on Chemistry of Soil Minerals, Parts 1 to 14, published in J. Chem. Soc. (London), from 1951 onwards. Results from some earlier papers from the author's laboratory are also included.

[b] Originally considered to be like erionites,[35] but recently shown to have a novel structure.[53]

as the bases.[44] Still later it was made from aluminosilicate gels with aqueous KOH+CsOH in particular proportions[45] and finally from TMA-aluminosilicate solutions with LiOH+CsOH.[41]

Knowledge of the synthesis of silica-rich forms of zeolite A using mixed TMAOH+NaOH bases was extended by Kacirek and Lechert.[43] In presence of seeds of Na-A the kinetic growth curves were similar to those in Fig. 7, described by Eqn (13), and the linear growth rate constants k were proportional to TMA/(TMA+Na). However pure N-A was not obtained because other species co-crystallized, important among which was faujasite (as zeolite N-Y). The term "N-A" means a zeolite A containing nitrogenous cations such as TMA.[42] In order to prepare pure N-A a particular procedure was required. Aluminosilicate gels were prepared from solutions of sodium aluminate and strongly alkaline sodium silicate, so that in each solution monomeric aluminate and silicate anions were likely to be dominant. The gel with Si/Al ~ 1·0 was washed to near-neutral pH and the wet gel was then treated with solutions of $(TMA)_8[Si_8O_{20}].69H_2O$ and 2–3% of Na-A seeds. At 88°C crystallization to pure N-A occurred, with the Si/Al ratios given in Table 7.[43] The above tetramethylammonium silicate introduces silica in the aqueous phase as cubic (i.e. double four-ring) anions. Dissolution of the alumino-silicate gel in the alkaline solution was considered in addition to re-create monomeric aluminate and silicate anions which, with the cubic anions, underwent condensation-polymerization to give zeolite A. Despite the presence of Na-A seeds the kinetics of crystallization no longer followed Eqn (13), but instead resembled the growth curves of Fig. 8 where nucleation is followed by deposition on the nuclei. The incubation period was attributed to a build-up of the dissolved monomeric anionic species from the gel. This work shows, whatever the detailed mechanism, not only the significance of the cations, here $TMA^+ + Na^+$, but also that of the way of performing the reaction, through the nature of the chosen reac-tants and the sequence of their mixing.

The results in Table 7 show that the higher the ratio TMA/(TMA+Na) the more siliceous the product becomes. The reason suggested for the unusually high Si/Al ratios in crystals of faujasites, zeolite A or other zeolites made in the presence of large organic cations is that because of their size a limited number only of such ions fills all the intracrystalline free volume. This imposes a limit to the anionic framework charge which in turn requires a high Si/Al ratio.[42] The higher the proportion in the reaction mixture of the small competing cation, here Na^+, the greater the proportion of these ions in the zeolite and so the more "normal" is its Si/Al ratio.

A further illustration of the influence of cations is provided by a study

TABLE 7
Silica-rich zeolites N-A from TMA-silicate and Na-aluminosilicate gel with
Si/Al = 1 [43]

Composition of mixture[a]	Si/Al in mixture	Si/Al in crystals	Time (days)
$NaAlO_2.SiO_2.0\cdot5TMASi.53H_2O$	1·50	1·50	7
$NaAlO_2.SiO_2.1\cdot0TMASi.70H_2O$	2·00	1·80	7
$NaAlO_2.SiO_2.1\cdot5TMASi.86H_2O$	2·50	2·10	8
$NaAlO_2.SiO_2.2\cdot0TMASi.103H_2O$	3·00	2·26	8
$NaAlO_2.SiO_2.3\cdot0TMASi.119H_2O$	4·00	2·33	8
$NaAlO_2.SiO_2.4\cdot0TMASi.200H_2O$	5·00	2·58	8

[a] TMASi denotes $(TMA)_8[Si_8O_{20}]/8$.

of crystallization sequences[46,47] at 175°, 162·5° and 150°C of compositions $10(TPA, Na, K)_2O.Al_2O_3.28SiO_2.750H_2O + 4KCl,$ 4NaCl   or
(2NaCl, 2KCl), where TPA denotes tetrapropylammonium. The sequences when $TPA^+ + Na^+$ were respectively the only cations present are
shown in Fig. 14.[47] TPA^+ favours the formation of zeolite ZSM-5,
whereas K^+, according to Table 6, favours the phillipsite-harmotome

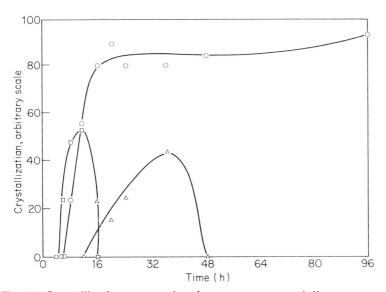

Fig. 14. Crystallization curves showing appearance and disappearance
of mordenite (□) and an analcime type phase (△) during growth of
zeolite   ZSM-5   (○)   from   the   parent   composition
$6(TPA)_2O\cdot4Na_2O.Al_2O_3.28SiO_2.750H_2O.$ [47]

zeolites and Na^+ favours gismondine type zeolite, mordenite and analcime. Figure 7, Chapter 5, shows these influences and the boundary lines very approximately show cation composition–crystal appearance zones after 4 days at 175°C.

5.4. *Organic Bases in Zeolite Synthesis*

In Table 6 there are various examples of zeolite formation in presence of tetramethylammonium (TMA) hydroxide. The use of organic bases as replacements for inorganic bases was initiated in 1961 by Barrer and Denny[42] and by Kerr and Kokotailo.[42a] The former workers used $N(CH_3)_4OH$, $N(CH_3)_3HOH$, $N(CH_3)_2H_2OH$, NCH_3H_3OH and NH_4OH with gels made from $Al(OH)_3$ freshly prepared and silica sols. Only adventitious traces of Na^+ were present. They obtained zeolites, mica and a montmorillonite-type compound as shown below:

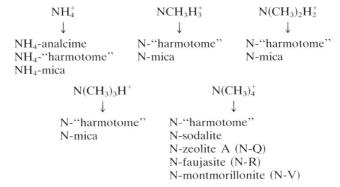

Also from oxide compositions $CaO.3TMAOH.Al_2O_3.nSiO_2 + aq$ Barrer and Denny[48] made (Ca, N)-sodalite (Ca, N)-"cubic harmotome" and a (Ca, N)-gmelinite type zeolite in addition to a montmorillonite type phase, the hexagonal dimorph of anorthite, boehmite (AlO.OH) and bayerite $(Al(OH)_3)$. The symbol N denotes as before "containing alkyl-ammonium and/or ammonium ions", since under the synthesis conditions alkylammonium ions may partially hydrolyse. As noted on p. 160, when large ions such as TMA are incorporated the zeolite A and faujasite produced will be more silica-rich than usual, because there is room for only a limited number of these big ions in the pore space available in the framework and therefore the anionic framework charge would necessarily have to be low (more Si, less Al). The use of TMA in forming silica-rich zeolite A was later examined in some detail by Kacirek and Lechert,[43] as already described in §5.3, and Baerlocher and Meier[49,50] subsequently made tetramethylammonium sodalite of unit cell composition

$(TMA)_2[Al_2Si_{10}O_{24}]$ in which one TMA cation occupied and filled each 14-hedral cavity (sodalite cage). This highly siliceous form can be compared with that of an ideal sodalite hydrate which is $Na_6[Al_6Si_6O_{24}]8H_2O$.

The work of Barrer and Denny and Kerr and Kokotailo resulted in extensive exploration of the use of organic bases in zeolite synthesis, with many interesting results, some of which are summarized in Table 8. The organic compounds have been successful in particular in producing extremely silica-rich zeolites such as the pentasils ZSM-5 and ZSM-11, and also zeolites Nu-1 and Fu-1. The table shows that ZSM-5 can be made in systems containing bases having a diversity of size and shape. This suggests a structure-directing influence which is surprisingly insensitive to the geometry of the organic base even though the base is required for successful synthesis. It is, however, to be recalled that in syntheses between 100 and 200°C, as is the case for most preparations of the pentasil zeolites,[68] ZSM-5 and -11, and of Nu-1 and Fu-1, there can be considerable hydrolysis of alkylammonium ions so that a number of cationic species can be present. When the organic compound is $NH_2(CH_2)_6NH_2$ synthesis of ZSM-5 at 100°C was reported,[61] which avoids the use of pressure vessels. The synthesis of the "Al-free" end-members of ZSM-5 and ZSM-11 termed respectively silicalite I and silicalite II[70] could be aided by partial hydrolysis yielding smaller cations down to NH_4^+, in view of the successful preparations of ZSM-5 and ZSM-11 from $(NH_4^+ + TPA)$ and $(NH_4^+ + TBA^+)$ mixtures respectively[62] (TBA = tetrabutylammonium).

In addition to analogues of many naturally occurring zeolites or of zeolites which so far have not been identified with natural counterparts but which can all be made with inorganic bases only, Table 8 shows that organic bases facilitate the formation of other novel structures. Some examples of both categories of novel zeolites are given below:

Made with inorganic bases	*Made in presence of organic bases*
Zeolite A[72] (*LTA*)	Zeolite N[81] (Z-21,[82] (Na, TMA)-V[51])
Zeolite L[73] (*LTL*)	Zeolite losod[58] (*LOS*)
Zeolite Li-A[74] (*ABW*)	Zeolite TMA-E[23] (*EAB*)
Species P and Q[37,21] (*KFI*)	Zeolite ZK-5[44] (*KFI*)
Zeolite U-l0[75]	Zeolite ZSM-5[60] (*MFI*)
Zeolite Li-H[74]	Zeolite ZSM-11[65] (*MEL*)
Zeolite Ba-J[76]	Zeolite ZSM-8[64]
Zeolite Ba-K[76]	Zeolite ZSM-12[67]
Zeolite Ba-N[76]	Zeolite Fu-1[55]
Zeolite *RHO*[77,40]	Zeolite Nu-1[54]
Zeolites ZSM-3[78] and -2[79,41]	
Zeolite K-I[80]	

TABLE 8
Some zeolite syntheses in presence of organic bases

Cations	Zeolite type	Cations	Zeolite type
TMA$^{+(49,50)}$	Sodalite Gismondine	Li$^+$ + Cs$^+$ + TMA$^{+(41)}$	Cancrinite Li-ABW Zeolite A Edingtonite (K-F) Analcime Offretite (O) Zeolite ZK-5 (KFI) Zeolite ZSM-2 Gismondine Sodalite Erionite
Na$^+$ + TMA$^{+(23,42,51,52,53,54,55)}$	Sodalite Cancrinite Gismondine Zeolite A (isotypes N-A, α, ZK-4) Faujasite Zeolite TMA-E (EAB) Mazzite (zeolites Ω, ZSM-4) Zeolite (Na, TMA)-V (isotypes N and Z-21) Zeolite Nu-1 Zeolite Fu-1	Na$^+$ + K$^+$ + benzyltrimethylammonium$^{+(57)}$ Na$^+$ + (58) Na$^+$ + (58)	Erionite Losod
Ca^{2+} + TMA$^{+(43)}$	Sodalite Gismondine Gmelinite	Na$^+$ + neopentyltrimethylammonium$^{+(58)}$	Losod

Template / cations	Zeolite product
$Ba^{2+} + TMA^{+}$ [56]	Erionite
$K^{+} + Na^{+} + TMA^{+}$ [23]	Zeolite L / Erionite / Offretite / Chabazite
$Na^{+} + NH_2(CH_2)_6NH_2$ [61]	Zeolite ZSM-5 (*MFI*)
NH_4^{+} + tetrapropylammonium^{+} [62]	Zeolite ZSM-5
Na^{+} + tetrapropylammonium^{+} [60,63]	Zeolite ZSM-5
Na^{+} + tetraethylammonium^{+} [64]	Zeolite ZSM-8
Na^{+} + tetrabutylammonium^{+} [65]	Zeolite ZSM-11 (*MEL*)
Na^{+} + benzyltriphenylammonium^{+} [65]	Zeolite ZSM-11
Na^{+} + tetrabutylphosphonium^{+} [65]	Zeolite ZSM-11
NH_4^{+} + tetrabutylammonium^{+} [62]	Zeolite ZSM-11
$Na^{+} + CH_3N^{+}$⟨bicyclic cage⟩$^{+}NCH_3$ [44,59]	Zeolite ZK-5 (*KFI*)
Na^{+} + primary n-alkylamines (C_2 to C_{12}) [60]	Levynite / Zeolite ZSM-5 (*MFI*)
Na^{+} + isopropylamine, dipropylamine or dibutylamine [60]	Zeolite ZSM-5
$K^{+} + CH_3N^{+}$⟨bicyclic cage⟩$^{+}NCH_3$ [66]	Zeolite ZSM-10
$Na^{+} + K^{+}$ + tetraethylammonium^{+} (with (Cr(OH)$_3$, Fe(OH)$_3$, Al(OH)$_3$) [67]	Zeolite ZSM-12

The zeolites with the framework topology of type *KFI* can be made either without an organic base or with the particular organic base of Table 8. This is also true of zeolite N, where the isotype Z-21 was made from Na-aluminosilicate gel while for the isotypes N and (Na, TMA)-V the mixed bases NaOH + TMAOH were used. Where adequate structure determinations have been made, the framework topologies are indicated by means of the three letters in brackets, following the designations of Meier and Olson (Chapter 1, Table 2).[83] The examples of novel zeolites given above are typical of the many possibilities for new porous tectosilicates based on the construction as models of numerous unrealized topologies.[84] This area of zeolite chemistry should continue to develop in interesting ways.

5.5. *Cation Templates*

There is in some instances direct evidence of the templating role of cations, as implied in §§5.3 and 5.4. The sodalite framework consists only of 14-hedra of type 1 (sodalite cages). Each cage is stacked in 8-fold co-ordination with respect to other like cages, by sharing one of its eight 6-ring faces with another cage. The only openings to cages are the 6-rings which are much too restricted to permit entry of large ions such as TMA^+ after formation. Nevertheless, it was noted in §5.4 that sodalites have been prepared[42,49,50] which have TMA^+ within the sodalite cages up to the limit of one per cage. Since TMA^+ cannot enter or leave the cages it must have been incorporated during growth. The water "icebergs" surrounding TMA^+ in solution must have been replaced by aluminosilicate sheaths, partially and then fully enclosing TMA in sodalite cages.

Offretite and zeolites Ω or ZSM-4 (two mazzite isotypes) are examples of zeolites the synthesis of which is strongly favoured by the presence of TMA^+ ions, as illustrated in Table 9.[85] Each zeolite has in its structure 14-hedral cages of type 2 (gmelinite cages), which again do not have faces open enough for entry or exit of TMA^+ after synthesis. Both zeolites retain close to one TMA^+ ion per gmelinite cage after synthesis and exhaustive ion exchange with Na^+ or K^+. There are no other cages large enough to hold TMA^+, and furthermore this ion is exchangeable from all the wide channels also present in each zeolite.[84] Accordingly, the non-exchangeable TMA^+ must be in the gmelinite cages and must have been incorporated during growth, whether as a precursor species in solution or at the surface of the crystals. Table 9 suggests that this templating process accounts for the ready appearance of offretite and the mazzite-type zeolite, ZSM-4 or Ω, when even small amounts of TMA^+ are present. The way in which yields of ZSM-4 depended upon the molar

TABLE 9
Dependence of zeolite type on presence or absence of TMA$^+$; mixtures with
$SiO_2/Al_2O_3 = 16$–20; $H_2O/SiO_2 = 14$–20. $T = 100°C$, static[85] conditions

Zeolite found	OH$^-$/SiO$_2$	Na$^+$/SiO$_2$	TMA$^+$/SiO$_2$	K$^+$/SiO$_2$	Ref.
Zeolite Y	0·8	0·8	0	0	(86)
ZSM-4^a	0·8	0·8	0·04	0	(87, 88)
Zeolite Ω^a	0·7	0·6	0·14	0	(86)
Zeolite L	0·8	0	0	0·8	(86)
TMA-O^b	0·9	0	0·09	0·8	(86)
TMA-offretite	1·1	0·4	0·10	0·7	(89)

^{a} Mazzite type zeolite.
^{b} Offretite type zeolite.

ratio $(TMA_2O/(SiO_2 + Al_2O_3 + Na_2O))$ is shown in Fig. 15.[90] When this
ratio was $\sim 1·5 \times 10^{-2}$, only ZSM-4 crystallized.

The use of cationic templates was extended[85,91] to linear polyelectrolytes made by reaction between 1,4-diazobicyclo[2.2.2]octane
(Dabco) and $Br(CH_2)_n Br$ where $n = 3, 4, 5, 6$ and 10. The polymers have
the formula

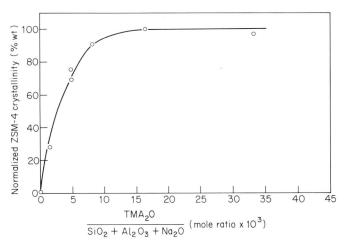

Fig. 15. The influence of the TMA$^+$ content of the reaction mixture
upon the content of mazzite type zeolite ZSM-4 in the product.[90]

They were designated according to the dibromide. Thus with 1,4-dibromobutane the polymer was termed Dab-4Br. The polyelectrolyte chains cannot because of their length be confined to individual cavities, as with TMA$^+$, but could only occupy wide channels within the zeolite. If they are incorporated they should cause the crystallization of zeolites having such channels, but not that of zeolites which lack wide channels. All the polymers were water-soluble, with molecular weights in the range 5000–15 000 (except Dab-3Br with ~2500).

Crystallizations of aluminosilicate gels with NaOH and the polyelectrolyte were effected at 85–90°C and also at 170–180°C, with the results summarized in Table 10.[91] The presence of very little Dab-4Br

TABLE 10

Zeolite crystallization in presence of polyelectrolytes.[91] $SiO_2/Al_2O_3 = 30$; $H_2O/SiO_2 = 21$; $OH/SiO_2 = 1.2$; $Na/SiO_2 = 12$, except where noted

Reaction mixture		Time	Product			
R^a	R/SiO$_2$	(days)	Zeolite^b	SiO$_2$/Al$_2$O$_3$	R$_2$O/Al$_2$O$_3$	Notes
(a) T = 90°C						
	0	4	Y + P	3·91		
	0	4	Y + P	3·61		Na/SiO$_2$ = 1·4
	0	11	Y + P	3·69		Na/SiO$_2$ = 1·8
	0	10	Y + P	4·41		Na/SiO$_2$ = 1·4
						H$_2$O/SiO$_2$ = 44
Dab-4Br	0·01	9	Gmelinite + Chabazite			
Dab-4Br	0·14	5	Gmelinite + Chabazite	5·15	0·15	
Dab-4Br	0·23	9	Gmelinite only	7·05	0·43	
Dab-4Br	0·43	60	Amorphous			
Dab-4Br	0·24	7	Gmelinite only	5·52	0·44	H$_2$O/SiO$_2$ = 44
Dab-4Br	0·11	3	Gmelinite only	5·78	0·33	SiO$_2$/Al$_2$O$_3$ = 15
Dab-4Br	0·23	13	Gmelinite only	5·69	0·25	SiO$_2$/Al$_2$O$_3$ = 15
Dab-4Br	0·11	13	Gmelinite + Chabazite			SiO$_2$/Al$_2$O$_3$ = 75
Dab-3Br	0·24	7	P	5·21	0·04	⎫
Dab-4Br	0·24	9	Gmelinite only	5·97	0·56	⎪
Dab-5Br	0·24	7	Gmelinite + Chabazite	5·29	0·24	⎬ H$_2$O/SiO$_2$ = 44
Dab-6Br	0·24	9	Gmelinite + Chabazite	4·69	0·12	⎪
Dab-10Br	0·24	9	Y + P			⎪
Dab-Pr$_2$	0·24	6	Y + P^c	4·51	0·03	⎪
Dab-Bu$_2$	0·24	6	Y + P^c	5·24	0·01	⎭
(b) T = 170–180°C						
	0	7	Analcime			⎫
Dab-4Br	0·11	7	Analcime	5·37	0·01	⎪
Dab-4Br	0·23	7	Analcime + Mordenite	8·46	0·16	⎬ Na/SiO$_2$ = 1·5
Dab-4Br	0·43	7	Mordenite only	20·8	0·40	⎪
	0	7	Analcime			⎭

^{a} R denotes the polyelectrolyte repeat unit.
^{b} Y denotes zeolite Y (faujasite), P denotes gismondine type zeolite.
^{c} Plus a trace of chabazite.

completely changed the products from zeolites $Y + P$ to gmelinite plus chabazite, and then with increasing amounts of polyelectrolyte up to a certain level to pure gmelinite, with a considerable uptake of the polyelectrolyte. The chains stretched out along the wide gmelinite channels should prevent the frequent stacking faults which turn natural and synthetic gmelinites into narrow pore sorbents. When the Dab-4Br was burnt out of the gmelinite it sorbed 7·5% by weight of cyclohexane, so that indeed the gmelinite was sufficiently free of stacking faults to show wide pore sorbent behaviour.

At 170–180°C the presence of Dab-4Br changed the zeolite grown from the dense zeolite, analcime, to a silica-rich wide pore mordenite, again with a considerable content of Dab-4Br in the wide channels.

The Dabco unit is cylindrical with ~6·1 Å diameter so that the polyelectrolyte chains would fit easily into the wide gmelinite channels (free diameter ~7 Å). The repeating units of Dab-3, -4, -5, -6 and -10Br have respectively the lengths 7·5, 8·7, 9·9, 11·0 and 14·5 Å. As Table 10 shows, all the polyelectrolytes yielding gmelinite, or this zeolite plus chabazite, had repeating units between 8·7 and 11 Å, to be compared with the 10 Å unit cell dimension of gmelinite in the channel direction. In mordenite this dimension is 7·5 Å, compared with 8·7 Å for Dab-4Br. The tendency to spatial correspondences may be relevant for the templating action of the polyelectrolytes.

The compositions of the gmelinites and mordenite are summarized in Table 11.[91] On average there are 2·3 cationic charges arising from the polyelectrolyte per channel in the unit cell of gmelinite and 1·7 for mordenite. There are two such charges per 8·7 Å along the Dab-4Br chain so there should be 2·3 charges per 10 Å channel length (the unit cell length) in gmelinite and 1·7 charges per 7·5 Å in mordenite. The

TABLE 11
Compositions of zeolites containing Dab-4Br[91]

Zeolite	SiO_2/Al_2O_3	Al/uc[a]	Na/uc	N/uc	C/N
Gmelinite	7·05	5·3	3·5	2·3	5·0
	5·52	6·4	4·2	2·8	6·3
	5·78	6·2	6·2	2·0	5·2
	5·69	6·2	5·4	1·6	6·1
	5·97	6·0	4·1	3·3	4·5
	6·71	5·5	3·3	2·1	5·2
	5·31	6·4	4·4	2·0	5·6
Mordenite	20·8	4·2	3·3	1·7	6·8

[a] uc denotes unit cell.

correspondence indicates channels filled in each zeolite by the poly-electrolyte chains. The C/N ratio for Dab-4Br is 5·07. The C/N ratios for gmelinites made at ∼90°C average 5·4 so that at this temperature there seems to be little hydrolysis of the polyelectrolyte. For mordenite made at 170–180°C the higher C/N ratio of 6·8 could indicate some polymer breakdown. The way in which the gmelinite and mordenite crystals form in association with and around the polyelectrolyte chains raises interesting stereochemical considerations. However, to some extent it has a parallel in the crystallization of urea and thiourea channel clathrates around hydrocarbon molecules.

5.6. *Some History-dependent Factors*

The way in which synthesis is performed has been found to influence the course of crystallization. This can be illustrated by an investigation by Freund[18] of factors controlling kinetics and yields in faujasite prepara-tion as Na-X. In one procedure sodium metasilicate pentahydrate solu-tion was mixed at the crystallization temperature, and normally in 5 seconds, with the requisite amounts of alkaline sodium aluminate solution to give the overall oxide composition $7·38Na_2O.Al_2O_2.5·03SiO_2.400H_2O$. Immediate formation of gel occurred and the mixture was then crystallized under a range of conditions. The quantity of Na-X formed at any time was determined by the extent of benzene adsorption on a sample of the solid product. From the ratio of amounts of Na-X and of the total solid, the fractional yield α was obtained. When $\alpha = 1$ all the solid is Na-X. Table 12 shows how the yields and the times for 50% crystallization were influenced by several reaction variables. As rate of stirring increased the S-shape of curves of yield against time was accentuated but the maximum yield decreased. For rates of 150 and 350 rev/min the yields passed through maxima followed by a slow decrease in which Na-X was progressively replaced by gis-mondine-like Na-P. Stirring promotes collision nucleation which appears to assist the competitive formation of Na-P. With the exception of etched glass, which appeared to aid nucleation of Na-P to some extent, the nature of the reactor wall (steel or nickel) did not much affect either the maximum yield or the half-life time, $t_{0·5}$. Where it is important to avoid contamination of the mixture by cations or silica leached from the glass reaction vessel, one can employ vessels of polythene, polypropylene or Teflon for temperatures at least up to about 100°C. Teflon can be employed at temperatures above 100°C, especially if the whole closed Teflon vessel is itself enclosed in a steel autoclave containing water to equalize the pressure inside and outside the sealed Teflon container. As

TABLE 12
Variables in the synthesis of zeolite Na-X[18]

Parameter varied	Crystallization temperature (°C)	Stirring speed (rev/min)	Reactor	Gel preparation time(s)	α_{max}	$t_{0.5}$ (h)
Stirring	98	0		5	1·00	3·60
		150		5	0·98	3·25
		150	stainless steel	5	0·99	3·00
		350		5	0·80	2·80
Reactor wall	98	150	stainless steel	5	0·96	3·35
			nickel	5	1·00	3·10
			etched glass[a]	5	0·91	3·40
			nickel	5, with nickel borate[b]	0·98	3·50
Temperature	82	0	stainless steel	5	1·00	7·00
	98	0		5	1·00	3·60
Gel preparation	98	150	nickel	5	0·86	3·30
				45	0·61	3·45

[a] Etched by 10 M NaOH at 150°C for 7 h.
[b] B_2O_3 concentration was $3·3 \times 10^{-2}$ mol dm^{-3}.

already discussed in §5.1 an increase in temperature increased the crystallization rate. Finally, slow mixing of the silicate and aluminate at ~98°C depressed the yield of Na-X and promoted that of Na-P.

In a second series of experiments the influence of silicate reagents was explored. Silicate solutions containing $1\cdot5$ mol SiO_2 dm^{-3} and with $Na_2O/SiO_2 = 1$ were prepared from various sources. Each was mixed with sodium aluminate solution with added NaOH to give an overall oxide composition $12\cdot10NaOH.Al_2O_3.17\cdot38SiO_2.500H_2O$, from which a homogeneous milky gel appeared after an induction time of about 20 s. The mixture was brought to 98°C and crystallized at that temperature. The silica sources could be divided into two groups, active and inactive for production of Na-X. The active group included hydrated sodium silicates and notably sodium metasilicate pentahydrate. The inactive group comprised mainly solid and colloidal silicas and yielded only Na-P. It was discovered that silica sources of the active group contained traces of Al. However intentional addition of Al_2O_3 to pure SiO_2 and Na_2CO_3 followed by fusion, crystallization and hydration did not give an active silicate. Presumably the active metasilicates through the presence of Al were able quickly to generate precursor species needed to nucleate Na-X or already contained these precursors as impurities. The division of silica sources into active and inactive categories had previously been observed by Whittam,[92,93] who did not at first trace the cause of the activity, but subsequently confirmed[94] the observations of Freund. Whittam made zeolites X, Y and A by the active silicate route.

Freund also reported that zeolite X could be crystallized from a clear solution with dissolved concentrations $2\cdot40$ mol dm^{-3} of NaOH, $0\cdot7$ mol dm^{-3} of SiO_2 and $0\cdot014$ mol dm^{-3} of Al_2O_3. The solution was prepared at -4°C. Precipitation at 20°C was observed only after an incubation time of 15 days. Samples of the solution were crystallized at 80°C after different maturation periods. The precipitate was crystalline from the start and was formed progressively. This behaviour complements the analogous preparation of analcime from clear solutions[15] referred to in Chapter 3 §2.5.

Some of the results obtained by Freund have more general application. Thus the beneficial effect on yield of an appropriate maturation period at temperatures below the subsequent crystallization temperature has been reported for zeolites Na-X,[18] the hexagonal polymorph of faujasite, ZSM-3,[78] and the mazzite-type zeolite ZSM-4.[95] It is also likely in producing zeolites other than faujasite and zeolite A that some sources of silica are more active than others. Thus, attention has already been drawn in §5.3 to the advantage in the synthesis of silica-rich varieties of zeolite A of using $(TMA)_8[Si_8O_{20}].69H_2O$ as one source of the silica.[43] It is of interest also that in syntheses of Na-A, unlike those of Na-X, stirring is no disadvantage.

The extent to which mixtures are homogenized before crystallization can also be important. For example, when a solution of water glass plus active silicate and one of sodium aluminate were thoroughly mixed at 25°C and a solution of aluminium sulphate was then added with vigorous stirring until a homogeneous mixture was assured, subsequent crystallization at 85°C for 22 h gave an excellent yield of faujasite. On the other hand, when the above procedure was followed without homogenization, crystallization at 85°C for 26 h yielded some amorphous material together with low yields of several zeolites including faujasite, a chabazite type, Na-P and analcime.[95]

The nature and treatment of one or more of the components of the reaction mixture can play a large part in determining the products obtained. Two examples can be given. In a partial exploration between 150 and 450°C of crystallization fields in the $CaO-Al_2O_3-SiO_2-H_2O$ system differences were observed when the silica was provided as powdered silica glass rather than as colloidal silica.[48] From oxide compositions $CaO.Al_2O_3.nSiO_2$ $(1 \leqslant n \leqslant 9)$ + aq a harmotome type zeolite formed with silica glass which was absent when colloidal silica was used. Also with silica glass, hydrogrossulars poor in silica, including the aluminium end member, appeared below 210°C in place of the aluminosilicates formed with colloidal silica.

Secondly, when the sources of silica and alumina were respectively kaolinite (oxide formula $Al_2O_3.2SiO_2.2H_2O$) and metakaolinite (oxide formula $Al_2O_3.2SiO_2$) and were reacted at 80° with aqueous $Ba(OH)_2$ the products were:[96]

From metakaolinite	*From kaolinite*
Ba-G, *L* (zeolite like Linde type L)	Ba-G, *L* (zeolite like Linde type L).
Hexagonal polymorph of celsian	Cymrite-type non-zeolite
Zeolite Ba-N (not identified)	Barium silicate hydrate
Zeolite Ba-T (not identified)	

Only one of the low-temperature products (i.e. Ba-G, *L*) obtained with the former corresponded with those derived from the latter. The metakaolinite is obtained from kaolinite by heating kaolinite at 500–550°C. These examples, for overall similar gross compositions, illustrate how the nature of the reactants can control the products.

5.7. *Crystallization in Presence of Dyestuffs*

It has long been known that additives in crystallizing solutions can result in habit modification and that dyestuffs are particularly active in producing this modification.[97,98] The dyestuff may be adsorbed selectively on certain crystal faces and can thereby change selectively the relative rates of growth of the faces. It was also shown that basic dyestuffs are often adsorbed on zeolite surfaces, as illustrated below for several zeolites:[99]

Gismondine type Na-P	crystal violet
	nuclear fast red
Chabazite types with Na or K	toluidine blue
exchange ions	methylene blue
Near phillipsite K-M	malachite green
	nuclear fast red
Faujasite	methylene blue
	malachite green

The above dyes were all strongly adsorbed on the respective zeolites; additionally other dyes were adsorbed, but less strongly, for example acriflavine on the faujasite zeolites X and Y. The adsorption behaviour depends both on the structure of the zeolite and on that of the dyestuff. In general the dye molecules are too large to permeate the crystals, the coloration of which is therefore primarily a surface effect.

In zeolite synthesis various kinds of germ nuclei can arise through fluctuation, which, if they become viable, result in co-crystallization of more than one zeolite. Therefore conditions are sought which favour the formation of viable nuclei of the desired kind only. If dyestuffs are differently adsorbed upon the germ nuclei as upon the crystals, then the right choice of dyestuff might suppress the appearance of one type of viable nucleus but not that of the type which it is desired to produce. That such effects are possible has been demonstrated in several instances. Thus, crystal violet and methyl red suppressed co-crystallization of gismondine type Na-P with the faujasite variant Na-X; and crystal violet, methyl red and nuclear fast red did the same with the faujasite variant Na-Y. Also crystal violet, methyl red, nuclear fast red and malachite green suppressed co-crystallization of phillipsite type K-M with zeolite L and its variants AG5 and AG4.[99] The molar ratio dyestuff/Al_2O_3 in the reaction mixtures was 5×10^{-3} in all cases.

6. Successive Transformations

According to Ostwald's rule, in the formation of polymorphs of a given element or compound the first polymorph to be formed from vapour, liquid or solution tends to be the least stable thermodynamically which is then in succession replaced by more and more stable polymorphs. The rule can apply to crystallizations from hydrothermal systems. For example, in Chapter 2 it was noted that in an aqueous system at 330–440°C and at high pressure the crystallization sequence from amorphous silica was[100]

$$\text{amorphous} \rightarrow \text{cristobalite} \rightarrow \text{keatite} \rightarrow \text{quartz}$$

In other instances also the initial appearance of cristobalite has been observed, which, when left at reaction temperature in more prolonged contact with the mother liquor, converts to quartz.

In zeolite syntheses similar product changes with time are found, examples of which were shown in Fig. 14. Syntheses of mordenite in strongly alkaline conditions are followed by its disproportionation into analcime and quartz.[38] If one mixes solutions of aluminate and silicate the first product, that which is thermodynamically least stable, is usually an aluminosilicate gel which is precipitated very rapidly. This metastable source is progressively consumed as crystalline phases appear. A typical reaction sequence under the appropriate synthesis conditions is

$$\text{amorphous} \rightarrow \text{faujasite} \rightarrow \text{gismondine type Na-P}$$

By adding tetramethylammonium to the reaction mixture this sequence can be changed to[90]

$$\text{amorphous} \rightarrow \text{faujasite} \rightarrow \text{mazzite type ZSM-4}$$

The progress of this sequence with time is shown in Fig. 16. The time-scale for sequences of the above type and the relative appearance times of the individual members can be altered by varying the reaction conditions. For example (§5.5), stirring the mixture promotes the rapid formation of zeolite Na-P at the expense of faujasite, so that crystallization under quiescent conditions is preferred in faujasite preparations. The great importance of these sequences, both in the laboratory and in large-scale manufacture of zeolites, is that once the desired product has appeared it can be isolated from the reaction medium and thereafter remains as an adequately stable product. It is no longer converted into other more stable products because there is no reaction medium to dissolve it and so allow nucleation of these other products; and because its change by a solid state reaction involves a large activation energy and hence a much higher temperature. Much of the success and richness of zeolite chemistry depends on one's ability to prepare and isolate metastable crystals.

Other instances of crystallization sequences, progressive with time, are considered in Chapter 5. Successive transformations are indeed very common.

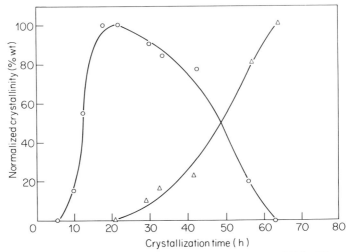

Fig. 16. Yields of crystals against time at about 100°C. The ratio $TMA^+/(TMA^+ + Na^+)$ is such as to yield ultimately only ZSM-4 (see Fig. 15). Faujasite (○) appears first and is then progressively replaced by ZSM-4.[90]

7. Growing Larger Zeolite Crystals

The synthesis of zeolites at temperatures around 100°C usually produces crystals in the size range 0·1–10 μm. In applications as catalysts, sorbents and ion exchangers this size range is fully suitable. The small sizes indicate that once nucleation begins there is rapidly formed a shower of numerous tiny particles all competing for the available chemical nutrients. At the end of crystal growth, as shown in §4, the first viable nuclei to appear in the shower are the largest because they have had the longest time in which to grow. Synthetic zeolite crystals large enough to mount in an X-ray camera must be about 100 μm or even more in diameter. In other situations one may be interested in still bigger crystals suitable for examination as molecule and ion-selective membranes or other electro-chemical uses.

Crystal size can be influenced by experimental factors which alter the numbers of viable nuclei which develop, so that if these numbers are reduced a greater share of the chemical nutrients is available for each growing crystal. The use of seed crystals can be similarly employed in situations where, as in growth of zeolite Na-Y on seeds of Na-X, little additional nucleation occurred.[10,11] The smaller the number of seeds involved the larger the average size of the final crystals, again because a larger amount of chemical nutrient is available for each crystal before these nutrients are exhausted. With initial mole fractions x_0 of seeds of Na-X (defined as in §3) of 0·044, 0·44 and 4·4% the distribution densities in the final states were as shown in Fig. 17,[10] the mean radii of these final distributions being respectively 2·7, 1·25 and 0·6 μm. The seed crystals had a mean radius of 0·23 μm. If one uses the mean radii of the final distributions with the mean radius of the seeds to calculate approximately the volumes of crystalline material added to the seeds one finds that about ten times more new crystal per crystal of seed was added when x_0 was reduced from 4·4% to 0·44% and again ten times more when x_0 was further reduced from 0·44% to 0·044%,

In §5.6 reference was made to the ease with which faujasites Na-X and Na-Y could be grown from reaction mixtures containing so-called active silicates. Freund[18] was able to influence the mean crystal dimensions by diluting the active silica source with an inactive one. With ratios of active silica to total silica of 0·005, 0·1 and 1 the maxima in the size distribution curves corresponded with crystal diameters of about 5·8, 2·8 and 1·8 μm respectively. This implies that more numerous nuclei were generated the larger the amount of the active silica.

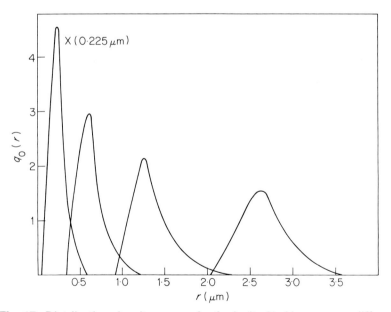

Fig. 17. Distribution density curves for faujasite Na-Y grown on different quantities of seed crystals.[10] $q_0(r)$ is the number fraction of crystals of radius r. The temperature was 88°C. The left hand curve is for the seed crystals.

With a view to producing still larger crystals suitable for X-ray single-crystal study the crystallization procedures have been differently modified by Ciric[101] and by Charnell.[102] Ciric employed a closed U-shaped tube as reaction vessel. It contained a firm gel of Carbo-pol in 1 M NaOH solution. In addition at one end of the U was a slurry of soluble sodium silicate and at the other a slurry of sodium aluminate. These reactants diffused slowly together and reacted. After two weeks at 80–90°C the reaction tube was sectioned, the products recovered and washed several times by stirring, sedimentation and decanting until only the crystalline fraction remained. This comprised in part polycrystals of the gismondine type zeolite Na-P up to 110 μm in diameter and also contained single-crystal cubes of Na-A of up to 60 μm edge.

Charnell improved on this by synthesizing crystals of zeolites Na-A or Na-X of about 100 and 140 μm dimensions respectively and in quantities of 40–50 g per preparation. The crystals were grown by reacting sodium metasilicate with sodium aluminate in the presence of triethanolamine. The latter appeared to reduce the rate of nucleation and so the number of viable nuclei formed. The triethanolamine was added to the sodium silicate solution and the sodium aluminate was stirred into the silicate with the usual gel formation. The reaction vessel was a two-quart poly-propylene covered jar and crystallization was usually completed at 75–85°C in 2–3 weeks for Na-A and in 3–5 weeks for Na-X. Similar crystals of more uniform size were obtained more quickly when the temperature was raised to 100°C. Various other water-soluble bases were also ex-amined (2-dimethylaminoethanol, ethanolamine, diethylenetriamine, 2-amino-2-methyl-1-propanol, morpholine and ethylene diamine) but tri-ethanolamine was the most satisfactory.

A significant influence of temperature upon crystal size has also been observed in certain instances. The silica end member of the ZSM-5 zeolite, when made in presence of NaF, was reported to give crystals of 2–15 μm at 100°C while at 200°C some of the crystals were rods exceeding 200 μm in length.[103] As observed in Chapter 2, §9, in the author's laboratory under approximately isothermal conditions at temper-atures above 300°C analcime crystals were made up to 0·5 mm diameter and nosean hydrate (i.e. the sodalite structure) in 1·5 mm blocks.[104] Bye and White[105] used the temperature gradient method to prepare sodalite crystals 1·2 cm across. This method has been very successful in growing large quartz and other crystals (Chapter 2, §9). A small seed of sodalite was supported near the colder top end of the autoclave at 450°C while the bottom of the autoclave was at 475°C. The reaction mixture consisted of 3·066 g Al_2O_3, 3·618 g SiO_2, 2·34 g NaCl and 7·2 g NaOH with water added to bring the volume to 50 ml. Mean linear growth rates on the seed of as much as 0·4 mm per day were achieved. While the temperature gradient was important it was not, contrary to the case with quartz, necessary to use crystals of the same species that was being grown in order to form the large crystals.

To obtain larger crystals by using higher temperatures, these temperatures must be within the crystallization field of the zeolite. Those zeolites which can be synthesized at higher temperatures tend to be less porous and to have a lower water content, such as analcime, rather than heavily hydrated structures like chabazite or faujasite. At elevated temperatures the water will have less and less structure and hydrated ions will retain their hydration shells less and less strongly. Template effects associated with the large units consisting of ions plus firmly held hydration shells, expected in low-temperature synthesis, will now be much less in evidence. Also at elevated temperatures aluminosilicate precursor species may be increasingly replaced by monomeric $Al(OH)_4^-$ and $Si(OH)_4$ and anions derived from $Si(OH)_4$. These factors should in turn favour the appearance of less open tectosilicates, as observed.

8. Compositional Zoning

Figure 12 in §5.2 shows that, in alkaline media and starting from silica-rich compositions, the zeolites formed (here chabazite and edingtonite type K-F) were more aluminous than the initial mixture. For zeolites which behave in this way the chemical nutrients become impoverished in Al more rapidly than silicon. It could then be anticipated that

(a) crystals nucleating later in the course of crystallization could have different Si/Al ratios from, and here be more siliceous than, crystals produced earlier; and

(b) as a crystal grows its core could be more aluminous than the outer layers.

The expectation (a) is however not necessarily the behaviour which has been found by experiment. Thus Ueda and Koizumi,[106] investigating analcime formation from extremely siliceous solutions of aluminosilicate in caustic soda (Chapter 3, §2.5) reported that the composition of the crystals formed late in the process did not differ from that of the crystals first precipitated. The analysis of the kinetics of crystallization by Zdhanov and Samulevich[12] (§4) is relevant in this connection. The nucleation rates peaked and then rapidly declined before any significant amount of crystal powder had been formed and therefore before any significant change in the Si/Al ratio of the chemical nutrients could have taken place. More investigations of the kind made by Ueda and Koizumi are required before generalizations relating to expectation (a) can be made.

As regards expectation (b) compositional zoning can be imposed. In §3 it has already been noted that zeolite Y can grow on seeds of zeolite X; and that silicalite I (Al-free ZSM-5) can grow around cores of ZSM-5 containing Al. Here there must be sharp discontinuities in chemical composition. What happens when there are no seeds is less clear. Compositional zoning has been reported for plagioclase felspars[107] but was not observed for natural faujasite, phillipsite and offretite.[108] In synthetic crystals of Na-X a weak zoning was reported in which the ratio Si/Al increased by a few per cent from core to rim,[109] in accord with expectation (b).

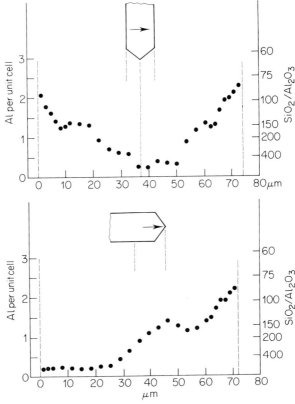

Fig. 18. Two perpendicular profiles of Al concentration across a section of a ZSM-5 crystal, determined by microprobe step scans, shown as the dotted lines.[110]

In the silica-rich zeolite ZSM-5 the zoning was the reverse of that suggested in (b). Crystals of this zeolite up to 200 μm are not difficult to make, and Ballmoos and Meier[110] examined the Al-content at different points in such crystals by electron microprobe scanning. The beam had a diameter of 0·8 μm, producing an analysis volume of about 2 μm^3. Figure 18 illustrates their results and shows very strong compositional zoning in which the crystal cores are much less aluminous than the outer parts in both longitudinal and lateral traverses. The explanation may be that nearly Al-free ZSM-5 nucleates more readily than ZSM-5 richer in Al, so that the nuclei from which growth proceeds tend to be very low in Al. Silica is then consumed preferentially to alumina as the Si/Al ratio in the chemical nutrients decreases and the crystalline layers laid down later are progressively more aluminous than those formed earlier. There is a need for additional definitive studies of this kind for synthetic zeolites which, using Charnell's[102] or other methods, can be made of adequate crystal size. Sodalite and cancrinite hydrates, analcime, faujasite and zeolite A are examples.

9. Concluding Remarks

Despite possibilities for diverse nucleation, conditions can usually be found which yield a particular zeolite from a particular kind of nucleus. The successful nucleation is kinetically determined and is characterized by an induction period sensitive to a number of factors (e.g. temperature, kinds of cations, nature and history of reactants, pH, presence or absence of additives, ageing prior to crystallization and presence or absence of stirring). Where a succession of crystalline products forms these products are expected to appear in order of increasing thermodynamic stability. Sigmoid curves of the amount of zeolite formed against time are characteristic for many zeolites where nucleation is followed by crystal growth. If crystal growth is limited to the surfaces of pre-formed seeds, the induction period is eliminated and sigmoid curves are no longer obtained. Linear rates of deposition of new crystalline material on growing crystals have been found to be constant during most of the growth and, among other factors, depend strongly upon temperature. There is evidence that the rate of nucleation increases rapidly with time and then passes through a maximum and declines, as the chemical nutrients start to be consumed competitively by the growing crystals.

Most crystallization studies appear to support the view that nucleation occurs from dissolved precursor species, and that crystal growth on viable nuclei involves a condensation-polymerization of dissolved species onto the growing surfaces. While nucleation can be in solution these nuclei can subsequently become mixed with hydrogel. It has also been suggested that nucleation can occur on or in hydrogel and even that crystal growth occurs by direct transformation of hydrogel.[106,107] Even if nucleation is within hydrogel it must be remembered that hydrogel is less compact than the crystals so that direct incorporation of gel onto the growing crystals would quickly leave gaps between crystal and ambient gel. The remaining gel would therefore need to dissolve to reach the crystal surfaces. The expected gaps have been observed during growth of zeolite crystals embedded in gel.[108] Thus, while nucleation could occur in solution, on hydrogel or within the gel it seems likely that crystal growth primarily involves deposition of dissolved species on the crystals.

References

1. P.H. Egli and S. Zerfoss, *Faraday Soc. Discussion No. 5* (1949) 61.
2. P.W.M. Jacobs and F.C. Tompkins, *in* "Chemistry of the Solid State" (Ed. W.E. Garner), Ch. 7, p. 184. Butterworth, London, 1955.
3. W. Meise and F.E. Schwochow, *in* "Molecular Sieves" (Eds. W.M. Meier and J.B. Uytterhoeven), p. 169. American Chemical Society Advances in Chemistry Series, No. 121, 1973.
4. W.J. Dunning, *in* "Chemistry of the Solid State" (Ed. W.E. Garner), Ch. 6, p. 159. Butterworth, London, 1955.
5. R. Becker, *Ann. Phys.* (1938) **32,** 128.
6. C.W. Bunn and H. Emmett, *Faraday Soc. Discussion No. 5* (1949) 119.
7. H.E. Buckley, *Faraday Soc. Discussion No. 5* (1949) 243.
8. R.M. Barrer and I.S. Kerr, *J. Chem. Soc.* (1963) 434.
8a. G.C. Edwards, D.E.W. Vaughan and E.W. Albers, to Grace Co., U.S.P. 4,175,059, 1979.
9. E.g., Dana's "System of Mineralogy", 7th Edn by Palache, Berman and Frendel, New York, 1944 and 1951.
10. H. Kacirek and H. Lechert, *J. Phys. Chem.* (1975) **79,** 1589.
11. H. Kacirek and H. Lechert, *J. Phys. Chem.* (1976) **80,** 1291.
12. S.P. Zdhanov and N.N. Samulevich, *in* "Proceedings of the 5th International Conference on Zeolites" (Ed. L.V.C. Rees), p. 75. Heyden, London, 1980.
13. E.g., C.V. McDaniel and P.K. Maher, to Grace Co., U.S.P. 4,166,099, 1979.
14. C.L. Kibby, A.J. Perrotta and F.E. Massoth, *J. Catal.* (1974) **35,** 256.
15. L.D. Rollmann, to Mobil Co., U.S.P. 4,088,605, 1978.
16. D. Dominé and J. Quobex, *in* "Molecular Sieves", p. 78. Society of Chemical Industry, 1968.
17. J. Ciric, *J. Colloid Interface Sci.* (1968) **28,** 315.
18. E.F. Freund, *J. Cryst. Growth* (1976) **34,** 11.
19. R.M. Barrer and J.F. Cole, *J. Chem. Soc. A* (1970) 1516.

20. D.T. Hayhurst and L.B. Sand, *in* "Molecular Sieves II" (Ed. J.R. Katzer), p. 219. American Chemical Society Symposium Series, No. 40, 1977.
21. R.M. Barrer and C. Marcilly, *J. Chem. Soc. A* (1970) 2735.
22. G.T. Kerr, *J. Phys. Chem.* (1968) **72,** 1385.
23. R. Aiello and R.M. Barrer, *J. Chem. Soc. A* (1970) 1470.
23a. H.J. Bosmans, E. Tambuyzer, J. Paenhuys, L. Ylen and J. Vancluysen, *in* "Molecular Sieves" (Ed. W.M. Meier and J.B. Uytterhoeven), p. 179. American Chemical Society Advances in Chemistry Series, No. 121, 1973.
24. D. Turnbull, *in* "Solid State Physics", Vol. 3, p. 252. (Ed. F. Seitz and D. Turnbull) Academic Press, London and New York, 1956.
25. R.A. Cournoyer, W.L. Kranich and L.B. Sand, *J. Phys. Chem.* (1975) **79,** 1578.
26. E. Bergman, *J. Appl. Chem,* (1963) **13,** 319.
27. Xu Qin-Hua, Bao Shu-Lin and Dong Jia-Lu, private communication.
27a. K.-J. Chao, T.C. Tasi, M.-S. Chen and I. Wang, *J. Chem. Soc. Faraday I* (1981) **77,** 465.
28. A. Culfas and L.B. Sand, *in* "Molecular Sieves" (Ed. W.M. Meier and J.B. Uytterhoeven), p. 140. American Chemical Society Advances in Chemistry Series, No. 121, 1973.
29. R. Aiello, C. Colella, D.G. Casey and L.B. Sand, *in* "Proceedings of the 5th International Conference on Zeolites" (Ed. L.V.C. Rees), p. 49. Heyden, London, 1980.
30. G.W. Morey, R.O. Fournier and J.J. Rowe, *Geochim. Cosmochim. Acta* (1962) **26,** 1029.
31. S.P. Zdhanov, *in* "Molecular Sieves", p. 70. Society of Chemical Industry, 1968.
32. R.M. Barrer and D.E. Mainwaring, *J. Chem. Soc. Dalton* (1972) 1254.
33. W. Lowenstein, *Amer. Mineral.* (1954) **39,** 92.
34. P. Caullet, J.-L. Guth and R. Wey, *C.R. Acad. Sci. Paris* (1979) **288,** Serie D-1.
35. "The hydrothermal chemistry of silicates", Parts 1 to 22 in *J. Chem. Soc.* (1951) to (1978).
36. "The chemistry of soil minerals", Parts 1 to 14, in *J. Chem. Soc.* (1965) to (1974).
37. R.M. Barrer, *J. Chem. Soc.* (1948) 127.
38. R.M. Barrer, *J. Chem. Soc.* (1948) 2158.
39. R.M. Barrer and D.J. Marshall, *Amer. Mineral.* (1965) **50,** 484.
40. R.M. Barrer, S. Barri and J. Klinowski, *in* "Proceedings of the 5th International Conference on Zeolites" (Ed. L.V.C. Rees), p. 20. Heyden, London, 1980.
41. R.M. Barrer and W. Sieber, *J. Chem. Soc. Dalton* (1977) 1020.
42. R.M. Barrer and P.J. Denny, *J. Chem. Soc.* (1961) 971.
42a. G.T. Kerr and G. Kokotailo, *J. Amer. Chem. Soc.* (1961) **83,** 4675.
43. H. Kacirek and H. Lechert, *in* "Molecular Sieves II" (Ed. J.R. Katzer), p. 244. American Chemical Society Symposium Series, No. 40, 1977.
44. G.T. Kerr, *Science,* (1963) **140,** 1412.
45. H.E. Robson, to Esso Co., U.S.P. 3,720,753, 1973.
46. A. Erdem and L.B. Sand, *in* "Proceedings of the 5th International Conference on Zeolites" (Ed. L.V.C. Rees), p. 64. Heyden, London, 1980.
47. A. Erdem and L.B. Sand, *J. Catal.* (1979) **60,** 241.

48. R.M. Barrer and P.J. Denny, *J. Chem. Soc.* (1961) 983.
49. C. Baerlocher and W.M. Meier, *Helv. Chim. Acta* (1969) **52,** 1853.
50. C. Baerlocher and W.M. Meier, *Helv. Chim. Acta* (1970) **53,** 1285.
51. R.M. Barrer and D.E. Mainwaring, *J. Chem. Soc. Dalton* (1972) 2534.
52. E.M. Flanigen, to Union Carbide, Dutch Patent 6,710,729, 1968.
53. R.L. Wadlinger, E.J. Rosinski and E.J. Plank, to Mobil Co., U.S.P. 3,375,203, 1968.
54. T.V. Whittam and B. Youll, to I.C.I. Ltd, U.S.P. 4,060,590, 1977.
55. T.V. Whittam, to I.C.I. Ltd, Ger. Patent 2,748,276, 1978.
56. R.M. Barrer and D.E. Mainwaring, *J. Chem. Soc. Dalton* (1972) 1259.
57. M.K. Rubin and E.J. Rosinski, to Mobil Co., U.S.P. 3,699,139, 1972.
58. W. Sieber and W.M. Meier, *Helv. Chim. Acta* (1974) **57,** 1533.
59. G.T. Kerr, to Mobil Co., U.S.P. 3,459,676, 1969.
60. M.K. Rubin, E.J. Rosinski and C.J. Plank, to Mobil Co., U.S.P. 4, 151, 189, 1979.
61. BASF AG, German Patent, 2,830,787, 1980.
62. D.M. Bibby, N.B. Milestone and L.P. Aldridge, *Nature* (1980) **285,** 30.
63. R.J. Argauer and G.R. Landolt, to Mobil Co., U.S.P. 3,702,886, 1972.
64. Mobil Co., Dutch Patent, 7,014,807, 1971.
65. P. Chu, to Mobil Co., U.S.P. 3,709,979, 1973.
66. J. Ciric, to Mobil Co., U.S.P. 3,692,470, 1971.
67. M.K. Rubin, C.J. Plank and E.J. Rosinski, to Mobil Co., Eur. Pat. Appl. 13,630, 1980.
68. G.T. Kokotailo and W.M. Meier, *Chem. Soc. Special Publ. 33* (1980) 133.
69. E.M. Flanigen, J.M. Bennett, R.W. Grose, J.P. Cohen, R.L. Patton, R.M. Kirchner and J.V. Smith, *Nature* (1978) **271,** 512.
70. D.M. Bibby, N.B. Milestone and L.P. Aldridge, *Nature* (1979) **280,** 664.
71. E.g. R.M. Barrer, "Zeolites and Clay Minerals as Sorbents and Molecular Sieves", Ch. 2. Academic Press, London and New York, 1978.
72. R.M. Milton, to Union Carbide, U.S.P. 2,882,244, 1959.
73. R.M. Milton, to Union Carbide, U.S.P. 3,030,181, 1962.
74. R.M. Barrer and E.A.D. White, *J. Chem. Soc.* (1951) 1267.
75. R.M. Barrer and W. Sieber, *J. Chem. Soc. Dalton* (1978) 598.
76. R.M. Barrer and D.J. Marshall, *J. Chem. Soc.* (1964) 2296.
77. H.E. Robson, D.P. Shoemaker, R.A. Ogilvie and P.C. Manor, *in* "Molecular Sieves" (Eds. W.M. Meier and J.B. Uytterhoeven), p. 106. American Chemical Society Advances in Chemistry Series, No. 121, 1973.
78. G.T. Kokotailo and J. Ciric, *in* "Molecular Sieves I" (Ed. R.F. Gould), p. 109. American Chemical Society Advances in Chemistry Series, No. 101, 1971.
79. J. Ciric, to Mobil Co., U.S.P. 3,411,874, 1968.
80. R.M. Barrer, J.F. Cole and H. Sticher, *J. Chem. Soc. A* (1968) 2475.
81. Union Carbide, U.S.P. 3,414,602, 1968.
82. H.C. Duecker, A. Weiss and C.R. Guerra, U.S.P. 3,567,372, 1971.
83. W.M. Meier and D.H. Olson, "Atlas of Zeolite Structure Types", published by the Structure Commision of the International Zeolite Association, 1978.
84. E.g., R.M. Barrer, "Zeolites and Clay Minerals as Sorbents and Molecular Sieves", Ch. 2. Academic Press, London and New York, 1978.
85. L.D. Rollemann, *in* "Inorganic Compounds with Unusual Properties", p. 387. American Chemical Society Advances in Chemistry Series, No. 173, 1979.

86. D.W. Breck, "Zeolite Molecular Sieves". Wiley, New York, 1974.
87. J. Ciric, to Mobil Co., French Patent 1,502,289, 1966.
88. C.J. Plank, E.J. Rosinski and M.K. Rubin, to Mobil Co., B.P. 1,117,568, 1968.
89. E.E. Jenkins, U.S.P. 3,578,398, 1971.
90. F.G. Dwyer and P. Chu, *J. Catal.* (1979) **50,** 263.
91. R.H. Daniels, G.T. Kerr and L.D. Rollmann, *J. Am. Chem. Soc.* (1978) **100,** 3097.
92. T.V. Whittam, to Peter Spence Ltd, B.P. 1,171,463, 1969.
93. T.V. Whittam, to I.C.I. Ltd, B.P. 1,450,411, 1976.
94. T.V. Whittam, communication at 5th International Zeolite Conference, Naples, June 2nd to 6th, 1980.
95. B.P. 1,117,463, 1968.
96. R.M. Barrer, R. Beaumont and C. Colella, *J. Chem. Soc. Dalton* (1974) 934.
97. H.E. Buckley, *Faraday Soc. Discussion No. 5* (1949) 243.
98. A. Butchart and J. Whetstone, *Faraday Soc. Discussion No. 5* (1949) 254.
99. T.V. Whittam, to I.C.I. Ltd, B.P. 1,453,115, 1976.
100. R.M. Carr and W.S. Fyfe, *Amer. Mineral.* (1958) **43,** 908.
101. J. Ciric, *Science* (1967) **155,** 689.
102. J.F. Charnell, *J. Cryst. Growth* (1971) **8,** 291.
103. E.M. Flanigen and R.L. Parton, to Union Carbide, U.S.P. 4,073,865, 1978.
104. R.M. Barrer, *Brit. Ceram. Soc.* (1957) **56,** 155.
105. K.L. Bye and E.A.D. White, *J. Cryst. Growth* (1970) **6,** 355.
106. S. Ueda and M. Koizumi, *Amer. Mineral.* (1979) **64,** 172.
107. J.V. Smith, "Felspar Minerals", Vol. 2, p. 206. Springer, Berlin, 1974.
108. R. Rinaldi, J.V. Smith and G. Jung, *Neues Jb. Mineralog. Mh.* (1975) 433.
109. T.J. Weeks and D.E. Passoja, *Clays and Clay Minerals* (1977) **25,** 211.
110. R. von Ballmoos and W.M. Meier, *Nature* (1981) **289,** 782.
111. D.W. Breck and E.M. Flanigen, *in* "Molecular Sieves", p. 47. Society of Chemical Industry, London, 1968.
112. B.D. McNicol, G.T. Pott, R.K. Loos and N. Mulder, *in* "Molecular Sieves" (Eds. W.M. Meier and J.B. Uytterhoeven), p. 152. American Chemical Society Advances in Chemistry Series, No. 121, 1973.
113. R. Aiello, R.M. Barrer and J.S. Kerr, *in* "Molecular Sieve Zeolites I", (Ed. R.F. Gould) p. 44. American Chemical Society Advances in Chemistry Series, No. 101, 1971.

Chapter 5

Representative Zeolite Syntheses, Crystallization Fields and Transformations

1. Introduction

In earlier chapters the pre-nucleation period in zeolite synthesis and special features of the post-nucleation stage, including the role of a number of the reaction variables, have been considered. Zeolite synthesis has become a substantial subject and in the present chapter a survey of some of the work accomplished and progress made will be given. Emphasis will be placed on mapping of crystallization fields under different conditions; on comparisons of several kinds of source materials for zeolite production (cf. Chapter 3, Table 1); and on transformation reactions involving zeolites.

2. Crystallization Fields

As emphasised in Chapter 4 the products of hydrothermal crystallization, especially at low temperatures, may be metastable, so that crystallization fields do not necessarily represent areas of final thermodynamic stability. They delineate instead regions of composition in which, under the selected experimental conditions, a particular product (or products) nucleates and can be isolated and hence be preserved as crystals. The importance of these crystallization fields is accordingly substantial. Their formal representation in terms of components, temperature and pressure can present difficulties. As a rule, the pressure is the autogenous pressure developed by the volatile components, water and air, in the closed system at the experimental temperatures. In crystallizations at or below 100°C open vessels can be employed, under reflux if desired, and the pressure is then that of the atmosphere. Temperature and chemical composition can be varied at will, and one may have a choice of compositions such as would not be found in Nature. Examples include the use of organic bases as described in Chapter 4.

The simplest compositions employed in growing porous crystals have been the three component ones, TPAOH–SiO_2–H_2O for making the silica end-member of the pentasil zeolite ZSM-5 and TBAOH–SiO_2–H_2O for the silica end-member of the pentasil zeolite ZSM-11. TPAOH here, as earlier, denotes tetrapropylammonium hydroxide and TBAOH denotes tetrabutylammonium hydroxide. A representation of the crystallization fields under constant or autogenous pressures would then employ a prism of triangular cross-section, with temperature measured normal to this cross-section and the apices of the triangle representing the pure organic base, pure silica and pure water respectively. The crystallization fields would be represented by specific volumes within the prisms and at any one temperature they would be areas within the triangular cross-section. However, four components are involved in the synthesis of homoionic zeolites, which are R_2O–Al_2O_3–SiO_2–H_2O, where R_2O is the base; and five or more components are involved when two or more bases are present in the reaction mixtures. With additives such as salts the number of components in the reaction mixture can be increased still further. For systems of many components, crystallization fields cannot be represented simply and at the same time in detail. Indeed, the number of experiments becomes so large for full exploration that in one laboratory detailed mapping would be impracticable within any reasonable period. This in turn could mean that in multicomponent systems at varied temperatures and pressures, novel zeolite species remain to be discovered in systems already subjected to considerable exploration.

At constant temperature T and pressure P, crystallization fields in four-component systems could be represented as volumes within a tetrahedron, each apex of which represents one pure component. At constant T, P, and water content the fields become areas in a triangular cross-section. In another simplified representation at constant or at autogenous pressures, one may use temperature as ordinate and the ratio $n = SiO_2/Al_2O_3$ as abscissa. This method and other ways of representing crystallization fields will be exemplified in what follows.

3. Hydrothermal Crystallization of some Four-component Mixtures

Crystallization has been investigated systematically for aqueous aluminosilicate gels and mixtures when the base was each one in turn of the series Li_2O,[1] Na_2O,[2,3] K_2O,[4] Rb_2O,[5] Cs_2O,[5] $(NH_4)_2O$,[6] $[NH_3CH_3]_2O$,[6] $[NH_2(CH_3)_2]_2O$,[6] $[NH(CH_3)_3]_2O$,[6] $[N(CH_3)_4]_2O$[6] CaO,[7] SrO[8] and BaO.[9] Some results are illustrated in Tables 1, 2 and 3 and in Figs 1 to 5 together with Fig. 16 of Chapter 2. The experiments were conducted with and without base in excess of the compositions $R_2O.Al_2O_3.nSiO_2 + aq$, and in part up to temperatures of about 450°C. The values of n were often in the range $1 \leqslant n \leqslant 12$. The following observations can be made:

1. The tables and especially Figs 1 to 5 and Fig. 16 of Chapter 2 show that the formation of particular species depends strongly upon temperature and upon the ratios SiO_2/Al_2O_3 and base/Al_2O_3.

2. The crystallization of non-zeolite tectosilicates (felspars and felspathoids) occurred as freely as that of zeolites, and crystallization fields of various species, both zeolite and non-zeolite, often overlapped. The appearance of layer silicates, recorded for several of the systems (Tables 1 to 3), suggests that conditions sometimes prevailed which were only mildly alkaline and so allowed nucleation and growth of both tecto- and phyllosilicates (cf. Chapter 2, Table 9).

3. Crystallization, especially from reaction mixtures with CaO as base, tended to be more sluggish with the alkaline earth bases than with the alkali metal hydroxides. This behaviour is likely to be related in part to the pH of the reaction mixture. As shown in Chapter 4 §5.2, the OH^- ion concentration plays a major mineralizing role. Co-crystallizations of two or more species were particularly noticeable with CaO as base. Despite the slowness of reaction when alkaline earth hydroxides were the bases, the diversity of species obtained was not less than with the alkali metal hydroxides. Zeolites were obtained with every one of the bases save Cs_2O; subsequently, however, even with this base, an edingtonite type zeolite, Cs-D, was readily made at 80°C from metakaolinite and CsOH aq.[11]

4. In many experiments measurements were made of the pH of the cold mother liquors after the hydrothermal reactions. With Li_2O, Na_2O and K_2O as the base R_2O, in excess of the oxide compositions $R_2O.Al_2O_3.nSiO_2$, all mother liquors were highly alkaline. With mixtures in which $R_2O/Al_2O_3 \sim 1$, the pH while still tending to be on the alkaline side was often close to the neutral value, especially when hydrothermal reactions were effected at the highest temperatures. This indicates virtually complete incorporation of the alkali into the reaction products. At the lowest reaction temperatures when CaO, SrO and BaO were the bases and for oxide compositions $RO.Al_2O_3.nSiO_2$ (i.e. no excess of base) the pH was sometimes but not always well on

the alkaline side, showing therefore that sometimes the base $R(OH)_2$ was not wholly incorporated in the products. However, as the reaction temperatures increased the cold mother liquors tended to have lower pH values which were sometimes even slightly on the acid side.

5. The effect of temperature on zeolite formation was of particular interest. Stringham[12] indicated 350°C as an approximate upper limit for zeolite formation (Chapter 2, Table 9). However, the results in Tables 1 and 3 show that this limit can be exceeded in particular instances (analcime; basic sodalite and cancrinite in which intracrystalline $NaOH + H_2O$ replaces the salts present in the felspathoid forms; mordenite, ferrierite; and yugawaralite).

The upper temperature limits may be considered in relation to the water contents of the zeolites and hence their intracrystalline porosity. Special circumstances apply in the case of analcime (Table 1), which is the least porous of all zeolites. Its formation at temperatures above 350°C reflects this fact. Its potassium-exchanged form is anhydrous and, as the felspathoid leucite ($KAlSi_2O_6$), can be made in the dry way as a high-temperature product having stability comparable with that of potassium felspar. Some crystallographic consequences of synthesis of analcime type phases or preparations of their ion-exchanged forms are illustrated in Table 4.[13] Among the hydrated forms, the unit cells increase in size with the Goldschmidt values of the cation radii up to the hydrated K-form. The much more usual anhydrous K-form (leucite) is, however, more compact and is tetragonal, as is true of the anhydrous NH_4-, Rb- and Tl-varieties. As the ionic radii increase from K^+ through to Cs^+ the tetragonal cells become more nearly cubic until with Cs-analcime (the mineral pollucite) the unit cell is again cubic and has the same cell dimension as Na-analcime. Thus, in its effect upon the framework, Cs^+ is equivalent to (Na^+, H_2O). Also, as the ion increases in size its mobility and ease of ion exchange within the rather tight mesh of the anionic framework decreases,[14,15] until with Cs^+ one must have recourse to temperatures of 200°C or more to effect ready cation exchange even in very small synthetic crystals. In consequence of the above behaviour, dependent upon the low intracrystalline porosity, the analcime topology is seen to share the properties of anhydrous felspathoid with those of hydrous zeolite, according to the cation size, so that the formation of this topology up to rather higher temperatures than those involved for other zeolites is to be expected.

The high-temperature appearance of basic cancrinite and basic sodalite (also termed cancrinite or sodalite hydrate) is likewise due to particular circumstances. In these frameworks the intracrystalline pore volume is

TABLE 1

Some zeolite and non-zeolite phases grown from 4-component systems nearly homoionic with respect to univalent cations

Type	Examples	Observed in range (°C)	Intracrystalline pore volume[a] (cm³ per cm³ of crystal)
(a) System Li₂O–Al₂O₃-SiO₂-H₂O[1] Temperature range 130–450°C (Fig. 1)			
Zeolites	Li-A(BW)[b] Li₄Al₄Si₄O₁₈.4H₂O	130–340	0·28
	Li-H[b] Li₂O.Al₂O₃.8SiO₂.5H₂O	150–300	–
Tectosilicate	α-eucryptite (Li-B)	300–450	–
non-zeolites	α-petalite (Li-C)	330–450	–
	α-spodumene (Li-J)	360	–
	β-spodumene (Li-I)	450	–
Silicate	metasilicate (Li-D)	190–275	–
(b) System Na₂O–Al₂O₃-SiO₂-H₂O[2] Temperature range 150–450°C (Fig. 2)			
Zeolites	analcime (Na-B)	150–390	0·18
	mordenite (Na-D)	240–330	0·26
	basic cancrinite (Na-C)	300–450	0·34
	basic sodalite and nosean (Na-I and Na-F)	150–450	0·34
Tectosilicates	albite (Na-A)	330–450	–
non-zeolites	nepheline (Na-E)	360–450	–
	nepheline hydrates I and II (Na-J and Na-K)	270–450	–
	cancrinite (Na-C) 3(Na₂O.Al₂O₃.2SiO₂)Na₂CO₃	150–450	–
	sodalite (Na-I) 3(Na₂O.Al₂O₃.2SiO₂)2NaCl	150–450	–
	nosean (Na-F) 3(Na₂O.Al₂O₃.2SiO₂)Na₂SO₄	150–450	–
	α-quartz (Na-H)	260–360	–
Layer silicate	paragonite mica (Na-G)	360–450	–
Silicate	metasilicate (Na-O), Na₂O.SiO₂	390	–
Unidentified	Na-L, Na₂O.Al₂O₃.6SiO₂.H₂O	330–400	–
	Na-M	450	–
	Na-N	330–420	–
(c) System Na₂O–Al₂O₃-SiO₂-H₂O[3] Temperature range 60–250°C (Fig. 3)			
Zeolites	Linde type A (Na-A)	60–110	0·47
	Analcime (Na-B)	150–250	0·18
	Gmelinite type (Na-S)	60–110	0·43
	Faujasite (Na-R)	60–110	0·53
	Gismondine type (Na-P)	60–250	0·47
	Basic nosean (Na-F)	110–250	0·34
	Basic sodalite (Na-I or G)	60–110	0·34
	mordenite (Na-D)	250	0·28
Tectsosilicate non-zeolite	Nepheline hydrate III(Na-T)	200	–
Unidentified	Na-U	250	–
(d) System K₂O–Al₂O-SiO₂-H₂O[4] Temperature range 60–450°C			
Zeolites	Analcime type (K-E)	350–450	0·18
	Edingtonite type (K-F)	100–150	0·40
	Chabazite types (K-G)	60–120	0·48
	Phillipsite type (K-M)	85–350	0·30
Tectosilicates	Sanidine (K-A)	150–450	–
non-zeolite	Leucite (K-C)	250–400	–
	Kaliophilite (K-D)	–	–
	Kalsilite (K-N)	300–450	–
	α-quartz (K-B)	150–450	–

TABLE 1 (continued)

Type	Examples	Observed in range (°C)	Intracrystalline pore volume^a (cm^3 per cm^3 of crystal)
Crystalline aluminas	Bayerite (K-H) $Al_2O_3.3H_2O$	60–85	–
	Boehmite (K-I) $Al_2O_3.H_2O$	250	–

(e) System Rb_2O–Al_2O_3-SiO_2-H_2O.[5]   Temperature range 160–450°C

Zeolites	Rb-D $Rb_2O.Al_2O_3.2SiO_2.H_2O$	165–180	–
	Rb-E^b $Rb_2O.Al_2O_3.6SiO_x.H_2O$	165–180	–
Tectosilicates	Rb-A $Rb_2O.Al_2O_3.2SiO_2$	180–450	–
non-zeolite	Rb-analcime I, $Rb_2O.Al_2O_3.4SiO_2$ (Rb-B)	200–450	–
	Rb-analcime II, $Rb_2O.Al_2O_3.4SiO_2$(Rb-H)	200–250	–
	Rb-felspar, $Rb_2O.Al_2O_3.6SiO_2$(Rb-C)	300–450	–
	Quartz	250–450	–

(f) System Cs_2O–Al_2O_3-SiO_2-H_2O[5]   Temperature range 160–450°C

Tectosilicates non-zeolite	Cs-F, $Cs_2O.Al_2O_3.2SiO_2$	160–450	–
	Pollucite, $Cs_2O.Al_2O_3.4SiO_2$(Cs-G)	160–450	–
	Quartz	250–450	–

(g) System $(NH_4)_2O$–Al_2O_3-SiO_2-H_2O[6]   Temperature range 200–450°C

Zeolites	NH_4-analcime (NH_4-D)	450	0·18
	Gismondine type (NH_4-P)^c	270–300	0·36
Tectosilicate non-zeolite	Cristobalite, SiO_2, (NH_4-H)	450	–
Layer silicate	NH_4-mica (NH_4-K)	250–450	–
Crystalline aluminas	Boehmite, $Al_2O_3.H_2O$ (NH_4-A)	250–400	–
	Corundum, Al_2O_3 (NH_4-B)	450	–

^{a} Measured as the volume of liquid water removable by heat and evacuation from the crystals.
^{b} Unidentified or no naturally occurring counterpart.
^{c} Originally termed harmotome-like and listed as N-L (or P) in Ref (6).

twice that in analcime so that their ready formation up to at least 450°C may seem unexpected. However, these high-temperature syntheses were effected in the presence of exceptionally large excesses of caustic soda. The intracrystalline pore volume is then shared both by intercalated NaOH (non-volatile) and by H_2O (volatile). The analogues of natural cancrinite, sodalite and nosean (the latter has the sodalite topology) were also made in the study by Barrer and White.[2] Here salts (Na_2CO_3, NaCl, Na_2SO_4) occupy much of the intracrystalline free volume and, being non-volatile, serve again to stabilize the structure. The thermodynamic basis[16] of this stabilization has been presented in Chapter 2 §4. It can in general be expected that even highly porous frameworks will be stabilized if the pore space is occupied by intercalated non-volatile species, so that in principle, if appropriate fillers could be found, very open structures similar to zeolites might be synthesized up to high temperatures.

TABLE 2

Phases grown from 4-component systems initially nearly homoionic with respect to organic cation. Systems $[NH_{(4-x)}(CH_3)_x]_2O$-Al_2O_3-SiO_2-H_2O, where $x = 1, 2, 3$ or 4 (Fig. 4)[6]

x (and temperature range (°C))	Type	Examples	Observed in range (°C)
x = 1 (125–300)	Zeolite	Harmotome-type (N-P)[a]	225–300
	Layer silicate	Mica (N-K)	225–300
	Crystalline alumina	Bayerite (N-C)	125–200
x = 2 (150–300)	Zeolite	Harmotome-type (N-P)[a]	175–300
	Layer silicate	Mica (N-K)	250–300
	Crystalline aluminas	Bayerite (N-C)	175
		Boehmite (N-A)	300
x = 3 (175–300)	Zeolite	Harmotome-type (N-P)[a]	250–300
	Tectosilicate	Sodalite (N-T)	250–300
	Layer silicate	Smectite (N-V)	250–300
	Crystalline alumina	Boehmite (N-A)	225–250
x = 4 (100–300)	Zeolites	Harmotome-type (N-P)[a]	250–300
		Faujasite-type (N-R)	100–150
		Linde type A (N-Q)	100–150
	Tectosilicate	Sodalite (N-T)	200–250
	Layer silicate	Smectite (N-V)	200–300
	Crystalline aluminas	Boehmite (N-A)	200–300

[a] Originally termed a harmotome type zeolite and listed as N-L (or P) in Ref. (6) but may have the gismondine topology.

Water, because of its volatility, is not such a filler and even at autogenous pressures highly porous zeolites tend to be replaced by less porous ones or by anhydrous crystals as the synthesis temperature increases, unless special factors such as those of the previous paragraph supervene (cf. Chapter 2, §8.4). This behaviour is shown by the relatively low upper temperature limits of formation of zeolite A, faujasite, chabazite and gmelinite as illustrated in Tables 1, 2 and 3, and the correspondingly higher upper temperature limits for the less porous zeolites, mordenite, ferrierite and yugawaralite. However, high intracrystalline porosity is not always the main objective in synthesizing zeolites. More important can be the size of the windows giving access to the intracrystalline channels and cavities, the steric configuration of these latter and the thermal and chemical stability of the zeolite. For these reasons among others there is much interest in mordenite and ferrierite and in other less porous structures such as the pentasil[17] zeolites ZSM-5 and -11.

Where organic bases are involved (Table 2 and Fig. 5) some hydrolysis can be expected under hydrothermal conditions:

$$[NR_{(4-x)}H_x]^+ + H_2O \rightleftharpoons [NR_{(3-x)}H_{(1+x)}]^+ + ROH \tag{1}$$

where R is the organic group and $x = 3, 2, 1$ or 0. Hence, as noted in

TABLE 3

Some zeolite and non-zeolite phases grown from 4-component systems nearly homionic with respect to divalent cations

Type	Examples	Observed in range (°C)	Intracrystalline pore volume[a] (cm³ per cm³ of crystal)
(a) System CaO–Al₂O₃-SiO₂-H₂O[7]	Temperature range 125–450°C (Fig. 5)		
Zeolites	Tetragonal analcime (Ca-D)	220–300	0·18
	Cubic analcime (Ca-E)	250–450	0·18
	Thomsonite (Ca-I)	225–245	0·32
	Epistilbite (Ca-J)	225–260	0·34
	Harmotome type (Ca-L)	200–300	0·36
	Mordenite (Ca-Q)	390	0·26
Tectosilicates non zeolite	Anorthite (Ca-F)	260–450	–
	Hexagonal dimorph of anorthite (Ca-P)	245–450	–
	α-cristobalite (Ca-H)	350–450	–
Silicates	Foshagite, 4CaO.3SiO₂.H₂O(Ca-R)	–	–
	Wollastonite, CaO.SiO₂ (Ca-S)	–	–
Aluminates	Ca-M, 4CaO.3Al₂O₃.3H₂O	180–285	–
	Ca-N, 3CaO.Al₂O₃.6H₂O	200	–
Hydrogrossular series	3CaO.Al₂O₃.6H₂O to 3CaO.Al₂O₃.3SiO₂ (Ca-K)	200–450	–
Crystalline aluminas	Boehmite (Ca-A)	200–440	–
	Corundum (Ca-B)	440–450	–
Unidentified	Ca-O	250–450	–
(b) System SrO–Al₂O₃-SiO₂-H₂O[8]	Temperature range 110–450°C		
Zeolite	Ferrierite (Sr-D)	340–440	0·24
	Analcime (Sr-I)	250–380	0·18
	Cancrinite hydrate (Sr-J)	150–200	0·34
	Mordenite (Sr-M)	250–300	0·26
	Yugawaralite (Sr-Q)	270–435	0·30
	Gmelinite (Sr-F)	200	0·43
	Chabazite type (Sr-G)	110–200	0·48
	Heulandite type (Sr-R)	250	0·35
Tectosilicates non-zeolite	Orthorhombic polymorph of felspar (Sr-C)	150–450	–
	Hexagonal polymorph of felspar (Sr-P)	150–450	–
	α-cristobalite (Sr-H)	300–435	–
Layer silicates	Sr-mica (Sr-L)	380–450	–
	Smectite (Sr-B)	405	–
Crystalline alumina	Boehmite (Sr-E)	150	–
Unidentified	Sr-K	400–435	–
	Sr-N	295–435	–
(c) System BaO–Al₂O₃-SiO₂-H₂O[9]	Temperature range 110 to 450°C		
Zeolites	Linde type L (Ba-G, L)[b]	110–250	0·28
	Harmotome type (Ba-M)	150–250	0·36
	Ba-J[c]	250–300	–
	Ba-K[c]	250–300	–
Tectosilicates non-zeolite	Celsian (Ba-C)	380–435	–
	Hexagonal polymorph of celsian (Ba-P)	150–435	–
	α-cristobalite (Ba-H)	340–435	–
Crystalline alumina	Boehmite (Ba-E)	150–250	–
Unidentified	Ba-A	250–300	–
	Ba-Z	300–340	–

[a] Estimated as the volume of liquid water removable by heat and evacuation from the crystals.
[b] Originally termed Ba-G.
[c] Unidentified and possibly novel zeolites. Good sorbents of O₂.[10]

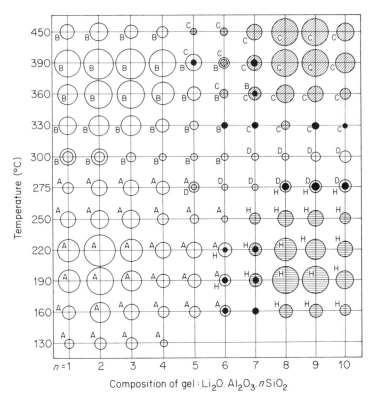

Fig. 1. Approximate areas of formation in the hydrothermal field of crystallization of aqueous gels $Li_2O.Al_2O_3.nSiO_2$.[1]

B = α-eucryptite	A = zeolite 1:1:2 (+4H₂O)
C = α-petalite	H = zeolite 1:1:8 (+5H₂O)
D = unidentified phase	

B = α-eucryptite   A = zeolite 1 : 1 : 2 ($+4H_2O$)
C = α-petalite   H = zeolite 1 : 1 : 8 ($+5H_2O$)
D = unidentified phase

Fig. 2. Crystallization fields of some aqueous hydrogels $Na_2O.Al_2O_3.n$-SiO_2, (*a*) with and (*b*) without additional alkali.[2] Where excess NaOH is added the % excess amount per Al_2O_3 is indicated on the diagram. In addition to compounds named on the diagrams:

NHI = nepheline hydrate I, $Na_2O.Al_2O_3.2SiO_2.H_2O$
NHII = nepheline hydrate II, $Na_2O.Al_2O_3.2SiO_2.0.5H_2O$
Species L = unidentified, monoclinic $a = 9.10$ Å, $b = 6.10$ Å, $c = 4.88$ Å, and $\beta = 105° 20'$. Composition $2Na_2O.Al_2O_3.6SiO_2.2H_2O$
Species M = unidentified, orthorhombic $a = 11.75$ Å, $b = 5.81$ Å, $c = 8.52$ Å
Species N = unidentified, probably hexagonal.

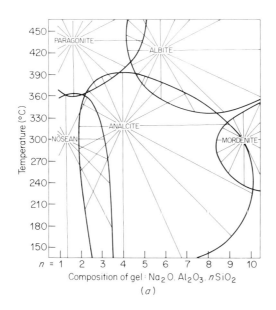

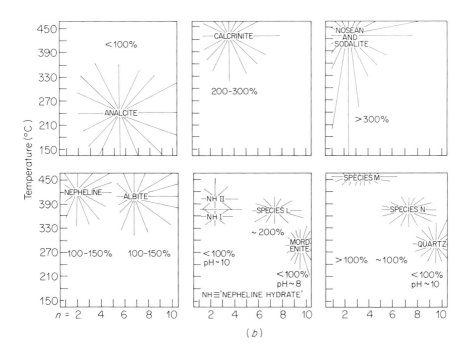

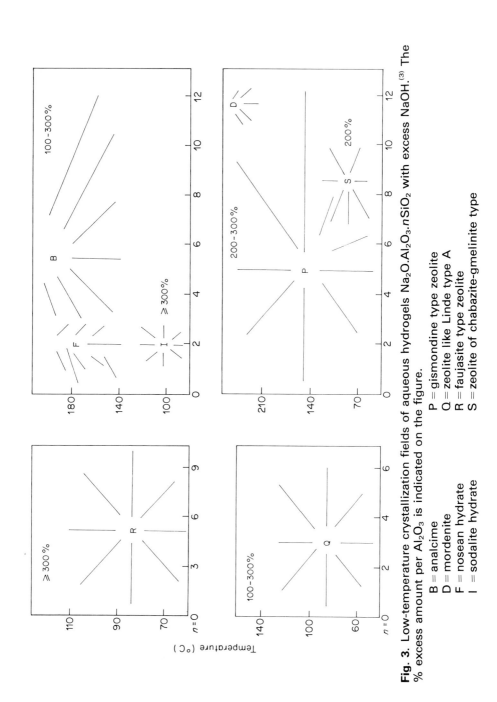

Fig. 3. Low-temperature crystallization fields of aqueous hydrogels $Na_2O.Al_2O_3.nSiO_2$ with excess NaOH.[3] The % excess amount per Al_2O_3 is indicated on the figure.

B = analcime
D = mordenite
F = nosean hydrate
I = sodalite hydrate

P = gismondine type zeolite
Q = zeolite like Linde type A
R = faujasite type zeolite
S = zeolite of chabazite-gmelinite type

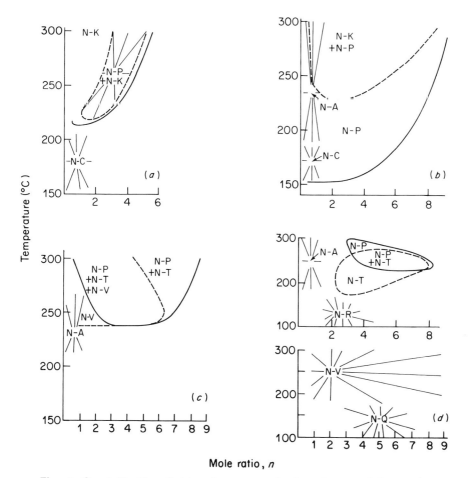

Fig. 4. Crystallization fields of aqueous hydrogels containing only organic bases[6] as follows: (a) monomethyl ammonium; (b) dimethylammonium; (c) trimethylammonium and (d) tetramethyl-ammonium

N-C = bayerite, $Al_2O_3.3H_2O$ N-V = montmorillonite
N-K = mica   N-Q = zeolite like Linde type A
N-P = harmotome type zeolite   N-R = faujasite type zeolite
N-T = sodalite   N-A = boehmite, $Al_2O_3.H_2O$
Temperatures are in °C.

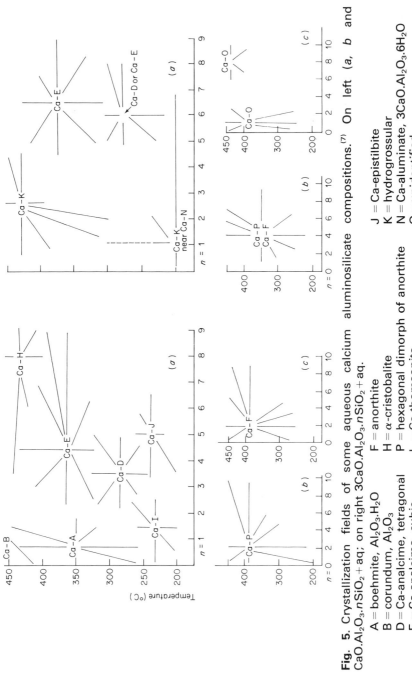

Fig. 5. Crystallization fields of some aqueous calcium aluminosilicate compositions.[7] On left (*a*, *b* and *c*) $CaO.Al_2O_3.nSiO_2 + aq$; on right $3CaO.Al_2O_3.nSiO_2 + aq$.

A = boehmite, $Al_2O_3.H_2O$
B = corundum, Al_2O_3
D = Ca-analcime, tetragonal
E = Ca-analcime, cubic
Temperatures are in °C.

F = anorthite
H = α-cristobalite
P = hexagonal dimorph of anorthite
I = Ca-thomsonite

J = Ca-epistilbite
K = hydrogrossular
N = Ca-aluminate, $3CaO.Al_2O_3.6H_2O$
O = unidentified
Q = Ca-mordenite

TABLE 4

Some properties of minerals having the analcime framework topology[13]

Cation	Radius (Å)	Unit cell[a] (Å)	Hydrated	Ion exchanger
Li^+	0·78	C; $a = 13·5$	yes	yes
Na^+	0·98	C; $a = 13·6_6$	yes	yes
Ca^{2+}	1·06	C; $a = 13·6_2$	yes	yes
		T; $a = 13·6_{2-8}$	yes	yes
		$b = 13·5_6$		
Ag^+	1·13	C; $a = 13·7$	yes	yes
Sr^{2+}	1·27	C; $a = 13·7_4$	yes	yes
K^+	1·33	C; $a = 13·7_9$	yes	yes
		T; $a = 12·9_2$	no	yes
		$b = 13·7_0$		
NH_4^+	1·43	T; $a = 13·1_7$	no	yes
		$b = 13·6_9$		
Rb^+	1·48	T_1; $a = 13·3_3$	no	yes
		$b = 13·6_4$		
		T_2; $a = 13·2$	no	yes
		$c = 13·6$		
Tl^+	1·49	T	no	yes
Cs^+	1·65	C; $a = 13·6_6$	no	no[b]

[a] C = cubic, T = tetragonal.
[b] Except at higher temperature (~200°C or more)

Chapter 4, the zeolitic products are referred to as N-zeolite where N can be ammonium, alklyammonium or a mixture of such ions. Because of the hydrolysis, the temperatures employed were not above 300°C. The synthesis of tetramethylammonium analogues of faujasite and Linde zeolite A was subsequently shown to have been facilitated by the presence of traces of Na^+ in the reaction mixtures.[18] The formation of N-P, originally considered as a harmotome variant[6] but probably a gismondine type,[19] was observed up to 300°C. The large size of the organic cations must limit the residual volume available to water so that the importance of water as a filler may be reduced. The Na-form of the gismondine type zeolite, Na-P, was obtained up to 250°C (Table 1).

The converse of the above considerations of the upper temperature limit for zeolite formation is the lower temperature limit for the appearance of synthetic anhydrous non-zeolite tectosilicates. Some examples are given in Table 5. In addition, the least porous of the zeolites, analcime, has been made at 100°C.[24] The formation of anhydrous tectosilicates can evidently proceed in certain instances at temperatures as low as those involved in the formation even of the most porous zeolites such as faujasite, chabazite, gmelinite or Linde type A.

TABLE 5

Examples of low-temperature synthesis of anhydrous tectosilicates

Type	Examples	Reaction mixture	Temperature (°C)	Reference
Felspar	Sanidine	Aqueous alkaline potassium aluminosilicate gel	150	(4)
	Albite	Aqueous alkaline Na-rich aluminosilicate gel	200	(20)
			175	(21)
Felspathoids	Hexagonal polymorph of celsian, Ba-P	Metakaolinite+aq. Ba(OH)$_2$	80	
	Sodalite	Kaolinite+aq. NaOH	80	(22, 23)
	Cancrinite	+salts		
	Kaliophilite	Kaolinite+aq. KOH	80	
	Rb-A (Rb$_2$O.Al$_2$O$_3$.2SiO$_2$)	Metakaolinite+aq. RbOH	80	(11)
	Cs-F (Cs$_2$O.Al$_2$O$_3$.2SiO$_2$)	Metakaolinite+aq. CsOH	80	
	Pollucite	Metakaolinite+aq. CsOH	80	
Crystalline SiO$_2$	α-quartz	Aqueous alkaline potassium aluminosilicate gel	150	(4)

Investigations summarized in Tables 1 to 3 have been supplemented by various authors at temperatures above 100°C and some of the results are given in Table 6. They confirm and extend the results of Tables 1 to 3. Thus to the zeolites of Tables 1 and 3 may be added Li-clinoptilolite,[26] Li-ferrierite[31] and bikitaite,[28] (Na, K)-clinoptilolite[30a] and Ca-, Sr- and Ba-clinoptilolites.[27] Some crystallization fields in the system $Na_2O–Al_2O_3–SiO_2–H_2O$ were mapped by Senderov[25] with the results shown in Fig. 6. As in Tables 1 to 3 the various authors observed the formation of many non-zeolites, mostly tectosilicates, with fields overlapping those of zeolites. The investigation of Hawkins[27] with Ca, Sr and Ba as the cations was particularly extensive and the results agreed well with those in Table 3. Sometimes zeolite formation was observed above the 350°C approximate upper limit suggested by Stringham[12] for the formation of natural zeolites. Analcime (up to 366°C), mordenite (to 430°C), clinoptilolite (to 370°C) and ferrierite (to 375°C) exemplify this behaviour. However, as in the instances quoted for Table 3, the yields tend to be low at these elevated temperatures.

To the hydrothermal zeolitizations recorded in Tables 1, 2, 3 and 6 one may add those of the pentasil zeolites ZSM-5, ZSM-8 and ZSM-11 and the silica end members of the ZSM-5 and ZSM-11 series, respectively termed silicalite 1 and silicalite 2. These have all been made in the range from 100 to 200°C. The formation of ZSM-5 and of silicalite 1 is favoured by tetrapropylammonium ions (TPA$^+$) as template;[32,33,34,35] ZSM-8 was made with tetraethylammonium ions (TEA$^+$);[36] ZSM-11 and silicalite 2 with tetrabutylammonium ions (TBA$^+$).[37,38,39] ZSM-5 was also made with hexamethylenediamine[40] and with primary amines[41] as the templates. Synthesis of a ZSM-5 type zeolite was reported without employing any organic template, for example by crystallizing a mixture of oxide composition $4·5Na_2O.Al_2O_3.80SiO_2.3196H_2O$ at 200°C for 3 days.[42] When the crystals of any of the above zeolites are prepared with the usual organic templates the organic species occupies sites within the structures. The products have been termed "precursors" of the final zeolites because to open the intracrystalline channels the organic templates which block them must be burned out by heating in air. To produce catalysts, charge-balancing Na$^+$ ions are replaced by NH$_4^+$ and the zeolite is then heated to eliminate NH$_3$, or the Na$^+$ is replaced by H$^+$ (as silanol OH) by direct treatment with acid:

TABLE 6

Some crystallizations under hydrothermal conditions

Author	Products			Reaction mixtures	Temp. range studied (°C)
	Zeolites	Non-zeolite tectosilicate	Other		
Senderov[25] (1963)	Mordenite Analcime Chabazite-type	Albite Quartz		Gels $Na_2O.Al_2O_3.$ $10SiO_2+NaOH+aq$ Excess Na_2O (as NaOH) from 0 to 60 mg/g H_2O	150–350
Ames[26] (1963)	Ferrierite Li-clinoptilolite			Gels $mLi_2O.Al_2O_3.nSiO_2$ $+aq$ ($m=0.5$–3; $n=8$–18)	300–350
Hawkins[27] (1967)	Analcimes (Sr, Ca) Mordenites (Ca, Sr) Clinoptilolites (Ca, Sr and Ba) Heulandite (Ca) Ferrierite (Ca, Sr) Yugawaralite (Sr)	Felspars (Ca, Sr, Ba, Na) Hexag. felspar (Ca) Paracelsian (Sr) Cymrite (Ba) Kalsilite (Ca, Sr, K) Nepheline (K) Nepheline hydrates I and III (Na) Quartz	Montmorillonite (Ca) Boehmite Sanbornite (Ba)	Glasses or sinters $RO.Al_2O_3.nSiO_2+aq$ ($R=$ Ca, Sr, Ba; $n=1$–10). Limited study with corresponding Na and K glasses	246–462
Drysdale[28] (1971)	Bikitaite Li-A (BW)	Eucryptite β-spodumene Quartz		Gels $mLi_2O.Al_2O_3.4SiO_2+$ aq ($m=1$–4)	300–350
Kibby, Perrotta and Massoth[29] (1974)	Ferrierite Mordenite Analcime			$NaAlO_2+SiO_2$ sol$+aq$	300–325
Winquist[30] (1976)	Ferrierites (Na, K)	Adularian felspar		Amorphous silica-alumina $+(Na$ or $K)_3PO_4$ or $KF+aq$	170–215
Goto[30a] (1977)	Mordenite Clinoptilolite Analcimes (Ca, Sr)			Gel $(Na_2, K_2)O.Al_2O_3.7SiO_2$ $+aq$ at pH ~7-9	200
Vaughan and Edwards[31] (1978)	Ferrierites (Li, Li+Na or K, Li+Ca, Sr or Ba)			Slurries $mM_2O.(1-m)RO.$ $Al_2O_3.10SiO_2+aq$ (125 moles per Al_2O_3). $M'=$ Li, Na, K; $R=$ Ca, Sr, Ba	200–250

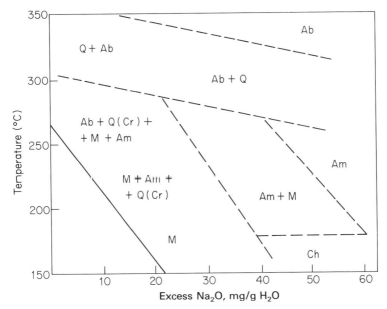

Fig. 6. Hydrothermal crystallization field of mixtures $(1 + x)Na_2O.Al_2O_3.10SiO_2 + aq.$[25]

Ab = albite	G = cristobalite
Am = analcime	M = mordenite
Q = quartz	Ch = chabazite

Bibby et al.[43] have shown that the separate steps of burning out the template, exchanging Na^+ by NH_4^+, and heating the NH_4-zeolite to eliminate NH_3 can be combined. The Na^+ and template in the reaction mixtures was replaced by NH_4^+ + template. The mixture containing NH_4^+ + TPA^+ yielded ZSM-5 and that containing NH_4^+ + TBA^+ gave ZSM-11 precursors which then gave the H-zeolites in the single step of heating in air.

Another siliceous zeolite, Nu-1, of unknown structure, has been reported,[44,45] in which the ratios SiO_2/Al_2O_3 ranged from about 20 to 150. The cationic content of the reaction mixtures was Na^+ + TMA^+ and temperatures of 170–180°C were used in synthesis. It was possible to make the zeolite with very little sodium in the structure and in which not all the intracrystalline TMA^+ neutralized framework charge. Some was presumably intercalated as TMAOH. The organic content could be burnt out in air, as with the pentasil zeolites referred to above. The water content of the zeolite free of TMA^+ approached 8%, so that the intracrystalline pore volume is low; a property shared with the pentasils (for ZSM-5 and -11 the water contents are ~9%).

A further siliceous zeolite-type phase, Fu-1, of unknown structure has been found in which the cations are Na^+ and TMA^+.[46,47] Typical products had SiO_2/Al_2O_3 between 15 and 35 and synthesis temperatures were again in the hydrothermal range ($\sim$180°C). The water content of a sample from which TMA had been burnt out was about 8·5%.

Other examples of highly siliceous zeolites are known[47a] (zeolites β, ZSM-12, ZSM-23, ZSM-48, zeta-1 and zeta-3). Zeolites having ratios SiO_2/Al_2O_3 much higher than those of naturally occurring siliceous zeolites like mordenite, svetlozarite, clinoptilolite and ferrierite are thus not uncommon in hydrothermal systems. For their synthesis an organic template base is needed; for zeolite ZSM-5 crystallization proceeds faster for low ratios OH^-/SiO_2 (0·05–0·15) and very low Al_2O_3 contents of the parent gels.[47a] The conditions of synthesis are accordingly distinctive when compared with those of the usual aluminous zeolites. These very

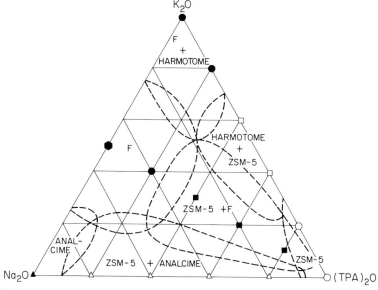

Fig. 7. Ternary composition diagram for crystallization fields in the Na_2O–K_2O–$(TPA)_2O$ cation system with salt additions.[21] Approximate stability areas are shown of the phases formed at 175°C after 4 days, from a reaction mixture:

$$10(TPA, Na, K)_2O.Al_2O_3.28SiO_2.750H_2O + 4xNaCl + 4(1-x)KCl$$

where

$x = Na/(Na + K)$ and TPA = tetrapropylammonium.
F = felspar.

siliceous zeolites have relatively low water contents which are comparable with those of analcime. On the other hand, for many of them the intracrystalline channels are much more accessible and of larger diameter than those of analcime so that they can be practical mole sieve sorbents and possess shape-selective catalytic properties. There is accordingly a lively interest in them and in the possibility of making others equally rich in silica. ZSM-5 has received most attention. Among other studies its crystallization field and transformations have been explored at 175°C by Erdem and Sand[21] as functions of the ratios $Na^+/(Na^+ + TPA^+)$ and $K^+/(K^+ + TPA^+)$. The approximate fields of formation of the phases observed after four days crystallization time are shown in Fig. 7. In the authors' notation F is felspar, which can only contain Na^+ and K^+ since organic cations are too large to be located within the small interstices of this tectosilicate framework. The reaction mixture contained Cl^- ions originally added as NaCl or KCl.

4. Low-Temperature Zeolite Formation

In §3 crystallization of a number of four-component reaction mixtures was considered, involving a single base and operating for the most part at temperatures above 100°C and up to 450°C (inorganic bases) and 300°C (organic bases). However, convenience in experimentation and in large-scale manufacture would favour operating near or below 100°C, a range in which numerous zeolitization reactions have been successful. A very extensive patent literature has developed in which low temperatures are used. Starting with compounds like soluble aluminates and silicates or silica sols which on mixing tend initially to give aluminosilicate gels, with various bases and mixtures of bases, with a range of additives and with different gel ripening and crystallization procedures, crystalline phases with diverse topologies have resulted. For the most part a given patent has specified and claimed conditions yielding a particular zeolite rather than exploring crystallization fields of all accessible species, whereas the open literature has usually attempted the latter. The claims of some patents have been more broadly drawn than is warranted by the experimental results presented, but nevertheless patents have recorded many valuable contributions to zeolite synthesis. Recent trends have been towards exploring the silica-rich ends of crystallization fields, as indicated in §3; towards low-temperature crystallization; and towards increasing the numbers of components to five or six by using mixtures of bases and by including salts and other non-bases in the reaction mixtures. Some results of these activities, for the most part at relatively low temperatures, will be considered.

Section 3 was concerned with zeolitization of gels under hydrothermal conditions. For this reason and because the number of researches using gels is so extensive and in part repetitive, attention will here be directed to a selection only of low-temperature syntheses from amorphous materials, mainly from alkaline aqueous hydrogels. The work of Regis et al.[20] in the sodium field yielded at 100°C most of the compounds previously made at low temperatures by Barrer et al.[3] (Table 1). Likewise Zdhanov and Ovsepyan[48] working at 90°C with alkaline potassium aluminosilicate gels made the chabazite (K-G) near-phillipsite (K-M) and edingtonite type zeolite (K-F) reported earlier by Barrer and Baynham[4] (Table 1) together with the novel but thermally unstable zeolite K-I ($K_2O.Al_2O_3.2SiO_2.3\cdot8H_2O$). Starting from soluble aluminates and silicates Bosmans et al.[49] explored the crystallization fields at 90°C of zeolites in the systems $K_2O-Al_2O_3-SiO_2-H_2O$, $Na_2O-Al_2O_3-SiO_2-H_2O$ and $K_2O-Na_2O-Al_2O_3-SiO_2-H_2O$. They obtained fourteen zeolites, all but two of which (K-J and (Na, K)-V) have previously been recorded in Table 1. The crystallization fields were mapped after selected time intervals and for different water contents on triangular diagrams of which the apices were SiO_2, Al_2O_3 and M_2O ($M_2O = Na_2O$, K_2O, or $(Na, K)_2O$). In the case of the mixed bases, the crystallization fields were mapped for different specified values of $K_2O/(K_2O + Na_2O)$. This detailed study showed that, in the sodium field, dilution with water caused sodalite hydrate to give way to Linde zeolite type A or to faujasites. In the potassium field dilution caused edingtonite type K-F to give way to the chabazite type K-G and to K-I. At longer times K-F was replaced by zeolite K-J (at 90 mole % H_2O) or by K-G and K-I (at 97 mole % H_2O). In the mixed base system with $K_2O/(K_2O + Na_2O) = 0\cdot5$ and 90 mole % H_2O a novel zeolite (K, Na)-V of unit cell oxide composition $3(Na, K)_2O.3Al_2O_3.6SiO_2.12H_2O$ was obtained. Crystallographic parameters of this and of the other zeolites, using letter designations assigned earlier were possible,† were given as shown on the opposite page.

The classification of zeolite type and the above tentative unit cell for K-M may both require revision because it has been suggested that it may have the merlinoite topology.[50] In §5 Table 15 it will also be seen that the K-F zeolite type can occur as a number of variants so that this designation represents a group of structurally related phases. Literature designations of a given zeolite type have been diverse: this is exemplified in Table 7 for some synthetic potassium phases which were discussed by Sherman.[51] Similar diversity arises in the designations in different laboratories of synthetic Na-, Li- and other zeolites.

---

† The notation of Bosmans et al.[49] is given in Table 7.

Sodalite hydrate, Na-T[a]	cubic body centred, $a = 8.97$ Å
Cancrinite hydrate, Na-C	hexagonal, $a = 12.6$ Å, $c = 5.14$ Å
Faujasites (X and Y)	cubic, face-centred, $a = 24.69–25.19$ Å
Zeolite A	cubic, primitive, $a = 12.28$ (24.56) Å
Zeolite Na-Pl	pseudo-cubic, gismondine type, $a = 10.03$ Å
Zeolite Na-P2	tetragonal $a = 10.12$ Å, $c = 9.84$ Å
Zeolite Na-G	chabazite-type, Na-form, hexagonal
	$a = 13.7$ Å, $c = 15.3$ Å
Edingtonite type K-F	tetragonal, body-centred,
	$a = 9.82$ Å, $c = 13.09$ Å
Zeolite K-M	pseudo-cubic, body centred,
(near phillipsite)	$a = 20.06$ Å (variable)
Zeolite L (Linde type L)	hexagonal, $a = 18.34$ Å, $c = 7.53$ Å
Zeolite-J	tetragonal, $a = 9.53$ Å, $c = 9.79$ Å
Zeolite K-I	hexagonal, $a = 13.48$ Å, $c = 13.38$ Å
Zeolite K-G	chabazite-type, hexagonal, K-form
	$a = 13.75$ Å, $c = 15.40$ Å (variable)
Zeolite (Na, K)-V	cubic primitive, $a = 9.43$ Å

[a] Sometimes termed Na-I (see Table 1).

Pereyron et al.[67] also made a study of the system $Na_2O–K_2O–Al_2O_3–SiO_2–H_2O$ in the low-temperature range from 30 to 95°C, with stirring of the reaction mixtures. The ratios SiO_2/Al_2O_3 were varied from 2.6 to 15; $K_2O/(K_2O + Na_2O)$ from 0 to 1; $(K_2O + Na_2O)/Al_2O_3$ from 1.6 to 12 and $H_2O/Al_2O_3 = 150$. Crystallization fields were presented similarly to those of Bosmans et al.,[49] and all the zeolites obtained are included among those made by the latter authors. Pereyron et al.[68] and also Borer and Meier[69] investigated zeolitization at 30–95°C and at 100°C, respectively, of gels in the system $Li_2O–Na_2O–Al_2O_3–SiO_2–H_2O$. The former authors again presented triangular diagrams of crystallization fields of a number of zeolites of types which are included among those made in the systems $Li_2O–Al_2O_3–SiO_2–H_2O$ and $Na_2O–Al_2O_3–SiO_2–H_2O$ and recorded in §3. The composition ranges were those given above with Li_2O replacing K_2O. In addition two polytypes of faujasite were reported but not characterized. Borer and Meier determined crystallization sequences for more than 400 compositions in the (Li, Na)-aluminosilicate hydrogel system with reaction times of up to 14 days at 100°C. The nine zeolitic species formed are again included among those recorded in §3. The total dry weight of constituents was about 0.65 mg per cm^3 of aqueous alkaline hydrogel and the molar ratios of components were such that always

$$n(Li_2O) + n(Na_2O) + n(Al_2O_3) + n(SiO_2) = 1$$

The crystallization sequences are shown in Table 8. They illustrate well the evolution of successive zeolites and show that such evolution depends

TABLE 7

Designations of some synthetic potassium zeolites and felspathoids in the system K_2O–Al_2O_3–SiO_2–H_2O

Related natural mineral	Barrer and Baynham (1956)[4]	Barrer et al. (1968)[22]	Barrer and Munday (1971)[52]	Barrer and Mainwaring (1972)[53]	Zdhanov and Ovsepyan (1964)[48]	Ovsepyan and Zdhanov (1965)[54]	Aiello and Franco (1968)[56]	Taylor and Roy (1964)[55]	Takahashi and Nishimura (1970)[57]	Bosmans et al. (1973)[49]	U.S. patent designations (Union Carbide only)
Edingtonite type	K-F	K-F	K-F	K-F	K-F	K-F	K-F		B	F	Z[58] and F[59]
Chabazite type (K-form)	K-G	K-G		K-G	K-chabazite	K-G	K-chabazite		C	H	K-G[60]; H[61] = K-G+K-I[51]
Linde type L				K-L						L	L[62]
"Phillipsite type"	K-M			K-M	K-phillipsite	K-M		K-H		W	W[63]
Analcime (K-form, hydrous)	K-E										
Unknown		K-I			K-I	K-I			A	Q	Q[64]
Unknown						K-II				J	J[65]
Kaliophilite	K-D	K-D		K-D							
Kalsilite	K-N										
F[59] + kaliophilite										M	M[66]

TABLE 8

Zeolite crystallization sequences at 100°C in the system $Li_2O–Na_2O–Al_2O_3\text{-}SiO_2$[69]

Composition of reaction mixture				Products at times given in hours[c]			
nLi_2O	nNa_2O	nAl_2O_3	$nSiO_2$	3	18	112	336
0·05	0·80	0·05	0·10	(Na, Li)-A	(Na, Li)-A	(Na, Li)-P2 and -T	(Na, Li)-P2
0·10	0·70	0·10	0·10	(Na, Li)-A	(Na, Li)-A	(Na, Li)-A	(Na, Li)-A and -T
0·30	0·50	0·10	0·10	(Na, Li)-A and -F	(Na, Li)-F and C	(Na, Li)-C	(Na, Li)-C
0·35	0·45	0·10	0·10	(Na, Li)-A and -F	(Na, Li)-F and -C	(Na, Li)-C	(Na, Li)-C and -A(BW)
0·45	0·35	0·10	0·15	(Na, Li)-F	(Na, Li)-F	(Na, Li)-A(BW)	(Na, Li)-A(BW)
0·40	0·40	0·05	0·15	(Na, Li)-F	(Na, Li)-F	(Na, Li)-A(BW) and -O	(Na, Li)-A(BW) and -O
0·20	0·55	0·10	0·15	(Na, Li)-A and -F	(Na, Li)-A and -F	(Na, Li)-A(BW)	(Na, Li)-A(BW)
0·60	0·15	0·05	0·20	—	—	(Na, Li)-O	(Na, Li)-O
0·35	0·35	0·10	0·20	(Na, Li)-C and -F	(Na, Li)-C and -F	(Na, Li)-C and -A(BW	(Na, Li)-A(BW)
0·25	0·45	0·10	0·20	(Na, Li)-A	(Na, Li)-A, -F and -C	(Na, Li)-C and -F	(Na, Li)-C
0·05	0·60	0·10	0·25	—	(Na, Li)-A and -P2	(Na, Li)-P2	(Na, Li)-P2
0·05	0·50	0·25	0·20	(Na, Li)-A	(Na, Li)-A	(Na, Li)-A	(Na, Li)-A
0·05	0·45	0·10	0·40	—	(Na, Li)-P1	(Na, Li)-P1	(Na, Li)-P and metasilicate
0·10	0·35	0·20	0·35	—	(Na, Li)-A	(Na, Li)-A and -C	(Na, Li)-A and -C
0·05	0·35	0·10	0·50	—	—	(Na, Li)-P1 and -P2	(Na, Li)-P2
0·15	0·25	0·15	0·45	—	(Na, Li)-P1	(Na, Li)-P1	(Na, Li)-A(BW) and -P1

[a] A = Linde type A; P1 = pseudo-cubic form of gismondine type zeolite; P2 = tetragonal form of the same; T = sodalite hydrate; C = cancrinite hydrate; F = edingtonite type zeolite; G = chabazite type; A(BW) = Li-zeolite of Barrer and White;[1] O = unidentified phase.

upon the composition of the reaction mixtures. They further illustrate the Ostwald rule of successive transformations (Chapter 4) in that the products found at the longest times represent phases which are thermodynamically more stable under the synthesis conditions than their predecessors. Table 8 illustrates also how at a given time one may have co-existing phases corresponding with overlap of crystallization fields (cf. §3).

Tetra-alkylammonium hydroxides can give clear aluminosilicate solutions which would be hydrogels if made with alkali metal or alkaline earth hydroxides. Such solutions may have dissolved species corresponding with formal concentrations of aluminate and silicate up to 2 mol per dm^3.[70] Advantage has been taken of this behaviour to synthesize zeolites by titrating CsCl and LiCl[71] or BaCl$_2$ and NaCl[72] in regulated amounts into tetramethylammonium aluminosilicate solutions followed by low-temperature (90°C) crystallization of the alkaline, aqueous hydrogels formed. From the mixtures containing TMA$^+$, Cs$^+$ or Na$^+$ and TMA$^+$ + Na$^+$ + Ba^{2+} the zeolites obtained were respectively those given in Table 9.

TABLE 9

Crystallization products from TMA-aluminosilicate solutions

Structural Type	Reference letters	Refs[a]
(a) With CsCl and LiCl[71]		
Zeolite, narrow pore	Li-A(BW)	(1)
Zeolite, Linde type A	N-A	(6)
Cancrinite hydrate	C	(2, 69)
Tectosilicate, unknown structure	Cs-F	(5)
Edingtonite type zeolite	K-F	(4, 74)
Pollucite	Cs-G	(5)
Offretite type zeolite	O	(75, 76)
Gismondine type zeolite	N-P	(6, 19)
Sodalite hydrate	N-T	(2, 6, 77)
Zeolite, known structure	ZK-5	(78, 73)
Zeolite, unknown structure	ZSM-2	(79)
(b) With BaCl$_2$ and NaCl[72]		
Zeolite, Linde type A	A	(80)
Zeolite, Linde type L	G, L	(81, 82)
Near-phillipsite	M	(4, 81, 82)
Gismondine type	P	(3, 19, 83)
Offretite type zeolite	O	(81, 75)
Zeolite, new type	U-10	
Unidentified	U-11	
(Ba, Na)-aluminate hydrate	BNA	(84)

[a] Refs. to papers originating the designations of column 2.

The appearance of zeolites ZSM-2, a wide-pore zeolite of unknown structure, and of ZK-5, of known structure[73] and of narrow-pore character, is particularly interesting. The sorption capacity and open-ness of ZSM-2 for N_2, $n-C_4H_{10}$, $iso-C_4H_{10}$ and $neo-C_5H_{12}$ was excellent, and the intracrystalline pore volume compared well with that of faujasite, the most porous zeolite known hitherto. The ZK-5 also had a good sorption capacity for N_2 and $n-C_4H_{10}$ but as expected from its structure excluded branched-chain paraffins. The syntheses of ZSM-2 and ZK-5 were often competitive and their production was examined in some detail using the mixtures and conditions given in Table 10. Al was provided as the isopropoxide and Si as $Si(OCH_3)_4$ in making up the tetramethylammonium silicate solutions, for all the experiments involved in Tables 9 and 10. The times t_1, t_2 and t_3 in Table 10 are, respectively, the time during which the tetramethylammonium aluminosilicate solution was kept at 90°C before adding the Cs and Li chlorides; the time for which the resulting mixture was kept at room temperature before heating at 90°C; and the crystallization time at 90°C. Table 10 shows that it is not strictly necessary to have Cs^+ in the reaction mixture in order to make either ZSM-2 or ZK-5. It also seems possible that a maturing period at room temperature (t_2) can assist the appearance of ZSM-2.

The crystallization process following the addition of $BaCl_2$ and NaCl to the tetramethylammonium aluminosilicate solution showed that small equivalent cation fractions of Ba^{2+} could exert a strong influence on the products obtained from otherwise well known zeolite synthesis mixtures. In the presence of TMA^+ and Na^+ the Ba^{2+} led to the replacement of Linde zeolite type A by offretite. If Ba^{2+} was present in mixtures which would otherwise yield only Linde A a new zeolite U-10 was formed, but as the amount of Ba^{2+} was increased U-10 was superseded by a zeolite of Linde type L. Also, at higher ratios Si/Al the gismondine type zeolite P gave way to a near-phillipsite. The new zeolite, U-10, had a hexagonal unit cell which was indexed to $a = 13.71 \pm 0.03$ Å and $c = 15.57 \pm 0.04$ Å. Two analysed samples gave the oxide formulae of $0.985Na_2O.0.026BaO.Al_2O_3.2.0SiO_2.4.12H_2O$; and $0.90K_2O.0.13BaO.Al_2O_3.2.20SiO_2.3.54H_2O$ when $BaCl_2$ and KCl were used in the reaction mixture instead of $BaCl_2$ and NaCl. Powder diffraction patterns of U-10, previously heated in steps to different temperatures, showed an irreversible change between 150° and 200°C.

This section may be concluded by tabulating some further examples of low-temperature syntheses covering the period from 1964 onwards (Table 11). The tabulation is illustrative and in no way exhaustive. Taken with the examples already given it emphasizes how rich and varied has been the harvest of porous tectosilicate crystals formed from gels. A great deal

TABLE 10
Syntheses of ZSM-2 and ZK-5[71]
$Si/Al = 2.5$; $(Cs+Li)/Al = 1.8$; $OH/Al = 3.5$; $OH = 1$; $H_2O/Al = 185$

Cs/(Cs+Li)	t_1	t_2 (hours)	t_3	Products
0^a	72	0	71	ZSM-2 + ZK-5
0^b	72	0	68	ZSM-2 + ZK-5
0^c	72	0	71	ZSM-2 + ZK-5 (trace)
0^d	72	0	70	ZSM-2
0·08	0	0	30	ZSM-2
$0·08^e$	0	0	30	ZSM-2 + ZK-5^f
$0·08^g$	0	0	70	ZSM-2
$0·08^g$	0	0	100	ZSM-2 + ZK-5 (trace)
$0·08^g$	5·2	0	94	ZK-5 + some ZSM-2
$0·08^g$	22	0	77	ZK-5^h
$0·08^g$	46	0	24	ZK-5^h
$0·08^g$	53	0	46	ZK-5
0·08	72	0	30	ZK-5
$0·08^e$	72	0	30	ZK-5
$0·08^a$	72	0	71	ZK-5 + some O^i
$0·08^b$	73	0	68	ZK-5 + some O^i
$0·08^c$	72	0	71	ZK-5 + 4% O^i
$0·08^d$	72	0	70	ZK-5 + 3% O^i
$0·08^e$	72	24	46	ZSM-2
$0·08^e$	72	24	46	ZSM-2 + ZK-5^f

^{a} Concentration, c, of TMAOH before adding Al(OPr^i)$_3$ was 1·56 mol kg^{-1}. Before adding Si(OMe)$_4$ the solution was diluted by a factor, f, of 2·34.
^{b} As for (a), but $f = 1·63$.
^{c} As for (a) but $f = 1·36$.
^{d} As for (a) but $f = 1·00$.
^{e} Seeded with 0·1 g of Na-exchanged Ba-P having the ZK-5 structure.
^{f} Unchanged seed crystals only.
^{g} Si(OMe)$_4$ added first, Al(OPr^i)$_3$ second.
^{h} This sample was free from Li$_2$SiO$_3$ by-product.
^{i} O = offretite type zeolite.

of crystallographic work remains to be done to determine the crystal architecture and place the zeolites in the most appropriate groups. Among the zeolites of §§3 and 4 there are many for which published structural studies are inadequate or absent. These include Li-H,[1] Li-K,[22] the Li-zeolite of Haden et al.,[85] (Na, K)-V,[49] Linde zeolite N[90] (with isotypes Z-21[93] and (Na, TMA)-V[11]), K-I,[22] K-J,[49] K-A(AF),[56] K-Z,[22] ZSM-2,[79] ZSM-3,[88] ZSM-10,[97] ZSM-12,[102] ZSM-21,[99] VK-2,[100] Fu-1,[46] Nu-1,[44] Ba-J,[9] Ba-K,[9] Ba-N,[81] Ba-T[81] and U-10.[72] The prospects are good for finding in Nature or synthesizing many more novel structures.

TABLE 11

Some examples of zeolite syntheses mainly at low temperatures

Authors	Designation of product	Example of reaction conditions	Comments
Haden et al.[85] (1964)	Li-zeolite	Slurry of metakaolinite + LiOH aq at $\sim$38°C. Stirring for 1 day	Structure unknown. Oxide composition $Li_2O.Al_2O_3.2SiO_2.4H_2O$
Kerr[86] (1966)	ZK-5	Aluminosilicate hydrogel with NaOH aq and T(OH)$_2$ at $\sim$100°C. $T = CH_3 \overset{+}{N} \Big\langle \Big\rangle \overset{+}{N} CH_3$	Narrow pore. Structure known[73]
Ciric[87] (1967)	ZSM-4	Na-aluminate in NaOH aq; Na silicate solution added then AlCl$_3$ solution and finally TMAOH solution. 100°C for 3 d.	Mazzite type zeolite
Ciric[79] (1968)	ZSM-2	Glass (2–6)Li$_2$O.Al$_2$O$_3$.(4–9-SiO$_2$ digested with excess water, initially at 20–30°C. Stirred at 55–60°C for 1–3 months	Wide pore. Indexed to tetragonal, $a = 27.4$ Å, $c = 28.1$ Å
Ciric[88] (1968); Kokotailo and Ciric[89] (1971)	ZSM-3	(Na, Li)-aluminosilicate hydrogels + aq at 60–100°C, 5 d to 16 h	Wide pore. Hexagonal polymorph of faujasite. $a = 17.5$ Å; $c \leqslant 129$ Å (a multiple of 14.3 Å); Si/Al = 1–3; L/(Li + Na) = 0.05–0.8
Acara[90] (1968)	Linde N	Metakaolin slurry with NaOH + TMAOH + aq. 1 d at $\sim$23 then shaken 2 h at $\sim$100°C	Narrow pore. Cubic unit cell, $a = 37.22$ Å. Oxide composition (0.83±0.05) Na$_2$O.(0.003± 0.001)TMA$_2$C.Al$_2$O$_3$.(2.0±0.2)SiO$_2$.3H$_2$O
Kerr[91] (1969)	ZK-20	Aqueous alkaline hydrogels from NaOH, TOH, aluminate and silicate 5 d at 80–100°C. T = $CH_3 \overset{+}{N} \Big\langle \Big\rangle \overset{+}{N}$	Levynite type. Narrow pore

(continued)

TABLE 11 (continued)

Authors	Designation of product	Example of reaction conditions	Comments
Aiello and Barrer[75] (1970)	0	Silica sol or gel added to solution of $Al(OH)_3$ in $NaOH+KOH+$ TMAOH.80°C for 7 d	Offretite type. Wide pore. Hexagonal $a = 13 \cdot 3$ Å; $c = 7 \cdot 6$ Å
	Ω	Similar to above	Wide pore, mazzite type zeolite. Hexagonal $a = 18 \cdot 15$ Å, $c = 7 \cdot 59$ Å
	(Na, TMA)-E	Similar to above	Novel zeolite.[91a] Hexagonal, $a = 13 \cdot 27$ Å $c = 15 \cdot 23$ Å
Jenkins[92] (1971)	(Na, K, TMA)-L Offretite type	Similar to above	Like Linde type L.[62] Wide pore. Wide pore
		Silica sol, Na-aluminate, NaOH, KOH, TMAOH+aq. 4d at 100°C	
Duecker et al.[93] (1971)	Z-21	Alkaline aqueous Na-aluminosilicate hydrogel at 95°C. Then refluxed ~1 h at ~100°C	Like Linde N. Small pore. Cubic unit cell, $a = 36 \cdot 7 \pm 0 \cdot 3$ Å. Oxide composition $Na_2O.Al_2O_3.2SiO_2.yH_2O$ ($1 \cdot 89 \leqslant y \leqslant 2 \cdot 2$)
Barrer and Mainwaring[11] (1972) and Barrer and Beaumont[94] (1974)	(Na, TMA)-V	Aqueous alkaline slurry of metakaolinite with NaOH+TMAOH. Cation fraction TMA=$0 \cdot 8$: 278 moles H_2O per $Al_2O_3.2SiO_2$. Stirred 6 d at 65–70°C	Like Linde N. Cubic unit cell. Small pore
Robson[95] (1972)	Erionite type	Aqueous alkaline (Na, Rb) aluminosilicate hydrogels at 100°C, 1–20 d	Small pore. Sorbs n-paraffins. Oxide composition $xNa_2O.(1-x)Rb_2O.Al_2O_3.(6 \cdot 5 \pm 1)SiO_2.yH_2O$
Rubin and Rosinski[96] (1972)	Erionite type	Colloidal SiO_2 added to solution of Na-aluminate, NaOH, KOH and TCl. 10 to 32 d at 100°C. T= Benzyltrimethyl ammonium	Small pore. Sorbs n-hexane. Oxide composition $xT_2O.yK_2O.zNa_2O.Al_2O_3.n.SiO_2$ if anhydrous. $x = 0 \cdot 1$–$0 \cdot 26$; $y = 0 \cdot 4$; $z = 0 \cdot 3$–$0 \cdot 6$; $n = 6 \cdot 9$–$9 \cdot 9$
Ciric[97] (1972)	ZSM-10	Alkaline aqueous hydrogels, from K-aluminate and K-silicate mixtures, with addition of $T(OH)_2$. 95–100°C, 10 d. $T = CH_3\overset{+}{N}$ ⬡ $\overset{+}{N}CH_3$	Wide pore. Hexagonal unit cell, $a = 18 \cdot 2$ Å, $c = 27 \cdot 8$ Å. Si/Al = $3 \cdot 5$–$3 \cdot 7$; ~13% by weight of water in hydrated crystals.

Reference	Zeolite	Synthesis conditions	Properties
H.E. Robson [97a] (1973)	RHO	NaAlO$_2$ with addition of CsOH to about 10% of the Na content. Silica sol added. 3–7 d incubation at 25°C, then 80–100°C for 2 to 4 d	Narrow pore. Cubic body-centred with a = 15·02 Å.
Robson [98] (1973)	ZK-5	(Cs, K)-aluminate solution from Al(OH)$_3$ + mixed CsOH and KOH. Combined with SiO$_2$ sol. 2–10d at 90–120°C.	Small pore. Structure known [98a]. Cubic cell. a = 18·7 Å
Sieber and Meier [70] (1974)	Losod	Compositions with T:Na:Al:Si: H$_2$O = 59:11:15:15:2000. Concentrations T = 1·35, Na = 0·252, Al = Si = 0·343 mol dm^{-3}. Stirred at 100°C. T = [cyclic quaternary ammonium structure]	Narrow pore. Structure determined by authors. Hexagonal cell, a = 12·91 Å, c = 10·54 Å. Cell content Na$_{12}$Al$_{12}$Si$_{12}$O$_{48}$.18H$_2$O
Plank et al. [99] (1977)	ZSM-21	Aluminosilicate hydrogel from aluminate, silicate, NaOH, alum and H$_2$SO$_4$. TOH added. 99°C for 70–83 d. T = (CH$_3$)$_3$NCH$_2$CH$_2$OH	Ferrierite type. Wide pore
Vaughan [100] (1978)	VK-2	KVO$_3$, KAlO$_2$ solution ripened with or without SiO$_2$. Mixture Na-silicate and Na-aluminate made and ripened. Then added to first in definite proportions. 60–100°C, 4–50 h	Narrow pore. Sorbs n-C$_4$H$_{10}$. Unit cell indexed as cubic with a = 30·1 Å SiO$_2$/Al$_2$O$_3$ = 3·1–2·3. Minimal vanadium content
Rubin et al. [101] (1978)	ZSM-34	KOH, NaOH, NaAlO$_2$ dissolved, choline chloride and then colloidal SiO$_2$ added. Gel cryst. 25 days at ~95°C.	Offretite-erionite type. Wide pore
Rubin et al. [102] (1980)	ZSM-12	Na silicate, chrome alum, water, H$_2$SO$_4$, Et$_4$NBr combined. Crystallized 195 d at 210°C	Wide pore, after organic content oxidized away. Sorbed 8% cyclohexane and 9·4% H$_2$O by weight. Very high silica content; Cr in structure.
Barrer et al. [102a] (1980)	RHO	Gel compositions (0·44–0·54)Cs$_2$O.(2·96–2·86)Na$_2$O.Al$_2$O$_3$.11·1SiO$_2$.110H$_2$O, several days at 90°C.	Narrow pore, body-centred cubic with a ~ 15 Å

5. Syntheses from Layer Silicates

Perhaps next to the number and variety of zeolite syntheses from gels come those from layer silicates such as kaolinite, $Al_2O_3.2SiO_2.2H_2O$, (often via its nearly amorphous dehydroxylation product metakaolinite, $Al_2O_3.2SiO_2$), halloysite, $Al_2O_3.2SiO_2.4H_2O$, or montmorillonite $0.67Na$ $[Al_{3.33}Mg_{0.67}(Si_8O_{20})(OH)_4]yH_2O$. The layer silicates may be fortified with extra silica if desired. The clay minerals can make the alumina and silica available in chemical combination and in particular molar ratios.

Individual aqueous bases reacting with kaolinite between 80 and 140°C have yielded the phases given in Table 12.[22] There are some differences between the products of this table and those of Table 3. Thus kaolinite served as a good starting compound for making basic sodalites and cancrinites (sodalite and cancrinite hydrates) but did not under the conditions employed yield zeolites like faujasite or Linde type A. Three unusual zeolitic phases are of interest: K-I, K-Z and Li-K. Of these K-I ($K_2O.Al_2O_3.2SiO_2.4H_2O$) was found to be hexagonal with $a = 13.51 \pm 0.04$ Å and $c = 13.50 \pm 0.4$ Å and to decompose at the low tem-

TABLE 12

Some low-temperature crystalline products obtained from kaolinite[a] with single aqueous bases[22]

Base	Product designation	Nature	Oxide composition
LiOH	Li-A(BW)[1]	Zeolite	$Li_2O.Al_2O_3.2SiO_2.4H_2O^e$
	Li-K[22]	Zeolite	–
	Li-L[22]	Ignition product of Li-K	–
	Li-O[22]	Ignition product of Li-exchanged Rb-D	–
NaOH	Basic sodalite[2]	Felspathoid-zeolite	$3(Na_2O.Al_2O_3.2SiO_2)yNaOH.xH_2O^d$
	Basic cancrinite[2]	Felspathoid-zeolite	$3(Na_2O.Al_2O_3.2SiO_2)yNaOH.xH_2O^d$
KOH	K-I[48]	Zeolite, unidentified	$K_2O.Al_2O_3.2SiO_2.3H_2O^b$
	K-G[4]	Chabazite type	$K_2O.Al_2O_3.xSiO_2.yH_2O$ $(2.3 \leqslant x \leqslant 4.2)^c$
	K-F[4]	Edingtonite type	$K_2O.Al_2O_3.2SiO_2.3H_2O^b$
	K-D[4]	Kaliophilite	$K_2O.Al_2O_3.2SiO_2$
	K-Z[22]	Basic zeolite	$K_2O.Al_2O_3.2SiO_2.KOH.3.5H_2O^b$
	K-Y[22]	Ignition product of K-Z	–
RbOH	Rb-D[5]	Edingtonite type	$Rb_2O.Al_2O_3.2SiO_2.2.6H_2O^b$
	Rb-A[5]	Ignition product of Rb-D	$Rb_2O.Al_2O_3.2SiO_2$
CsOH	Cs-D[22]	Edingtonite type	$Cs_2O.Al_2O_3.2SiO_2.2.4H_2O^b$
	Cs-F[5]	Ignition product of Cs-D	$Cs_2O.Al_2O_3.2SiO_2$

[a] The reference numbers in column 2 indicate syntheses from gels from which the letter designations originated. In Ref. (1) the zeolite Li-A(BW) was designated as Li-A. The initial of each author is added to differentiate the zeolite from Linde type A.
[b] Results of complete analyses.
[c] Analytical results of Ref. (4).
[d] x and y interdependent: $0 \leqslant x \leqslant 8$; $y = 2 - x/4$.
[e] Analytical results of Ref. (1).

perature of ~168°C in a heating X-ray camera. It appeared to be the same as K-I made by Zdhanov and Ovsepyan.[48] K-Z ($K_2O.Al_2O_3.2SiO_2.KOH.3.5H_2O$) formed only by reaction between kaolinite and saturated KOH solution and contained additional alkali, probably intercalated, like NaOH in basic sodalite. It remained stable on heating to 400°C. Li-K appeared between 80 and 140°C, but tended in longer runs to be replaced by Li-A(BW). None of these three phases could be identified with known structures.

Metakaolinite, usually obtained by dehydroxylating kaolinite from 500 to 700°C, has been a prolific source of zeolites, examples of which are given in Table 13. Many of the products appeared at temperatures as low as 80°C, as exemplified in the crystallization fields[11] shown in Figs 8, 9 and 10 in which the bases were respectively LiOH; NaOH; and RbOH and CsOH. The lettering in these figures corresponds with that in Table 13, save that the chemical symbol of the cation is not prefixed. The symbol "Am" denotes amorphous. All the sodium minerals in Table 13 have also been made from gels (Table 1). Unlike kaolinite, metakaolinite readily yielded faujasites, Linde zeolite A and the gmelinite type zeolite, Na-S. Basic sodalite and cancrinite formed as readily as from gels or from kaolinite. Also the edingtonite type zeolite appeared when KOH, RbOH and CsOH were the bases, again as from gels or from kaolinite. TlOH likewise produced this type of zeolite from metakaolinite. The diversity of Na-zeolites formed at low temperatures from metakaolinites (Table 13) contrasts with the behaviour of kaolinite (Table 12). In general metakaolinite appears to be rather more reactive (of higher free energy) than kaolinite. The synthesis of the zeolite Ba-G,L, a barium form of the zeolite (Na, K)-L (Linde type L) merits attention because the barium form is much richer in alumina than the usual (Na, K)-form. A barium form of zeolite L was also made from gels (Table 3). The two other zeolitic phases, Ba-N and Ba-T could not be identified with known zeolites. They extend the number of Ba zeolites already recorded in Tables 3 and 6 from five to seven. Of this number four have not been identified and are of unknown structure (Ba-J, Ba-K, Ba-N and Ba-T), and four (Ba-mordenite, Ba-G,L, Ba-J and Ba-K) are good sorbents of permanent gases. In the Li field the synthesis of Li-H, first made from gels and of unknown structure, is of interest. It co-crystallized with the zeolite Li-A(BW) (indicated by the symbol A in Fig. 8) at temperatures between 110° and 170°C from metakaolinite compositions fortified with 4 and 8 moles of SiO_2 per "mole" of metakaolinite ($Al_2O_3.2SiO_2$). Metakaolinite with and without added silica yielded six potassium zeolites of which five match those obtained by Bosmans et al.[49] from potassium aluminosilicate gels at 90°C and four are the same as were prepared from

TABLE 13
Reactions of metakaolinite, with and without added silica, with single aqueous bases[11,53,81,82]

Base	Product designation[a]	Nature
LiOH	Li-A(BW)[1]	Zeolite, $Li_2O.Al_2O_3.2SiO_2.4H_2O$
	Li-H[1]	Zeolite, $Li_2O.Al_2O_3.8SiO_2.5H_2O$
	Li-D[1]	Metasilicate, Li_2SiO_3
NaOH	Na-B[2]	Analcime, $Na_2O.Al_2O_3.4SiO_2.2H_2O$
	Na-C[2]	Cancrinite hydrate, $3(Na_2O.Al_2O_3 2SiO_2.yNaOH.xH_2O)$
	Na-J[2]	Nepheline hydrate I $(Na_2O.Al_2O_3.2SiO_2.H_2O)$
	Na-P1[3]	Gismondine-type, pseudocubic
	Na-P2[3]	Gismondine-type, pseudo-tetragonal
	Na-P3[3]	Gismondine-type, pseudo-orthorhombic
	Na-Q[3]	Zeolite, Linde type A
	Na-R[3]	Faujasite
	Na-S[3]	Gmelinite type
	Na-T[3]	Hydroxy sodalite, $3(Na_2O.Al_2O_3.2SiO_2)yNaOHxH_2O$
KOH	K-F[4]	Edingtonite type
	K-G[4]	Chabazite type
	K-I[48]	Zeolite $K_2O.Al_2O_3.2SiO_2.yH_2O$
	K-L[62]	Zeolite, like Linde type L[62]
	K-M[4]	Near-phillipsite
	K-Z[22]	Basic zeolite $K_2O.Al_2O_3.2SiO_2.KOH.yH_2O$
	K-D[4]	Kaliophilite, $K_2O.Al_2O_3.2SiO_2$
	K-N[4]	Kalsilite, $K_2O.Al_2O_3.2SiO_2$
RbOH	Rb-A[5]	Non-zeolite, $Rb_2O.Al_2O_3.2SiO_2$
	Rb-D[5]	Edingtonite-type zeolite, like K-M
	Rb-M	Zeolite, like K-M[4]
CsOH	Cs-D	Edingtonite-type zeolite, like K-F[4]
	Cs-F[5]	Non-zeolite, like Rb-A, $Cs_2O.Al_2O_3.2SiO_2$
	Cs-D[5]	Pollucite, $Cs_2O.Al_2O_3.4SiO_2$
TlOH	Tl-A	Unidentified, minor yield
	Tl-B	Unidentified, minor yield
	Tl-C	Unidentified, minor yield
	Tl-F	Edingtonite-type zeolite, like K-F[4]
Ba(OH)$_2$	Ba-G, L[9]	Zeolite, like Linde type L,[62] $1·1BaO.Al_2O_3.2·5SiO_2.5·1H_2O$
	Ba-N[81]	Zeolite, $BaO.Al_2O_3.2SiO_2.2·8H_2O$
	Ba-P[9]	Hexagonal polymorph of celsian
	Ba-T[81]	Zeolite, $BaO.Al_2O_3.2SiO_2.1·2Ba(OH)_2.2H_2O$

[a] The references in column 2 are to earlier syntheses, usually from gels, in which the designations originated. Ba-G, L was originally termed Ba-G,[9] but the letter L was subsequently added[81] to differentiate this zeolite from the chabazite type zeolite K-G[4] and to indicate that it has the topology of Linde zeolite L, as shown by Baerlocher and Barrer.[103]

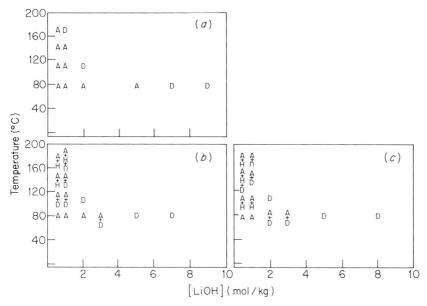

Fig. 8. Crystallizations from the compositions (a) 1MTK + (2·5–45)LiOH + 275H$_2$O; (b) 1MTK + 4SiO$_2$ + (2·5–45)LiOH + 275H$_2$O; and (c) 1MTK + 8SiO$_2$ + (2·5–45)LiOH + 275H$_2$O, showing dependence upon temperature and alkalinity.[11]

        A = Li-A(BW)   MTK denotes metakaolinite.
        H = zeolite of ref. 1   D = Li$_2$SiO$_3$

kaolinite. Linde zeolite type L, near-chabazites, -phillipsites, -edingtonites and zeolite K-I were of most common occurrence.

In further exploration of low-temperature zeolitization involving metakaolinite, aqueous mixed bases were employed.[11,82] The metakaolinite was used, as in Table 13, both with and without added silica. Results are summarized in Table 14, in which the reference letters are those in Table 13 for the same kind of zeolite. The ease with which diverse zeolites form is well illustrated by both tables. Where comparisons can be made between the results of Tables 13, for each of two single bases, with those of Table 14 for the same two bases in admixture, the mixtures yield most of the zeolites formed by the single bases acting separately. Some differences also arise: for example considering zeolitization by LiOH and NaOH (Table 13) and by LiOH + NaOH (Table 14), Li-H (zeolite), Li-D (silicate) and Na-B (analcime) were not found with the mixed bases which however yielded (Na, Li)-F (edingtonite type zeolite) and (Na, Li)-U, a layer silicate like hectorite, both absent from Table 13. The zeolite Linde

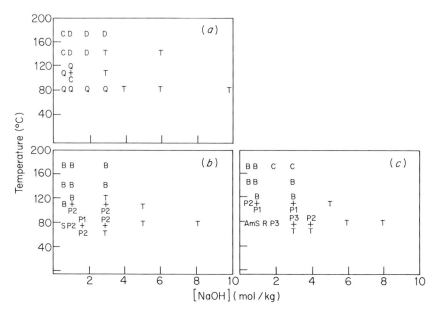

Fig. 9. Crystallizations from the compositions[11] (a) 1MTK + (1–50)NaOH + 275H$_2$O; (b) 1MTK + 4SiO$_2$ + (2·5–50)NaOH + 275H$_2$O; and (c) 1MTK + 8SiO$_2$ + (2·5–50)NaOH + 275H$_2$O. MTK denotes meta-kaolinite.

B = analcime	R = faujasite type
C = cancrinite hydrate	S = gmelinite type
J = nepheline hydrate I	T = sodalite hydrate
P1 = pseudo-cubic near gismondine	Am = amorphous
P2 = pseudo-tetragonal near gismondine	
Q = zeolite like Linde type A	

type A is known to be favoured by a sodic environment (Chapter 4, Table 6). It can also form, as Table 14 shows, when sodium is diluted by the larger potassium or tetramethylammonium ions but was not observed when the sodium was diluted with the smaller Li$^+$ ion. On the other hand, faujasite, the formation of which is also favoured by a sodic environment, was not observed when sodium was diluted by potassium but did crystallize when the sodium was diluted by lithium. It also formed when both sodium and tetramethylammonium were present. Because such results must depend on the relative ease of nucleation of potentially competing species they must be considered as typical only for the particular compositions and conditions employed.

The ways in which the equivalent cation fraction of a given base (as ordinate) and total alkalinity (as abscissa) controlled the crystallization fields was investigated[11,82] for all the pairs $Ba(OH)_2 + LiOH$, $Ba(OH)_2 + NaOH$, $NaOH + KOH$, $NaOH + LiOH + LiOH$, $KOH + LiOH$ and $NaOH + N(CH_3)_4OH$. Examples of these crystallization fields are given in Figs 11, 12 and 13. The compositions of the reaction mixtures, except for the relative proportions of the two bases, were held constant at values indicated in the captions. Certain zeolites are seen from Tables 13 and 14

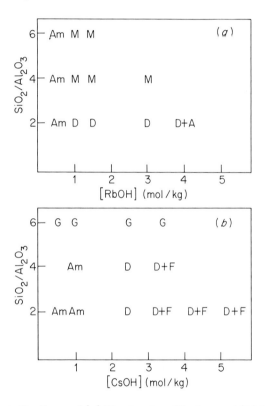

Fig. 10. Crystallizations of (a) Rb-aluminosilicates and (b) Cs-aluminosilicates at 80°C as functions of the ratio SiO_2/Al_2O_3 and of alkali concentration.[11]

Rb-A = tectosilicate like Cs-F[5]
Rb-D = edingtonite type zeolite like K-F[4]
Rb-M = zeolite like K-M[4]
Cs-D = edingtonite type zeolite like K-F[4]
Cs-F = tectosilicate[5]
Cs-G = pollucite
Am   = amorphous

TABLE 14

Zeolitization of metakaolinite with and without extra silica, at low temperatures and using aqueous mixed bases[11,82]

Bases (and temperature, °C)	Crystal designations	Crystal types
NaOH + KOH (80)	(Na, K)-D, Kal	Kaliophilite, $K_2O.Al_2O_3.2SiO_2$
	(Na, K)-F	Edingtonite type, like K-F[4]
	(Na, K)-G	Chabazite type
	(Na, K)-M	Like zeolite K-M[4]
	(Na, K)-Pl	Pseudo-cubic gismondine type
	(Na, K)-P2	Pseudo-tetragonal gismondine type
	(Na, K)-Q	Zeolite like Linde type A
	(Na, K)-S	Gmelinite type
	(Na, K)-T	Sodalite hydrate
NaOH + LiOH (80)	(Na, Li)-A(BW)	Like zeolite Li-A(BW)[1]
	(Na, Li)-C	Cancrinite hydrate
	(Na, Li)-D	Metasilicate like Li-D[1]
	(Na, Li)-F	Edingtonite type, like K-F
	(Na, Li)-P1	Pseudo-cubic gismondine type
	(Na, Li)-P2	Pseudo-tetragonal gismondine type
	(Na, Li)-R	Faujasite type
	(Na, Li)-S	Gmelinite type
	(Na, Li)-T	Sodalite hydrate
	(Na, Li)-U	Layer silicate like hectorite
KOH + LiOH (80)	(K, Li)-A(BW)	Like zeolite Li-A(BW)[1]
	(K, Li)-D	Metasilicate
	(K, Li)-D, Kal	Kaliophilite
	(K, Li)-F	Edingtonite type, like K-F[4]
	(K, Li)-G	Chabazite type
	(K, Li)-M	Like zeolite K-M[4]
	(K, Li)-S	Gmelinite type
	(K, Li)-U	Layer silicate like hectorite
NaOH + TMAOH (85)	(Na, TMA)-C	Cancrinite hydrate
	(Na, TMA)-P1	Pseudo-cubic gismondine type
	(Na, TMA)-Q	Like Linde zeolite type A
	(Na, TMA)-R	Faujasite type
	(Na, TMA)-T	Sodalite hydrate
	(Na, TMA)-V	Zeolite like Linde type N[90]
Ba(OH) + LiOH (80)	(Li, Ba)-A(BW)	Zeolite like Li-A(BW)[1]
	(Ba, Li)-E	Edingtonite
	(Ba, Li)-F	Edingtonite type like K-F[4]
	(Ba, Li)-G, L	Zeolite like Linde type L[62]
	(Ba, Li)-M	Zeolite like K-M[4]
Ba(OH)$_2$ + NaOH (80)	(Ba, Na)-G, L	Zeolite like Linde type L[62]
	(Ba, Na)-M	Zeolite like K-M[4]
	(Na, Ba)-P, *GIS*	Gismondine type like Na-P[3]
	(Ba, Na)-P*	Like cymrite
	(Na, Ba)-S	Gmelinite type, like Na-S[3]
	(Ba, Na)-T	Zeolite, like Ba-T[81]
Ba(OH)$_2$ + KOH (80)	(K, Ba)-G, *CHA*	Chabazite type zeolite
	(K, Ba)-G, L	Zeolite like Linde type L[62]
	Ba-P	Hexagonal polymorph of celsian
	(Ba, K)-T	Zeolite like Ba-T[81]
Ba(OH)$_2$ + TlOH (80)	(Ba, Tl)-B	Like Tl-B[82]
	(Ba, Tl)-G, L	Zeolite like Linde type L[62]
	(Tl, Ba)-F	Edingtonite type zeolite like K-F[4]
Ba(OH)$_2$ + TMAOH (85)	(Ba, TMA)-E	Erionite type zeolite
	(Ba, TMA)-G, L	Zeolite like Ba-G, L[81]
	Ba-P	Hexagonal polymorph of celsian
	(Ba, TMA)-T	Zeolite like Ba-T[81]

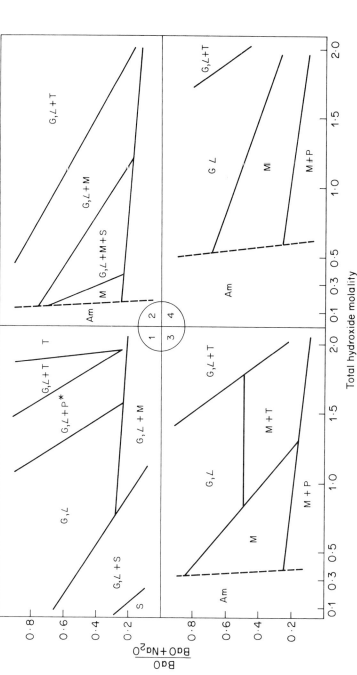

Fig. 11. Crystallization fields at 80°C in the system[82] Ba(OH)$_2$–NaOH–MTK–SiO$_2$–H$_2$O. Quadrant 1, no SiO$_2$ added; quadrant 2, SiO$_2$/MTK = 1; quadrant 3, SiO$_2$/MTK = 2; quadrant 4, SiO$_2$/MTK = 4. MTK denotes meta-kaolinite.

G, L = zeolite (Ba, Na)-G, L, like Linde type L[62]
M = zeolite (Ba, Na)-M, like K-M[4]
P = zeolite (Ba, Na)-P, gismondine type like Na-P[3]

P* = (Ba, Na)-P*, non zeolite like cymrite
S = zeolite (Ba, Na)-S, like Na-S[3]
T = (Ba, Na)-T, zeolite like Ba-T[81]

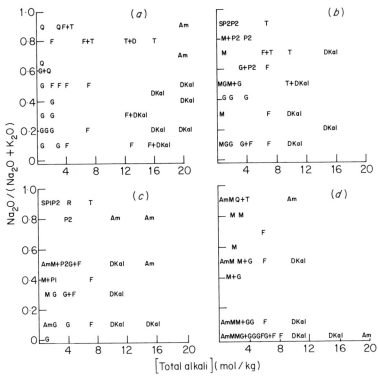

Fig. 12. Crystallization of aluminosilicates at 80°C with mixed NaOH + KOH solutions,[11] from metakaolinite (MTK) with and without added SiO_2.

(a) $1MTK + (2 \cdot 8-75)(Na, K)OH + 275H_2O$
(b) $1MTK + 4SiO_2 + (2 \cdot 8-75)(Na, K)OH + 275H_2O$
(c) $1MTK + 6SiO_2 + (2 \cdot 8-75)(Na, K)OH + 275H_2O$
(d) $1MTK + 8SiO_2 + (2 \cdot 8-75)(Na, K)OH + 275H_2O$

D, Kal	= (Na, K)-kaliophilite
F	= (Na, K)-F, like K-F[4]
G	= (Na, K)-chabazite
M	= (Na, K)-M, zeolite like K-M[4]
P1	= pseudo-cubic gismondine type
P2	= pseudo-tetragonal gismondine type
Q	= zeolite like Linde type A
S	= gmelinite type zeolite
T	= sodalite hydrate
Am	= amorphous

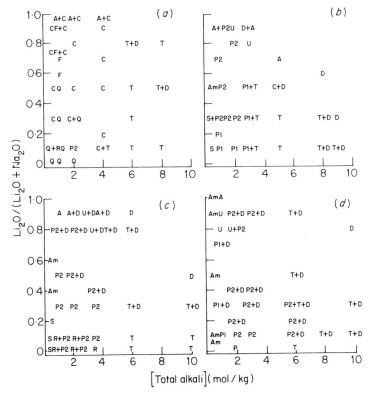

Fig. 13. Crystallization of aluminosilicates at 80°C with mixed NaOH + LiOH solutions[11] from metakaolinite (MTK) with and without added SiO₂.

(a) $1MTK + (2·5–40)(Na, Li)OH + 275H_2O$
(b) $1MTK + 4SiO_2 + (2·5–40)(Na, Li)OH + 275H_2O$
(c) $1MTK + 6SiO_2 + (2·5–40)(Na, Li)OH + 275H_2O$
(d) $1MTK + 8SiO_2 + (2·5–40)(Na, Li)OH + 275H_2O$

A	= zeolite like Li-A(BW)[1]
C	= cancrinite hydrate
D	= metasilicate[1]
F	= zeolite like K-F[4]
P1	= pseudo-cubic gismondine type
P2	= pseudo-tetragonal gismondine type
R	= faujasite type
S	= gmelinite type
T	= sodalite hydrate
U	= layer silicate like hectorite
Am	= amorphous

TABLE 15

Syntheses of zeolites having edingtonite topologies[11]

Designations of variant	Oxide compositions	Synthesis conditions
K-F	$K_2O.Al_2O_3.2SiO_2.3H_2O$	<150°C, K-aluminosilicate gels[3]
	$K_2O.Al_2O_3.2SiO_2.3H_2O$	80°C, KOH aq + kaolinite[22]
	$K_2O.Al_2O_3.2SiO_2.3H_2O$	80°C, KOH aq + metakaolinite[53]
Rb-D	$Rb_2O.Al_2O_3.2SiO_2.H_2O$	165°C, Rb-aluminosilicate gels[5]
	$Rb_2O.Al_2O_3.2SiO_2.2·6H_2O$	80°C, RbOH aq + kaolinite[22]
	$Rb_2O.Al_2O_3.2SiO_2.2·5H_2O$	80°C, RbOH aq + metakaolinite[11]
Cs-D	$Cs_2O.Al_2O_3.2SiO_2.2·4H_2O$	80°C, CsOH aq + kaolinite[22]
		80°C, CsOH aq + metakaolinite[11]
(Na, Li)-F	$0·8Li_2O.0·2Na_2O.Al_2O_3.2SiO_2.3H_2O$	80°C, NaOH + LiOH aq + metakaolinite[11]
(Na, Li)-F	$0·75Li_2O.0·25Na_2O.Al_2O_3.2SiO_2.3·2H_2O$	100°C, (Na, Li)-aluminosilicate gel[69]
(Na, K)-F	$(Na, K)_2O.Al_2O_3.xSiO_2.yH_2O$[a]	80°C, NaOH + KOH aq + metakaolinite[11]
(K, Li)-F	$(Li, K)_2O.Al_2O_3.xSiO_2.yH_2O$[a]	80°C, KOH + LiOH aq + metakaolinite[11]
Species N	$K_2O.Al_2O_3.4SiO_2.yKCl$[a]	200–450°C, analcime + excess KCl[104]
K-F (KCl)	$K_2O.Al_2O_3.2·54SiO_2.0·8KC.10·5H_2O$	100°C, K-aluminosilicate gel + excess KCl[105]
K-F (KCl)	$K_2O.Al_2O_3.2SiO_2.0·3KCl.2·5H_2O$	80°C, KOH aq, excess KCl + kaolinite[22]
Species O	$K_2O.Al_2O_3.4SiO_2.yKBr^+$	200–450°C, analcime + excess KBr[104]
K-F (KBr)	$K_2O.Al_2O_3.2·53SiO_2.0·72KBr.0·4H_2O$	200°C, K-aluminosilicate gel + excess KBr[105]
K-F (KBr)	$K_2O.Al_2O_3.2SiO_2.0·2KBr.2·5H_2O$	80°C, KOH aq, excess KBr + kaolinite[22]
K-F (KI)	$K_2O.Al_2O_3.2·5SiO_2.0·45KI.0·5H_2O$	200°C, K-aluminosilicate gel + excess KI[105]
(Ba, Li)-F	$(Ba, Li_2)O.Al_2O_3.xSiO_2.yH_2O$[a]	80°C, Ba(OH)$_2$ + LiOH aq + metakaolinite[82]

[a] Values of x and y not determined

and the figures to crystallize from a variety of cationic environments (cf. Chapter 4, §5.3). These include zeolites having the framework topologies of edingtonite, Linde zeolite type L,[62] sodalite and cancrinite. Table 15 gives examples of syntheses from gels, analcime, kaolinite and metakaolinite of variants of the edingtonite topology. In addition to frameworks containing only water molecules and charge-neutralizing cations, the edingtonite structure may, like sodalite and cancrinite,[23] form with salt molecules (KCl, KBr, and KI) included,[22,105] displacing some zeolitic water. The salt molecules, once intercalated, cannot be extracted from this zeolite without decomposing it. The silica contents given in Table 15 for Species O and N are those of the parent analcime and the actual silica content may be less than this. The X-ray powder patterns of preparations of (Na, K)-F showed that the unit cell size changed with the cation fraction of Na^+. For certain values of this fraction two patterns with slightly different d-spacings indicated two very similar co-existent phases. In agreement with this, exchange isotherms for $Na^+ \rightleftarrows K$ and $Na^+ \rightleftarrows Li^+$ have both shown miscibility gaps between the end members of the exchange pairs.[52] Changes in the unit cell type were observed among certain of the preparations, as shown below:[11]

Edingtonite[106]	Orthorhombic, $P2_12_12$	$a = 9.54$ Å; $b = 9.65$ Å; $c = 6.50$ Å
K-edingtonite[107]	Orthorhombic	$a = 19.46$ Å; $b = 20.04$ Å; $c = 13.36$ Å
K-F[108]	Orthorhombic	$a = 14.02$ Å; $b = 13.92$ Å; $c = 13.14$ Å
Rb-D[11]	Tetragonal, body-centred	$a = 9.98$ Å; $c = 13.23$ Å
K-F (KCl)[105]	Tetragonal, body-centred	$a = 9.83$ Å; $c = 13.12$ Å
K-F (KBr)[105]	Tetragonal, primitive	$a = 9.79$ Å; $c = 6.54$ Å
K-F (KI)[105]	Tetragonal, primitive	$a = 9.81$ Å; $c = 6.59$ Å

These results indicate that there can be synthetic variants of a given framework topology, such as are found also in some naturally occurring zeolites (e.g. stilbite, stellerite and barrerite; or heulandite and clinoptilolite). The edingtonite topology appears to provide an unusual range of such variants.

Interesting differences in composition which were sometimes linked with changes in sorption properties were demonstrated for the Ba-containing phases of Tables 13 and 14 having the topology of Linde zeolite type L.[62] The observed behaviour is shown in Table 16.[81] From $Ba(OH)_2$ aq or $Ba(OH)_2 + KOH$ aq and kaolinite or metakaolinite highly aluminous forms of zeolite L can be synthesized as seen in the table. The zeolite is normally a wide-pore sorbent which takes up neo-pentane, but one preparation differentiated between iso-butane (sorbed) and neo-pentane (not sorbed at 0°C). This difference could arise if there were in the preparation which failed to sorb neo-pentane some intracrystalline detrital material restricting the openings giving access to the interior of

TABLE 16

Approximate saturation capacities[81] of different zeolites having the framework of Linde zeolite type L[62]

	SiO_2/Al_2O_3		Capacities (cm^3 at stp per g)				Lattice
Cations present	Before synthesis	In product	O_2 (78K)	n-C_4H_{10} (273K)	iso-C_4H_{10} (273K)	neo-C_5H_{12} (273K)	stable at °C
Ba	2	2·50	87	12·2	9·7	nil	
Ba, K	2	2·08	82	23·5	19·4	13·0	>800
Ba	5	4·80	72	9·5			
Ba, K	6	5·40	74	11·0	10·9		>800
K, Na	15–28[62]	5–7		25	20	10	

the crystals. Additional examination was made of samples (numbered 1, 2 and 3) of (Ba, TMA)-G,L heated in air at 540, 590 and 700°C respectively to oxidize away the organic moiety. After subsequent equilibration with the atmosphere and then out-gassing at temperatures rising to 300°C they lost water in amounts, in order, 15·3, 16·1 and 21·7% of their dry weights. They were all wide-pore sorbents which freely imbibed cyclohexane and n-hexane at 22°C. Analyses of samples 1 and 3 were made, and from the BaO/Al_2O_3 ratios and the assumption that $(BaO + (TMA)_2O)/Al_2O_3 = 1$ the water free oxide compositions were:

Sample 1: $0·78BaO.0·22(TMA)_2O.Al_2O_3.2·35SiO_2$

Sample 2: $0·84BaO.0·16(TMA)_2O.Al_2O_3.2·23SiO_2$

TABLE 17

Transformations of halloysite and montmorillonite

		NaOH aq	KOH aq
(a) Halloysite[109]	Temp. (°C)		
	~200	Analcime, Na-phillipsite type	Phillipsite type
	~340	Analcime	Phillipsite type
	~270	Analcime	Phillipsite type
	~300	Analcime, nepheline hydrate	Phillipsite type, leucite, kaliophilite
(b) Halloysite[56]	Conc.		
(80°C)	M/4	Linde type A	Zeolite K-A(AF)
	M/2	Linde type A	K-chabazite
	M	Linde type A, sodalite hydrate	K-chabazite
	2 M	Sodalite hydrate	K-chabazite
(c) Montmorillonite[56]	M/4	Gismondine type Na-P	Zeolite K-F
(80°C)	M/2	Na-P, sodalite hydrate	Zeolite K-F
	M	Sodalite hydrate	Zeolite K-F
	2 M	Sodalite hydrate	Zeolite K-F

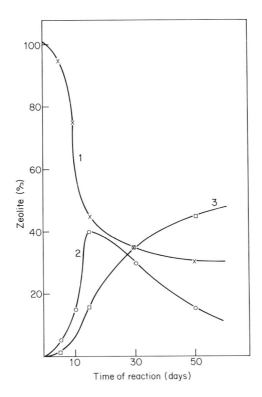

Fig. 14. Conversion of halloysite to Linde zeolite A and to faujasite.[110] The ratio Na_2O/Al_2O_3 was 7·35 and the NaOH solution was 1 M. Curve 1 shows the amount of halloysite remaining; curve 2, the formation of zeolite A and its later conversion to faujasite; and curve 3, the yield of faujasite.

Evidently thermally stable, aluminous and so cation-rich variants of Linde zeolite type L[62] can be made as Ba-containing forms which, like the original siliceous (Na, K)-L, are usually wide pore sorbents.

Clay minerals other than kaolinite and its ignition product metakaolinite have been examined as sources of zeolites, although less extensively. Franco and Aiello,[109] using a ratio $M_2O/Al_2O_3 = 2$ in their reaction mixtures (M = Na or K), and with a volume of aqueous solution of 50 cm³, studied the hydrothermal transformation of 2 g portions of halloysite. Aiello and Franco[56] made a parallel study of the low-temperature reactions of halloysite and montmorillonite, using aqueous solution of NaOH and KOH of various concentrations. The results of both investigations are summarized in Table 17. In addition to a number of the zeolites formed from gels, kaolinite and metakaolinite, an apparently new zeolite,† K-A(AF), was reported which had the oxide composition $K_2O.Al_2O_3.2SiO_2.4\cdot25H_2O$. It was indexed to a cubic unit cell with $a = 15\cdot85$ Å. At 300°C several felspathoids co-crystallized with the zeolites (nepheline hydrate ($Na_2O.Al_2O_3.SiO_2.H_2O$) with analcime; and leucite ($K_2O.Al_2O_3.4SiO_2$) and kaliophilite ($K_2O.Al_2O_3.2SiO_2$) with phillipsite). When montmorillonite was zeolitized at 80°C Na-P replaced the Linde type A from halloysite and the edingtonite type K-F replaced the K-A(AF) and chabazite obtained from halloysite. Fernandez et al.[110] extended zeolite synthesis from halloysite to the unusually low temperature of 35°C. They used aqueous sodium hydroxide solutions between M/4 and 5 M. Faujasite and Linde zeolite type A formed when the solutions were M to 3 M. Various ratios Na_2O/Al_2O_3 were successfully employed and the amounts of each zeolite and of the remaining halloysite were monitored as functions of time in a number of the experiments. The yield of zeolite A went through a maximum and then declined while that of faujasite increased steadily over the reaction time. This behaviour is shown in Fig. 14 for $Na_2O/Al_2O_3 = 7\cdot35$ and a concentration of NaOH equal to 1 M.

6. Syntheses from Natural and Synthetic Glasses

The formation of zeolites from natural and synthetic glasses has also been systematically investigated. In the high temperature hydrothermal range (~235 to ~375°C) using 1% and 3% contact solutions of NaOH and KOH the phases obtained from natural glasses are given in Table 18.[111] The glassy reactants yielded the felspar, orthoclase; the felspathoids, leucite and kaliophilite; and the zeolites herschelite, chabazite and analcime. Herschelite and chabazite formed only when the mineralizing solution was squeous KOH. Analcime was a dominant reaction product from all the glasses with sodic contact solutions and from the alkali-rich glasses (nepheline-leucite and basaltic) even when the contact solution was potassium hydroxide. From the viewpoint of zeolite use only the chabazite type zeolites are of interest and attention may therefore be directed to the zeolitization of glasses at lower temperatures.

With sodium hydroxide contact solutions and a natural volcanic glass in the form of finely ground rhyolitic pumice the crystallization field after 60 hours at 65°C is shown in Fig. 15.[113] The zeolites derived are the

---

† The authors[56] termed this zeolite K-A. However, to differentiate it from Li-A(BW) and Linde type A, the authors' initials (A and F) have been added.

TABLE 18

Phases[a] obtained from some natural glasses under hydrothermal conditions[111]

Conditions: Temperature (°C), Pressure (kg cm^{-2})

Glasses	T ~235°C, P ~30 (kg cm^{-2})				T ~280°C, P ~60 (kg cm^{-2})			T ~330°C, P ~120 (kg cm^{-2})		T ~375°C, P ~200 (kg cm^{-2})			
	KOH aq		NaOH aq		KOH aq		NaOH aq	KOH aq	NaOH aq	KOH aq		NaOH aq	
	1%	3%	1%	3%	1%	3%	1%	1%	1%	1%	3%	1%	3%
Rhyolitic	orthoclase	orthoclase	analcime + orthoclase	analcime + orthoclase	orthoclase	–	analcime + orthoclase	orthoclase	analcime + orthoclase	orthoclase	–	orthoclase + analcime	–
Alkali-trachitic	herschelite	–	analcime	–	herschelite	–	analcime + orthoclase	analcime + orthoclase	analcime + orthoclase	–	orthoclase	–	analcime + orthoclase
trachi-andesitic	–	–	–	–	chabazite	leucite	analcime + (orthoclase)	analcime + orthoclase	analcime + orthoclase	analcime + orthoclase	–	analcime + orthoclase	–
Leuco-phonolitic	–	–	analcime	–	herschelite	–	analcime	leucite	analcime + orthoclase	leucite	–	analcime + orthoclase	–
Leuco-tephritic	(herschelite)	–	analcime	–	herschelite	–	analcime	leucite + analcime	–	–	leucite + kaliophilite	–	analcime
Leucitic	(chabazite)	leucite + kaliophilite + (chabazite)	–	–	leucite + (kaliophilite + chabazite)	–	analcime	leucite + kaliophilite	–	leucite + kaliophilite	–	analcime	–
Nepheline-leucitic	–	–	–	–	analcime	–	analcime	analcime	analcime	analcime	–	analcime	–
Basaltic	–	–	–	–	analcime	–	analcime	analcime	analcime	–	–	–	analcime

[a] Products given in parentheses denote formation only in minor amount.

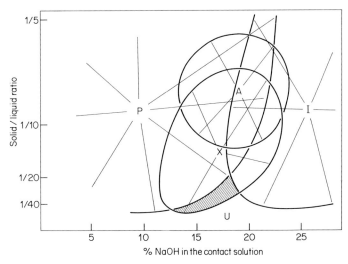

Fig. 15. Formation areas of products crystallized from Lipari pumices at 65°C for 60 hours with various solid/liquid ratios and NaOH concentrations.[113]

A = zeolite Linde type A   P = gismondine type zeolite
I = sodalite hydrate   X = faujasite
U = amorphous

gismondine type P, Linde type A, faujasite (X) and sodalite hydrate (I). The symbol U denotes "amorphous". There is considerable overlap in the crystallization fields, but zeolite P begins to appear at the lowest and sodalite hydrate at the highest alkali concentrations. The kinetics of formation of zeolite X were of the usual sigmoid form (Chapter 4 §4). The rhyolitic pumice employed was from Isola di Lipari, Italy, and had the composition

SiO_2	70·85%	MgO	0·55%	H_2O	3·71%
Al_2O_3	12·83	FeO	1·35		
Na_2O	4·46	Fe_2O_3	1·02		
K_2O	4·70	TiO_2	0·15		
CaO	0·83	MnO	0·11		

Thus the molar ratio $SiO_2/Al_2O_3 = 9·4$ and the glass was highly siliceous.

Subsequently, using a second rhyolitic ash from the western United States having $SiO_2/Al_2O_3 = 9·3$, the zeolitization processes were reinvestigated[114] with contact solutions of caustic soda to which various amounts of salts (NaF or NaCl) had been added. This enables one to vary the ratio cation/OH^- as desired but also introduces an additional anion. For values of cation/$OH^- > 1$, crystallization fields were obtained at 60, 90 and 120°C, in which the weight ratio glass/water was the ordinate and the OH^- concentration (with different ratios anion/OH^-) the abscissa. When this latter ratio was unity zeolites obtained were those given below:

NaOH aq	(NaOH + NaF) aq	(NaOH + NaCl) aq
Gismondine type Na-Pl (90, 120°C)	Na-Pl (60°C)	Na-Pl (60, 90°C)
Analcime (120°C)	Analcime (120°C)	Sodalite hydrate (120°C)
Sodalite hydrate (120°C)	Erionite (120°C)	Erionite (120°C)
Phillipsite (90, 120°C)	Phillipsite (90, 120°C)	Phillipsite (60, 90, 120°C)
Mordenite (120°C)	Chabazite (90°C)	Faujasite (60, 90°C)
Faujasite (60°C)	Faujasite (60, 90°C)	

The additions of salts modified not only the approximate boundaries of the crystallization fields but also changed certain of the zeolite types which formed.

In an earlier study[115] of zeolite formation from the rhyolitic pumice having $SiO_2/Al_2O_3 = 9.4$, using aqueous caustic soda as contact solution a zeolite Na-V was obtained (isotypes are Linde N,[62] Z-21[93] and (Na, TMA)-V[11]). However, the ratio SiO_2/Al_2O_3 of the Na-V was only ~2 and so the siliceous natural glass was later replaced as the solid reactant by synthetic glasses[116] having the compositions (in molar proportions):

	(a)	(b)
Na_2O	0·98	0·98
Al_2O_3	1·00	1·00
SiO_2	1·00	2·10

From glass (a) Na-V, Linde zeolite A and sodalite hydrate (I) were formed, and from (b) Na-V, zeolite A, faujasite (X) and sodalite hydrate were obtained. Some of the experiments yielding Na-V from glass (b) under conditions aiming to optimize the yield as against competing species are given in Table 19. Na-V is cubic, with a very large unit cell ($a \sim 37$ Å) which may vary a little according to the exchange cation and the Si/Al ratio.[94] It is a narrow-pore zeolite which does not sorb even permanent gases at low temperatures. It undergoes ion exchange reactions freely and has an intracrystalline pore volume of about 0·35 cm³ per cm³ of crystal, similar to that of sodalite hydrate. Like the latter[117] it might serve as a good trap for room temperature storage of permanent or inert gases like Ar and Kr, sorbed into it at high pressure and temperature. No organic cation template was required for its syntheses from glass, unlike its isotypes Linde N and (Na, TMA)-V. As Z-21 it was also made from alkaline aluminosilicate hydrogels in absence of the organic template tetramethylammonium (TMA).

TABLE 19

Some conditions giving cubic zeolite Na-V from glass (b)[116]

Temperature (°C)	Ratio Solid/liquid	% NaOH in contact solution	Time (hours)	Zeolites formed[a]
80	1/5	28	24	V + (I)
80	1/20	28	24	V + I
60	1/10	28	24	V + (I)
60	1/20	22	24	V + I + (A)
60	1/40	23	24	V + I
50	1/5	20	48	X + (V) + (I)
50	1/10	26	48	V
50	1/10	27	48	V
50	1/20	29	48	V
50	1/30	28	48	V + (I)
45	1/10	28	95	V
45	1/5	27	120	V + (I)
45	1/10	28	120	V + (I)

[a] A = Linde type A; X = Linde type X; I = sodalite hydrate. Parentheses indicate minor yield.

Synthetic and semi-synthetic glasses have also served to give a number of other zeolites. Compositions of some synthetic glasses used for this purpose are given in Fig. 16.[113] Thus the a to f series had $Na_2O/Al_2O_3 = 1$ with different ratios SiO_2/Al_2O_3 while the series d_1 to d_4 had $SiO_2/Al_2O_3 = 4$ but different ratios Na_2O/Al_2O_3. Crystallization fields of the a to f series are indicated in the top of Fig. 17 for 36 hour runs at 80 and 120°C. The ratio of solid to liquid was 1/20. Similarly the bottom parts of Fig. 17 show crystallization fields for the series d_1 to d_4 at 80°C after 12 and 36 hours. In the diagram A = Linde zeolite type A, B = amalcime, P = gismondine type Na-P and X = faujasite. U as before denotes amorphous. The solid/liquid ratio was maintained at 1/20. The overlap in crystallization fields is also considerable, as in Fig. 15. The gismondine type zeolite P tended to be dominant and at 120°C faujasite was not seen although zeolite A still formed at low ratios SiO_2/Al_2O_3. Increased alkali content in the glass favoured the appearance of faujasite, the yields of which were influenced not only by the alkali content of the glass but also by the concentration of NaOH in the contact solutions, and by the reaction time. The yields of faujasite at a given time passed through maxima when plotted against the concentration of caustic soda, but the maxima for d_2, d_3 and d_4 glasses were between 80% and 100%.

Leucite fused with sodium carbonate has been the source of semi-synthetic glasses having the compositions given below, in molar proportions:

Na$_2$O	K$_2$O	Al$_2$O$_3$	SiO$_2$	Symbol
0·97	0·98	1	3·98	l$_1$
1·98	0·97	1	4·05	l$_2$
2·95	0·94	1	4·03	l$_3$
3·94	0·95	1	4·02	l$_4$

These glasses were then used in an attempt to optimise the yields of faujasite.[118] The crystallization fields after 36 hours at 65°C with solid/liquid ratios as ordinate and % NaOH in the contact solutions as abscissa showed areas in which faujasite was the only product. Faujasite prepared also from the rhyolitic glass having SiO$_2$/Al$_2$O$_3$ – 9·4, referred to earlier, showed very good sorption of toluene, cyclohexane and n-pentane, and good catalytic cracking properties towards n-cetane.[115] Natural glasses usually contain iron which remains associated with the crystalline products. It was found by examining layers of a bed sedimented from suspension in a column that the amount of Fe$_2$O$_3$ varied with depth of the layer, thus showing that at least part of the iron oxide was physically mixed with the zeolite rather than combined with it.[115]

Investigations have also been made using aqueous potash at 60–140°C as mineralizing solution. The same rhyolitic glass with SiO$_2$/Al$_2$O$_3$ = 9·4 was employed as solid powdered reactant and yielded the edingtonite type zeolite, K-F, chabazite, K-G, the hexagonal zeolite K-I of Table 12, Linde zeolite type L, the near-phillipsite K-M and a gmelinite type

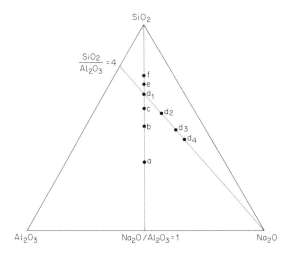

Fig. 16. Molecular compositions of two series of glasses (a to f and d$_1$ to d$_4$) used in zeolitization studies.[113]

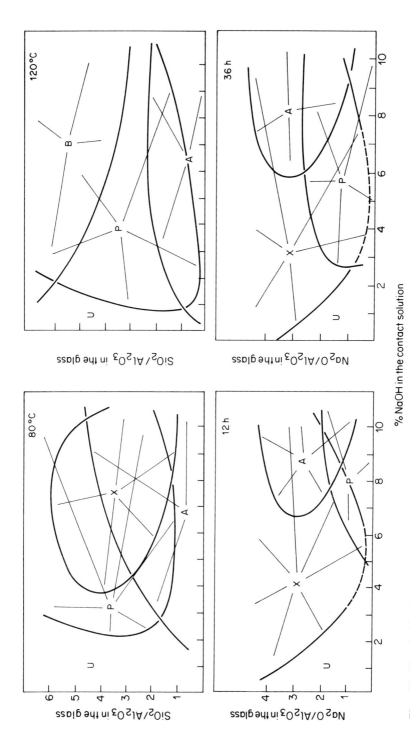

Fig. 17. *Top half:* Formation areas of products from glasses a to f of Fig. 16 with solid/liquid ratios 1/20, and NaOH solutions of varied concentration, at 80°C and 120°C.

Bottom half: areas of formation of products after 12 and 36 hours from glasses d_1 to d_4 of Fig. 16 with solid/liquid ratio 1/20 using NaOH solutions of varied concentration at 80°C.[113]

zeolite, K-S. In addition a cubic phase considered to be a hydrogrossularite, K-T, and an unidentified phase, K-Q, were reported. Crystallization fields were mapped as with the sodic media.[119] In a further example, a rhyolitic glass having $SiO_2/Al_2O_3 = 10\cdot4$ was zeolitized with 2 M mixed Na_2CO_3 and K_2CO_3 solution, principally in the range 130–150°C.[120] At pressures of 1 kbar in 13- to 15-day runs, clinoptilolite was obtained as a single phase when in the contact solution $1\cdot5 \leqslant K/Na \leqslant 0\cdot67$. In solutions richer in K, phillipsite became dominant; in those richer in Na, mordenite became the major phase together with less clinoptilolite and minor phillipsite. Above 150°C with $K/Na = 1$ mordenite was the important product, while in K-rich solutions below 130°C phillipsite formed preterentially.

In natural zeolite occurrences zeolitization often occurs in a temperature profile, with zoning of the zeolites formed in preferred temperature ranges (Chapter 1 §2). In addition to providing some confirmation of the above work of Hawkins *et al.*,[120] experiments of Kirov *et al.*[121] are of considerable interest because they zeolitized volcanic glasses under conditions simulating the above temperature zoning. To give a steady thermal gradient and eliminate convection, the top part of a long autoclave was heated giving the gradient shown in Fig. 18. The autoclave contained a column of the fragmented glass with contact solutions of 1 N and of 0·2 N NaOH or its mixtures with KOH filling the interstices, and a ratio by weight of solid/liquid = 1·3. The runs lasted, in various experiments, between 9 and 288 hours, and after reaction the column was removed and divided into nine sections, corresponding to the different temperatures, as indicated on the ordinate of Fig. 18. The glassy phases used included a perlite having a molar ratio $SiO_2/Al_2O_3 = 9\cdot6$ and a basaltic glass in which SiO_2/Al_2O_3 was 5·6. When the temperature range covered in the profile was 220 down to 110°C, the basaltic glass yielded analcime giving way to the gismondine type zeolite Na-P as temperature along the profile decreased. For the perlite, the sequence for descending temperatures tended to be: analcime, mordenite, clinoptilolite and phillipsite, with some overlap. Clinoptilolite yield was maximized between 180 and 135°C when $Na/(Na+K)$ was 0·5 or 0·2. Qualitatively the results follow the same pattern as those of Hawkins *et al.*[120]

In large-scale applications, zeolites need to be pelleted with an appropriate binder since they are prepared only as fine powders. In one investigation, however, mordenite was prepared in pebble form.[122] Powdered pumice and aqueous water glass were mixed and reaction occurred with constant rocking of the autoclave to yield crystalline pebble-like masses. Thus it may be possible to combine zeolite synthesis and bonding in certain instances.

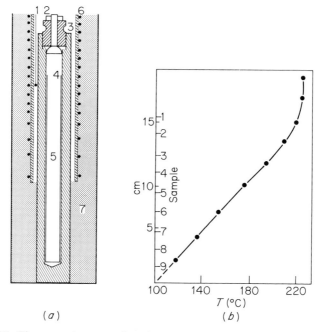

Fig. 18. The autoclave used and temperature gradient involved in the zoning experiment of Kirov *et al.*[121] In (*a*), 1 = thermocouple, 2 = autoclave seal; 3 = autoclave; 4 = copper cylinder; 5 = starting material; 6 = heater; and 7 = thermal insulation. In (*b*), the temperature distribution along the autoclave and the heights at which the phases crystallized were investigated.

The examples summarized show that many of the zeolites prepared from gels, layer silicates or metakaolinite can equally well be made from natural or synthetic glasses. Although zeolite formation using alkalis other than NaOH or KOH has been less explored for glasses there is, from the results available, reason to believe that glasses are potentially just as good sources of zeolites as are gels and clay minerals.

7. Syntheses from Other Tectosilicates

Non-zeolite tectosilicates can co-crystallize with zeolites (Tables 1, 3, 5, 6 and 18), but they can also be converted into zeolites. Some examples of these processes are summarized in Table 20[123] in which various reactions yielding analcime are given and also transformations of analcime to other species. The table shows that analcime can result from transformation of nepheline, adularia and microcline. Since 1959 many additional syntheses of analcime have been reported, but the table shows adequately how readily alkaline sodic contact solutions yield this zeolite from varied starting materials. Some further examples are referred to below or have already been mentioned in earlier sections.

From nepheline ($Na_2O.Al_2O_3.2SiO_2$) and a nepheline-rich volcanic rock (nephelinite) six zeolites have been made (analcime, wairakite, natrolite, phillipsite, chabazite and thomsonite). The zeolitization was performed under conditions which, by periodically changing the mineralizing solutions, sought to imitate the open reaction systems found in Nature.[136] Reaction temperatures were between 100 and 250°C while the contact solutions were 0.01 equiv dm^{-3} NaOH, KOH, NaCl, $CaCl_2$, HCl or H_2SO_4. Only alkaline solutions readily yielded zeolites with the possible exception of analcime. This zeolite was an initial product even with dilute HCl or H_2SO_4, probably because the nepheline provided an adequate reservoir of alkali. However, over longer periods the analcime was replaced by the layer silicates illite and kaolinite and by boehmite. The nearly neutral NaCl solutions or pure water also initially gave analcime, replaced in part, at longer times, by illite. Dilute NaOH and nepheline gave analcime (and some natrolite at 150°C). With nephelinite at 100°C a chabazite type zeolite was reported which was replaced in part by phillipsite and analcime at 150°C and fully by analcime at 200 and 250°C. Dilute KOH and nepheline gave phillipsite at 150°C which was replaced by analcime and leucite at 250°C. The dilute $CaCl_2$ and nepheline yielded analcime and thomsonite at 150°C, and at 250°C gave the same zeolites together with wairakite, hexagonal anorthite and illite. Although much less study has been devoted to the formation of zeolites from non-zeolite tectosilicates than from gels, clay minerals and glasses, the examples given suggest that many zeolites could be made from felspars and felspathoids under hydrothermal or even low-temperature conditions using appropriate alkaline contact solutions.

8. Some Zeolite Transformations

Like other silicates zeolites can themselves be transformed into different species, both zeolite and non-zeolite. Table 20 illustrates the versatility of analcime in the hydrothermal production of felspars, felspathoids, other zeolites, clay minerals and even quartz and hieratite (K_2SiF_6). The zeolites formed from analcime were cancrinite and sodalite (or nosean) hydrates, Li-A(BW),[1] salt-bearing ZK-5 (as Species P and Q) and

TABLE 20
Formation and transformations of analcime[123]

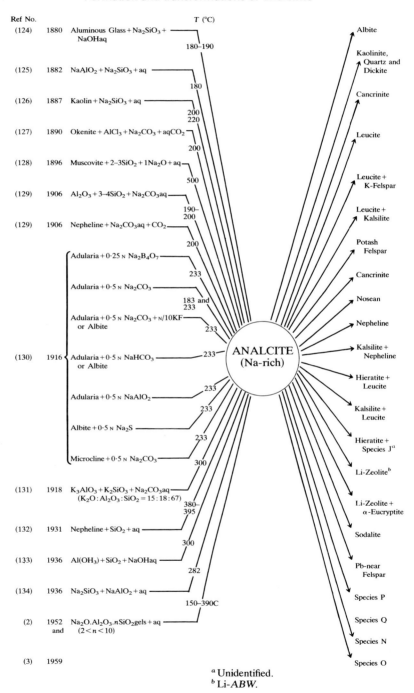

Ref No.			T (°C)

(124) 1880 Aluminous Glass + Na₂SiO₃ + NaOHaq — 180–190

(125) 1882 NaAlO₂ + Na₂SiO₃ + aq — 180

(126) 1887 Kaolin + Na₂SiO₃ + aq — 200 220

(127) 1890 Okenite + AlCl₃ + Na₂CO₃ + aqCO₂ — 200

(128) 1896 Muscovite + 2–3SiO₂ + 1Na₂O + aq — 500

(129) 1906 Al₂O₃ + 3–4SiO₂ + Na₂CO₃aq — 190–200

(129) 1906 Nepheline + Na₂CO₃aq + CO₂ — 200

(130) 1916
- Adularia + 0·25 N Na₂B₄O₇ — 233
- Adularia + 0·5 N Na₂CO₃ — 183 and 233
- Adularia + 0·5 N Na₂CO₃ + N/10KF or Albite — 233
- Adularia + 0·5 N NaHCO₃ or Albite — 233
- Adularia + 0·5 N NaAlO₂ — 233
- Albite + 0·5 N Na₂S — 233
- Microcline + 0·5 N Na₂CO₃ — 300

(131) 1918 K₃AlO₃ + K₂SiO₃ + Na₂CO₃aq (K₂O : Al₂O₃ : SiO₂ = 15 : 18 : 67) — 380–395

(132) 1931 Nepheline + SiO₂ + aq — 300

(133) 1936 Al(OH)₃ + SiO₂ + NaOHaq — 282

(134) 1936 Na₂SiO₃ + NaAlO₂ + aq — 150–390C

(2) 1952 and Na₂O.Al₂O₃.nSiO₂gels + aq (2 < n < 10)

(3) 1959

ANALCITE (Na-rich)

Albite

Kaolinite, Quartz and Dickite

Cancrinite

Leucite

Leucite + K-Felspar

Leucite + Kalsilite

Potash Felspar

Cancrinite

Nosean

Nepheline

Kalsilite + Nepheline

Hieratite + Leucite

Kalsilite + Leucite

Hieratite + Species Jᵃ

Li-Zeoliteᵇ

Li-Zeolite + α-Eucryptite

Sodalite

Pb-near Felspar

Species P

Species Q

Species N

Species O

ᵃ Unidentified.
ᵇ Li-*ABW*.

3 days at 395°C with SiO_2 + aq	1931	(132)
Aqueous CO_2 at 300°C	1941	(135)

4 days at 260° to 410°C with 2 N Na_2CO_3; 2 days at 360°C with excess Na_2SeO_4 + NaOH aq; or 4 days at 450°C with 8 N Na_2CO_3

1 day at 280°C with 5 N K_2CO_3; 2 days with excess KNO_3 at 250°C; or 3 days with excess K_2SO_4 at 400°C

1 day at 360°C with 5 N K_2CO_3

2 days at 220°C with excess K_2CO_3 + KOH

16 hr at 195°C with sat. K_2CO_3 + Na_2CO_3

2 to 4 days at 330° to 450°C with 3 N NaOH

4 days at 450°C with 5 N NaOH

2 days at 450°C with NaOH,KOH + aq

2 days at 450°C with KOH,NaOH + aq

2 days at 250°C with KF aq   1953   (104)

2 days at 350°C with KF aq

14 h–1 day at 110° to 200°C with excess (K,Na)F + aq

$2\frac{1}{2}$–4 days at 220° to 270°C with excess LiCl or $LiNO_3$ + aq

6 days at 310°C with excess $LiNO_3$ + aq

3–4 days at 400° to 450°C with excess NaCl or Na_2S + aq

7 days at 400°C with excess $PbCl_2$ + aq

2 days at 225° to 450°C with excess $BaCl_2$ + aq

2–3 days at 200° to 450°C with excess $BaBr_2$ + aq

4–2 days at 360° to 450°C with excess KCl + aq

4–1 days at 200° to 450°C with excess KBr + aq

salt-bearing edingtonite type zeolites (Species N and O). The salts displace water from within the frameworks for obvious spatial reasons.

The investigation of salt-bearing forms of zeolites with the ZK-5 and the edingtonite framework topologies was extended by Barrer and Marcilly,[105] who transformed analcime, chabazite, faujasite (as zeolites X and Y), felspar, pyrophyllite and kaolinite (with added silica) to Species P and Q (i.e. to salt-bearing forms of zeolite ZK-5); and also transformed analcime, sodalite, faujasite (as zeolite X) and Linde zeolite type A to the salt-bearing forms of the edingtonite type zeolite K-F. Instances of successful reactions are given in Table 21. The specific role of the salts in favouring the ZK-5 topology ($BaCl_2$ and $BaBr_2$) or that of edingtonite (KCl, KBr and KI) is evident, although in experiments not listed in Table 21 other phases co-crystallized. The transformation of analcime to Species P and Q followed the sigmoid form of reaction kinetics discussed in Chapter 4 §4, which is typical of nucleation followed by crystal growth. The ZK-5 structures termed P' and Q' in Table 21 which were grown from zeolite X showed some variations in the X-ray pattern and had a larger cubic unit cell, possibly in part because of their preparation from a zeolite with a lower ratio SiO_2/Al_2O_3 (cf. Chapter 7, Table 8). The cubic unit cell edges of some of the preparations were:

		$a/Å$
P	(from analcime)	$18·59_3$
P	(from zeolite Y)	$18·62_2$
Q	(from analcime)	$18·63_8$
Q	(from zeolite Y)	$18·68_6$
P'	(from zeolite X)	$18·93_6$
Q'	(from zeolite X)	$18·94_3$
P	(from analcime, $BaCl_2$ extracted)	$18·86_2$
ZK-5		$18·72_0$

The kind of salt (chloride or bromide) and the relative proportions of water and salt within the lattice should also contribute to changes in the unit cell dimensions.

Natural heulandite and clinoptilolite have been re-crystallized to analcime in three-week runs at 100°C using 0·1 M NaOH and 0·1 or 0·01 M Na_2CO_3 as contact solutions,[137] and analcime was also formed from mordenite if the reaction medium from which the mordenite was initially grown was too highly alkaline.[138,139] Clinoptilolite and mordenite have been recrystallized to faujasite with contact solutions of caustic soda at 100°C, often with the addition of salts.[140] The conversion to faujasite of natural clinoptilolites of sedimentary origin was reported later[141] to be limited to Hector clinoptilolite and indeed to depend upon the chloride

TABLE 21

Examples of syntheses[105] of salt-bearing forms[a] (Species P and Q) having the topology of zeolite ZK-5 and of the edingtonite type zeolite K-F

Parent silicate	Parent silicate (g)	salt (g)	H_2O (cm^3)	Time (days)	Temp. (°C)	Products
Analcime	2	10 BaCl$_2$	7	2	230	v. good P
	2	13 BaBr$_2$	7	2	230	v. good Q
Chabazite	0.5	10 BaCl$_2$	6	4	250	good P
	0.5	13 BaBr$_2$	6	4	250	good Q
Felspar	0.5	6 BaCl$_2$	6	4	400	v. good P, trace of felspar
	0.5	8·2 BaBr$_2$	6	4	400	v. good Q, trace of felspar
Faujasite (Y)	0.5	6 BaCl$_2$	6	4	250	v. good P
Faujasite (X)	0.5	8·2 BaBr$_2$	6	4	250	v. good Q
	0.5	6 BaCl$_2$	6	4	250	P′, traces of Ba-P*
	0.5	8·2 BaBr$_2$	6	4	250	Q′, traces of Ba-P*
Pyrophyllite	0·72	6 BaCl$_2$ + 0·63 Ba(OH)$_2$	6	4	400	P, with pyrophyllite
	0·72	8·2 BaBr$_2$ + 0·63 Ba(OH)$_2$	6	4	230	traces of Q
Kaolinite + 0·84 cm^3 Syton 2X^b	0·43	6 BaCl$_2$ + 0·525 Ba(OH)$_2$	6	4	230	good P
Kaolinite + 0·84 cm^3 Syton 2X^b	0·43	8·2 BaBr$_2$ + 0·525 Ba(OH)$_2$	6	4	230	moderate Q
Analcime	1	6 KCl	6	4	400	Leucite, traces K-F(Cl) and sanidine
	1	10 KBr	6	4	400	Leucite, traces K-F(Br) and sanidine
	1	14 KI	6	4	400	Leucite, traces K-F(I) and sanidine
Faujasite (X)	0·5	6 KCl	6	4	260	K-F(Cl)
	0·5	10 KBr	6	4	260	K-F(Br)
	0·5	14 KI	6	4	300	K-F(I)
Linde A	0·5	6 KCl	6	4	260	K-F(Cl), kaliophilite
	0·5	10 KBr	6	4	240	K-F(Br), kaliophilite

^{a} Ba-P* is the hexagonal polymorph of the Ba-felspar, celsian; P′ and Q′ are aluminous varieties of P and Q, all with the ZK-5 topology (Chapter 7, Table 8).
^{b} Syton-2X is a colloidal silica with 30 wt % of SiO_2.

impurity content and the Na^+ and K^+ contents of the material. High ratios of Na^+ relative to K^+ favoured transformation to faujasite but this conversion was not complete.

Zeolites have been converted to layer silicates[142] and silicates[143] under hydrothermal conditions. Erionite, mordenite and analcime were first decationated as far as possible with $0 \cdot 1$ M HCl at 25°C and then the products were heated for 10 days with $0 \cdot 1$ M HCl at 175 and 230°C, with additions of $AlCl_3$ sufficient to bring the overall Al/Si ratio to unity. Boehmite was formed in addition to kaolinite.[142] When clinoptilolite was digested hydrothermally and for different times with lime at temperatures from 90 to 280°C various products were obtained according to temperature, time and the ratio $CaO/(Al_2O_3 + SiO_2)$ in the mixtures.[143] When this ratio was $0 \cdot 8$ calcium silicate hydrate and finally 11 Å tobermorite, $Ca_5(OH)_2[Si_6O_{16}]4H_2O$, were formed at 90°C, while at 170 and 280°C only tobermorite resulted. With $CaO/(Al_2O_3 + SiO_2) = 1$ the products were tobermorite at 140, 180 and 280°C, together with xonotlite, $Ca_6(OH)_2[Si_6O_{17}]$ at the highest temperature. When the ratio was 2 the products were α-dicalcium silicate hydrate at 140 and 180°C and tricalcium silicate hydrate at 280°C, in each case with minor amounts of hydrogarnet.

In the dry way zeolites can also be heated and thereby converted into other tectosilicates. Because heating drives off zeolitic water which is a stabilizer of the open zeolite structures (Chapter 2 §4) one expects sintering reactions of zeolites to yield more compact tectosilicates. The nature of the resultant tectosilicate is in part dependent upon the SiO_2/Al_2O_3 ratio of the parent compound and very much upon the cation present. Because various nearly homoionic ion-exchanged forms of the zeolite can be produced before sintering, a range of tectosilicates can be made in the subsequent sintering process. This is illustrated in Table 22,[144] for a series of chabazite type zeolites with different SiO_2/Al_2O_3 ratios and different exchange cations. The influence of the SiO_2/Al_2O_3 ratio can be seen for example in the formation of nosean (variant of sodalite) and nepheline which takes place only from the two most aluminous Na-chabazites. Likewise, kaliophilite appears as final product of sintering the two most aluminous and leucite appears on sintering the two most siliceous K-chabazites. The role of the cations is equally evident: the Li-chabazites, whatever the SiO_2/Al_2O_3 ratio, always yielded β-eucryptite; Na-chabazites yielded quartz and nepheline; K-chabazites gave leucite and kaliophilite; and Ca-chabazites gave quartz and Ca-felspar. Thus the kind of tectosilicate frameworks which formed were largely specific to the cations present.

TABLE 22

Products derived by sintering exchange forms of chabazite type zeolites[144]

SiO_2/Al_2O_3	Cation form	Products and heating temperatures (°C)
5·05	Li	Li-Ch[a] (200, 400 and 600°C), quartz (850°C), β-eucryptite (1070°C)
	Na	Na-Ch (200, 400 and 600°C), modified Na-Ch (820°C), quartz (1070°C)
	K	Modified Ch (200, 500 and 700°C), unidentified (1000°C), leucite (1100°C)
	NH₄	Modified Ch (240, 410°C), H-Ch (560 and 650°C), quartz (800 and 1095°C)
	Ca	Ca-Ch (240 and 500°C), modified Ca-Ch (810°C), quartz (1090°C)
4·15	Li	Li-Ch (200°C), modified Li-Ch (550°C), mainly amorphous (820°C), β-eucryptite 1070°C)
	Na	Na-Ch (160, 500, 700°C), amorphous (920 and 1070°C)
	K	Modified K-Ch (200 and 550°C), modified K-Ch and amorphous (700°C), amorphous (970°C), leucite (1100°C)
	Ca	Ca-Ch (200°C), modified Ca-Ch (500 and 700°C), amorphous (940°C), Ca-felspar (1090°C)
2·65	Li	Modified Li-Ch (200°C), modified Li-Ch and amorphous (550°C), amorphous (550 and 820°C), β-eucryptite (1060°C)
	Na	Na-Ch (160°C), modified Na-Ch and some nosean (550 and 700°C), nosean (940°C), nepheline (1080°C)
	K	Modified K-Ch (200°C), modified K-Ch and amorphous (500 and 700°C), amorphous (980°C), kaliophilite (1080°C)
	Ca	Ca-Ch (200°C), modified Ca-Ch and some hauyne (550°C), amorphous (940°C), Ca-felspar (1090°C)
2·30	Li	Li-Ch (200°C), amorphous (550°C), β-eucryptite (900 and 1080°C)
	Na	Na-Ch (170°C), nosean (300, 500 and 840°C), α-carnegeite (950°C), nepheline (1080°C)
	K	modified K-Ch (200°C), largely amorphous (500°C), amorphous (700 and 950°C), kaliophilite (1090°C)
	Ca	modified Ca-Ch (200°C), hauyne (500 and 820°C) amorphous (900°C), Ca-felspar (1080°C)

[a] Ch denotes the parent chabazite type phase.

Because felspars and felspathoids can yield zeolites, the products of sintering reactions such as those in Table 22 become in their turn potential sources of other zeolites, as in the case of nepheline (§7), and so can be intermediates in the conversion of chabazite to other zeolites. Such a method supplements the direct transformations illustrated earlier.

9. Concluding Remarks

From §§7 and 8 and from other results in this chapter and in Chapters 2 and 4, chemical pathways are seen to exist allowing one to transform one mineral species to another and readily to reverse this process. The examples given show how one may go from felspars or felspathoids to zeolites, from zeolites to clay minerals from clay minerals to felspars, felspathoids or zeolites and from zeolites to felspathoids or felspars. Indeed, forward and reverse processes seem to be available for each combination of pairs of mineral types. Moreover, with proper attention to conditions it is frequently possible to make the forward and reverse chemical pathways result in pure or almost pure preparations. There is a noteworthy versatility in this area of chemistry, dependent upon hydrothermal operating conditions.

Under given zeolitization conditions the thermodynamic instability of the reactant mixture could vary appreciably as between gels, glasses, metakaolinite, clay minerals and non-zeolite tectosilicates. Thus in a given alkaline contact solution the concentrations of dissolved species derived from these reactants could differ, in this way influencing the chances of nucleation of potentially competing reaction products. It is therefore of interest that, although differences certainly do emerge among the kinds of resultants using different starting materials of similar gross composition, these differences are offset by many broad similarities. Such similarities tend to support the view, favoured by other evidence given in Chapter 4, that nucleation frequently involves dissolved species and that growth upon nuclei also involves deposition of dissolved species. The alternative view that nucleation and growth takes place by rearrangement and ordering of Si–O–Si and Al–O–Si bonds within gels[145,146] is not tenable within the compact frameworks of glasses, metakaolinite, or non-zeolite tectosilicates at low temperatures. According to this alternative, gels should behave very differently from glasses or metakaolinite, which the foregoing account indicates is on the whole not the case. Instead compact phases undergo gelatinization and solution through attack by the aqueous alkalis as a prelude to crystallization as zeolites.

References

1. R.M. Barrer and E.A.D. White, *J. Chem. Soc.* (1951) 1269.
2. R.M. Barrer E.A.D. White, *J. Chem. Soc.* (1952) 1561.
3. R.M. Barrer, J.W. Baynham, F.W. Bultitude and W.M. Meier, *J. Chem. Soc.* (1959) 195.
4. R.M. Barrer and J.W. Baynham, *J. Chem. Soc.* (1956) 2882.
5. R.M. Barrer and N. McCallum, *J. Chem. Soc.* (1953) 4029.
6. R.M. Barrer and P.J. Denny, *J. Chem. Soc.* (1961) 971.
7. R.M. Barrer and P.J. Denny, *J. Chem. Soc.* (1961) 983.
8. R.M. Barrer and D.J. Marshall, *J. Chem. Soc.* (1964) 485.
9. R.M. Barrer and D.J. Marshall, *J. Chem. Soc.* (1964) 2296.
10. R.M. Barrer and J.A. Lee, unpublished measurements.
11. R.M. Barrer and D.E. Mainwaring, *J. Chem. Soc. Dalton* (1972) 2534.
12. B. Stringham, *Econ. Geol.* (1952) **47,** 661.
13. R.M. Barrer, *Chemistry in Britain* (1966) **2,** 380.
14. R.M. Barrer and J.D. Falconer, *Proc. Roy. Soc.* (1956) **A236,** 227.
15. R.M. Barrer and L.V.C. Rees, *Trans. Faraday Soc.* (1960) **56,** 709.
16. R.M. Barrer, *J. Phys. Chem. Solids* (1960) **16,** 84.
17. G.T. Kokotailo and W.M. Meier, *in* "Properties and Applications of Zeolites" (Ed. R.P. Townsend), p. 133. Special Pub. No. 33, The Chemical Society, London, 1980.
18. R.M. Barrer, P.J. Denny and E.M. Flanigen, U.S.P. 3,306,922, 1967.
19. Ch. Baerlocher and W.M. Meier, *Helv. Chim. Acta* (1970) **53,** 1285.
20. A.J. Regis, L.B. Sand, C. Calmon and M.E. Gilwood, *J. Phys. Chem.,* (1960) **64,** 1567.
21. A. Erdem and L.B. Sand, *in* "Proceedings of the 5th International Conference on Zeolites" (Ed. L.V.C. Rees) p. 64. Heyden, London, 1980.
22. R.M. Barrer, J.F. Cole and H. Sticher, *J. Chem. Soc. A* (1968) 2475.
23. R.M. Barrer and J.F. Cole, *J. Chem. Soc. A* (1970) 1516.
24. S. Ueda and M. Koizumi, *Amer. Mineral.* (1979) **64,** 172.
25. E.E. Senderov, *Geochem.* (1963) No. 9, 848.
26. L. Ames, *Amer. Mineral.* (1963) **48,** 1374.
27. D.B. Hawkins, *Mat. Res. Bull.* (1967) **2,** 951.
28. D.J. Drysdale, *Amer. Mineral.* (1971) **56,** 1718.
29. C.L. Kibby, A.J. Perrotta and F.E. Massoth, *J. Catalysis* (1974) **35,** 256.
30. B.H.C. Winquist, 1976, U.S.P. 3,933,974.
30a. Y. Goto, *Amer. Mineral.* (1977) **62,** 330.
31. D.E.W. Vaughan and G.C. Edwards, 1978, U.S.P., 4,088,739.
32. R.J. Argauer and G.R. Landolt, 1972, U.S.P. 3,702,886.
33. F.G. Dwyer and E.E. Jenkins, 1976, U.S.P. 3,941,871.
34. R.W. Grose and E.M. Flanigen, 1977, U.S.P. 4,061,724.
35. E.M. Flanigen, J.M. Bennett, R.W. Grose, J.P. Cohen, R.L. Patton, R.M. Kirchner and J.V. Smith, *Nature* (1978) **271,** 512.
36. Mobil Oil Corp., 1971, Dutch Pat. 7,014,807.
37. P. Chu, 1973, U.S.P. 3,709,979.
38. G.T. Kokotailo, P. Chu, S.L. Lawton and W.M. Meier, *Nature* (1978) **275,** 119.
39. D.M. Bibby, N.B. Milestone and L.P. Aldridge, *Nature* (1979) **280,** 664.
40. L. Maros, J. Stabenow and M. Schwarzmann, 1980, German Pat. 2,830,787.

41. M.K. Rubin, E.J. Rosinki and C.J. Plank, 1979, U.S.P. 4,151,189.
42. R.W. Grose and E.M. Flanigen, 1977, Belgian Pat. 7,701,115.
43. D.M. Bibby, N.B. Milestone and L.P. Aldridge, *Nature* (1980) **285,** 30.
44. T.V. Whittam and B. Youll, 1977, U.S.P. 4,060,590.
45. M.S. Spencer and T.V. Whittam, *Acta Phys. Chem.* (1978) **24,** 307.
46. T.V. Whittam, 1978, German Pat. Appl., 2,748,276.
47. M.S. Spencer and T.V. Whittam, *in* "Properties and Applications of Zeolites" (Ed. R.P. Townsend), p. 342. Special Pub. No. 33, The Chemical Society, London.
47a. H.E. Robson, Pittsburgh Catalysis Society Symp., 28th May 1981.
48. S.P. Zdhanov and M.E. Ovsepyan, *Dok. Akad. Nauk, S.S.R.* (1964) **157,** 913.
49. H.J. Bosmans, E. Tambuyzer, J. Paenhuys, L. Ylen and J. Vancluysen, *in* "Molecular Sieves" (Ed. W.M. Meier and J.B.Uytterhoeven), p. 179. American Chemical Society, Advances in Chemistry Series, No. 121, 1973.
50. E. Passiglia, D. Pongiluppi and R. Rinaldi, *Neues Jb. Mineral. Mh* (1977) **8,** 355.
51. J.D. Sherman, *in* "Molecular Sieves II" (Ed. J. R. Katzer) p. 30. American Chemical Society Symposium Series, No. 40, 1977.
52. R.M. Barrer and B. Munday, *J. Chem. Soc. A* (1971) 2914.
53. R.M. Barrer and D.E. Mainwaring, *J. Chem. Soc. Dalton* (1972) 1254.
54. M.E. Ovsepyan and S.P. Zdhanov, *in* "Zeolites. Their Synthesis, Properties and Applications" (Ed. M.M. Dubinin and T.G. Plachenov) pp. 2–10 through 2–15. Transl. by International Information, Inc., 1964.
55. A.M. Taylor and R. Roy, *Amer. Mineral.* (1964) **49,** 656.
56. R. Aiello and E. Franco, *Rend. Accad. Sci. Fis. Mat., Naples* (1968) **35,** 1.
57. H. Takahashi and Y. Nishimura, *Nippon Kagaku Zasshi* (1967) **88,** 528; (1968) **89,** 378; and (1970) **91,** 23.
58. R.M. Barrer and J.W. Baynham, 1961, U.S.P. 2,972,516.
59. R.M. Milton, 1961, U.S.P. 2,996, 358.
60. R.M. Barrer and J.W. Baynham, 1962, U.S.P. 3,056,654.
61. R.M. Milton, 1961, U.S.P. 3,010,789.
62. D.W. Breck, 1965, U.S.P. 3,216,789.
63. R.M. Milton, 1961, U.S.P. 3,012,853.
64. D.W. Breck and N.A. Acara, 1961, U.S.P. 2,971,151.
65. D.W. Breck and N.A. Acara, 1961, U.S.P. 3,011,869.
66. D.W. Breck and N.A. Acara, 1961, U.S.P. 2,995,423.
67. A. Pereyron, J.-L. Guth and R. Wey, *Comp. Rend.* (1971) **272,** 1331.
68. A. Pereyron, J.-L. Guth and R. Wey, *Compt. Rend* (1971) **272,** 181.
69. H. Borer and W.M. Meier, *in* "Molecular Sieve Zeolites II" (Ed. W.M. Meier and J.B. Uytterhoeven), p. 122. American Chemical Society, Advances in Chemistry Series, No. 101, 1971.
70. W. Sieber and W.M. Meier, *Helv. Chim. Acta* (1974) **57,** 1533.
71. R.M. Barrer and W. Sieber, *J. Chem. Soc. Dalton* (1977) 1020.
72. R.M. Barrer and W. Sieber, *J. Chem. Soc. Dalton* (1978) 598.
73. W.M. Meier and G. Kokotailo, *Zeit. Krist.* (1972) **135,** 374.
74. Ch. Baerlocher and R.M. Barrer, *Zeit. Krist.* (1974) **140,** 10.
75. R. Aiello and R.M. Barrer, *J. Chem. Soc. A* (1970) 1470.
76. M.K. Rubin, 1969, German Pat. 1,806,154.
77. Ch. Baerlocher and W.M. Meier, *Helv. Chim. Acta* (1969) **52,** 1853.
78. G.T. Kerr, *Inorg. Chem.* (1966) **5,** 1539.

79. J. Ciric, 1968, U.S.P. 3,411,874.
80. D.W. Breck, W.G. Eversole and R.M. Milton, *J. Amer. Chem. Soc.* (1956) **78,** 2338.
81. R.M. Barrer and D.E. Mainwaring, *J. Chem. Soc. Dalton* (1972) 1259.
82. R.M. Barrer, R. Beaumont and C. Colella, *J. Chem. Soc. Dalton* (1974) 934.
83. Ch. Baerlocher and W.M. Meier, *Zeit. Krist.* (1972) **135,** 339.
84. E.T. Carlson and L.S. Wells, *Bur. Stds, J. Res.* (1948) **41,** 103.
85. W.L. Haden and F.J. Dzieranowski, 1964, U.S.P. 3,123,441.
86. G.T. Kerr, 1966, U.S.P. 3,247,195.
87. J. Ciric, 1967, French Pat. 1,502,289.
88. J. Ciric, 1968, U.S.P. 3,415,736.
89. G. Kokotailo and J. Ciric, *in* "Molecular Sieve Zeolites I" (Ed. R.F. Gould), p. 109. American Chemical Society, Advances in Chemistry Series, No. 101, 1971.
90. N.A. Acara, 1968, U.S.P. 3,414,602.
91. G.T. Kerr, 1969, U.S.P. 3,459,676.
91a. W.M. Meier and M. Groner, *J. Solid State Chem* (1981) **37,** 204.
92. E.E. Jenkins, 1971, U.S.P. 3,578,398.
93. H.C. Duecker, A. Weiss and C.R. Guerra, 1971, U.S.P. 3,567,372.
94. R.M. Barrer and R. Beaumont, *J. Chem. Soc. Dalton* (1974) 405.
95. H.E. Robson, 1972, U.S.P. 3,674,425.
96. M.K. Rubin and E.J. Rosinski, 1972, U.S.P. 3,699,139.
97. J. Ciric, 1972, U.S.P. 3,692,470.
97a. H.E. Robson, D.P. Shoemaker, R.A. Ogilvie and P.C. Manor, *in* "Molecular Sieves" (Ed. W.M. Meier and J.B. Uytterhoeven), p. 106. American Chemical Society, Advances in Chemistry Series, No. 121, 1973.
98. H.E. Robson, 1973, U.S.P. 3,720,753.
98a. W.M. Meier and G. Kokotailo, *Zeit. Krist.* (1965) **121,** 211.
99. C.J. Plank, J. Rosinski and M.K. Rubin, 1977, U.S.P. 4,046,859.
100. D.E.W. Vaughan, 1978, U.S.P. 4,091,079.
101. M.K. Rubin, E.J. Rosinski and C.J. Plank, 1978, U.S.P. 4,086,186.
102. M.K. Rubin, C.J. Plank and E.J. Rosinski, 1980, Eur. Pat. Appl. 13630.
102a. R.M. Barrer, S. Barri and J. Klinowski, *in* "Proceedings of the 5th International Conference on Zeolites" (Ed. L.V.C. Rees), p. 20. Heyden, London, 1980.
103. Ch. Baerlocher and R.M. Barrer, *Zeit. Krist.* (1972) **136,** 245.
104. R.M. Barrer, L. Hinds and E.A. White, *J. Chem. Soc.* (1953) 1466.
105. R.M. Barrer and C. Marcilly, *J. Chem. Soc. A* (1971) 2914.
106. W.M. Meier and D.H. Olson, *in* "Molecular Sieve Zeolites I" (Ed. R.F. Gould), p. 155. American Chemical Society, Advances in Chemistry Series, No. 101, 1971.
107. W.H. Taylor, *Min. Mag.* (1935) **24,** 208.
108. Ch. Baerlocher, private communication.
109. E. Franco and R. Aiello, *Rend. Soc. Ital. Mineral. Petrolog.* (1968) **24,** 3.
110. I. Fernandez, J.L.M. Vivaldi and A. Pozzuoli, *Boletin Geol. Minero.* (1974) **84–85,** 442.
111. R. Sersale and R. Aiello, *in* "Chimica delle Alte Temperature e delle Alte Pressioni". Consiglio Nazionale delle Richerche, Rome, 1967.
112. D. Breck, "Zeolite Molecular Sieves", p. 110. Wiley, New York 1974.
113. R. Aiello, C. Colella and R. Sersale, *in* "Molecular Sieve Zeolites I" (Ed.

R.F. Gould), p. 151. American Chemical Society, Advances in Chemistry Series, No. 101, 1971.
114. R. Aiello, C. Colella, D.G. Casey and L.B.Sand, *in* "Proceedings of the 5th International Conference on Zeolites" (Ed. L.V.C. Rees), p. 49. Heyden, London, 1980.
115. C. Colella and R. Aiello, *Annali di Chimica* (1971) **61,** 721.
116. R. Aiello and C. Colella, *Chimica e L'Industria* (1973) **55,** 692.
117. R.M. Barrer and D.E.W. Vaughan, *J. Phys. Chem. Solids* (1971) **32,** 731.
118. C. Colella and R. Aiello, *Annali di chimica* (1970) **60,** 587.
119. C. Colella and R. Aiello, *Rend. Accad. Sci. Fis. e Mat. Soc. Nazionale di Sci., Lettre ed Arti, Naples* (1971) **38,** 243.
120. D. B. Hawkins, R. A. Sheppard and A. J. Gude, 3rd, *in* "Natural Zeolites" (Ed. L.B. Sand and F.A. Mumpton), p. 337. Pergamon Press, Oxford, 1978.
121. G.N. Kirov, V. Pechigargov and E. Landzheva, *Chem. Geol.* (1979) **26,** 17.
122. I.M. Keen, W.J. King and R. Walls, *Nature* (1968) **217,** 1045.
123. R.M. Barrer *in* "Transactions of the 7th International Ceramics Congress", p. 379. Clowes and Sons, London, 1960.
124. A. de Schulten, *Bull. Soc. Fr. Mineral.* (1880) **3,** 150.
125. A. de Schulten, *Bull. Soc. Fr. Mineral.* (1882) **5,** 7.
126. J. Lemberg, *Zeit. Deut. Geol. Ges.* (1887) **39,** 559.
127. C. Doelter, *Neues Jb. Mineral.* (1890) **1,** 118.
128. G. Friedel, *Bull. Soc. Fr. Mineral.* (1896) **19,** 5.
129. C. Doelter, *Tsch. Min. Petr. Mitt.* (1906) **25,** 79.
130. E.A. Stephenson, *J. Geol.* (1916) **24,** 180.
131. J. Konigsberger and W.J. Muller, *Zeit. Anorg. Chem.* (1918) **104,** 1.
132. C.J. v. Nieuwenberg and H.B. Blumendahl, *Rec. Trav. Chim. Pays-Bas* (1931) **50,** 989.
133. W. Noll, *Chemie der Erde* (1936) **10,** 129.
134. F. G. Straub, *Ind. Eng. Chem.* (1936) **28,** 113.
135. F. H. Norton, *Amer. Mineral.* (1941) **26,** 1.
136. H. Höller and U. Wirsching, *in* "Proceedings of the 5th International Conference on Zeolites", (Ed. L.V.C. Rees), p. 164. Heyden, London, 1980.
136a. W.M. Meier and G. Kokotailo, *Z. Krist.* (1965) **121,** 211.
137. J.R. Boles, *Amer. Mineral.* (1971) **56,** 1724.
138. R.M. Barrer, *J. Chem. Soc.* (1948) 2158.
139. D. Dominé and J. Quobex, *in* "Molecular Sieves", p. 78. Society of Chemical Industry, London, 1968.
140. H.E. Robson, 1973, U.S.P. 3,733,390.
141. H.E. Robson, K.L. Riley and D.D. Maness *in* "Molecular Sieves II" (Ed. J.R. Katzer), p. 233. American Chemical Society Symposium Series, No. 40, 1977.
142. C.R. de Kimpe and J.J. Fripiat, *Amer. Mineral.* (1968) **53,** 216.
143. T. Mitsuda, *Mineral. J.* (1970) **6,** 143.
144. R.M. Barrer and D.A. Langley, *J. Chem. Soc.* (1958) 3811.
145. B.D. McNicol, G.T. Pott, K.R. Loos and N. Mulder, *in* "Molecular Sieves" (Eds. W.M. Meier and J.B. Uytterhoeven), p. 152. American Chemical Society, Advances in Chemistry Series, No. 121, 1973.
146. D.W. Breck and E.M. Flanigen, *in* "Molecular Sieves", p. 47. Society of Chemical Industry, London, 1968.

Chapter 6

Isomorphous Replacements in the Frameworks of Zeolites and Other Tectosilicates

1. Introduction

Isomorphous replacements involving tectosilicates can include the following chemical processes:

(i)   Cation exchange. This can usually be performed directly.
(ii) Replacements in the anionic frameworks of Si by such elements as Al, Ga, Ge, Be, B, Fe, Cr, P, Mg effected if at all only during synthesis.
(iii) Replacement of framework oxygen.
(iv) In zeolites and some felspathoids, isomorphous replacements of intracrystalline salts or molecular water by other species.

The study of cation exchange in tectosilicates forms an extensive subject best considered separately but briefly described in Chapter 1 §6. Salt-bearing tectosilicates will be discussed in the next chapter so that the present account will be confined to anionic framework substitutions. Aluminosilicate tectosilicates are frequently non-stoichiometric in that simple numerical ratios between Al and Si or between cations such as Na^+ and K^+ are not always maintained. Electrical neutrality is of course preserved with the result that the number of equivalents of charge-balancing cations equals the number of Al atoms in the anionic framework. Each Al^{III} replacing Si^{IV} introduces a negative charge on the framework which would otherwise, as in quartz, carry no nett charge. Also, because each TO_4 tetrahedron ($T = Al$ or Si) shares each of its apical oxygens with one of four other tetrahedra one must always have $(Al + Si)/O = 1/2$.

251

2. Replacement of Si by Al

The number of stably occupiable cation sites has an important bearing on the replacement of Si by Al. Thus in the felspar anorthite, $CaAl_2Si_2O_8$, all stable cation sites are occupied by Ca^{2+} ions. It is therefore impossible to replace each Ca^{2+} by $2Na^+$ to give $Na_2Al_2Si_2O_8$ because there would then be twice as many cations as there were sites for them to occupy. Accordingly, isomorphous substitutions by appropriate synthesis would have to be of the type

$$Ca, Al \rightleftarrows Na, Si$$

This ensures that the number of cations never exceeds the number of sites available for stable occupation. The end member of such substitution is the felspar albite, $NaAlSi_3O_8$. There is the further possibility of miscibility gaps between the end members of an isomorphous substitution series, but this does not alter the requirement that over the accessible range of substitution replacement must be of the foregoing type.

In zeolites, on the other hand, there are normally more, and sometimes many more, cation sites available than the number of cations needed for neutralizing the negative charge on the anionic framework, so that isomorphous substitution can occur in ways that allow the total number of cations to vary. Framework substitutions can therefore be of types

$$Ca, Al \rightleftarrows Na, Si$$
$$Na, Al \rightleftarrows Si$$

The greater flexibility of this situation sometimes means that one may have large variations in the relative proportions of Al to Si according to the conditions of formation of a particular zeolite. Some examples of this are given in Table 1(a). The lower limits tend to the value 2·0, which is that expected according to Lowenstein's[8] Ål–O–Ål avoidance rule. This rule has recently been questioned, as described later (§4), but where valid it requires that no AlO_4 tetrahedron shares an apical oxygen with another AlO_4 tetrahedron. It is of interest that in some samples of Linde zeolite A, values of the ratio SiO_2/Al_2O_3 less than 2 (e.g. 1·85) have been reported but were attributed to intercalated non-framework $NaAlO_2$.[9] It may be possible by direct synthesis to push the upper limits of the examples given in Table 1(a) to still more silica-rich phases. The upper limit given in Table 1(a) for chabazite is from a naturally occurring sample[2] and the lower limit for a synthetic material;[1] upper limits for the other compounds are for synthetically prepared zeolites, often employing special methods of synthesis, some of which have already been referred to in Chapters 4 and 5. Thus the silica-rich forms of zeolite A and of sodalite were obtained by replacing some or all of the sodium in the alkaline reaction mixtures by tetramethylammonium and in the case of zeolite A also introducing phosphate to complex Al (§8). As noted in Chapter 4 §5.4 there is not sufficient room in the intracrystalline pore space of zeolite frameworks to accommodate many of the large organic cations and thus, if the zeolite forms, the anionic framework must have low negative charge, so requiring few cations to neutralize this charge and being silica-rich. In the 14-hedral cavities of sodalite there is room for just one tetramethylammonium ion per cavity and in accordance with this steric requirement the sodalite grown from sodium-free alkaline tetra-methylammonium aluminosilicate hydrogel had the silica-rich composition of Table 1(a). When grown from alkaline (Na, K)-aluminosilicate hydrogels zeolite L is rather siliceous.[6] In order to make aluminous forms of this zeolite the Na and K were replaced by Ba and the aluminous

TABLE 1(a)
Variations in ratio SiO_2/Al_2O_3 in some zeolites

| Compound | SiO_2/Al_2O_3 | | Ref. |
	Lower limit	Upper limit	
Chabazite	~2·16	~7·8	(1, 2)
Faujasite	~2·2	~6·8	(3)
Linde zeolite A	~2·0	~6·8	(4)
Linde zeolite L	~2·08	~7·0	(5, 6)
Sodalite	~2·0	~10·0	(7)

TABLE 1(b)
SiO_2/Al_2O_3 ratios of some very siliceous synthetic zeolites[14a]

Zeolite	Approx. range in ratios SiO_2/Al_2O_3
Beta	30–75
ZSM-5	25–1000
ZSM-11	78–1000
ZSM-12	45–160
ZSM-23	55–217
ZSM-48	870–1340
Nu-1	40–120
Fu-1	20–40
Zeta-1	25–32
Zeta-3	60–74

variety of the zeolite could then be made,[5] with the low SiO_2/Al_2O_3 ratio of the table. The double charge on the Ba^{2+} ion is critical in this case in allowing the L structure to form with a larger anionic framework charge. Thus the nature of the cations can be critical in determining the extent of isomorphous replacement of Si by Al.

The silica-rich faujasite of Table 1(a) was obtained by a classical procedure in which zeolite Y was deposited upon seeds of the more aluminous variety zeolite X (Chapter 4, Table 1), using as source material alkaline hydrogel with SiO_2/Al_2O_3 as high as 28. The rate of deposition upon the seeds occurred more slowly the greater the ratio SiO_2/Al_2O_3 in the faujasite (Chapter 4, Fig. 6), thus setting a practical limit to this method of enriching faujasite in silica. Factors which play a part in determining the silica/alumina ratio in the crystals are this ratio in the reactant mixture and the concentration and excess amount of the alkali. The influence of these factors in the crystallization of chabazite type zeolites (K-G), and of the edingtonite type zeolite, K-F, was shown in Figure 12 of Chapter 4. For a given concentration of KOH, the larger the ratio SiO_2/Al_2O_3 in the initial hydrogel, the larger its value in the crystalline products; and for a given ratio SiO_2/Al_2O_3 in the parent mixture the greater the concentration of KOH the less siliceous the crystals, until SiO_2/Al_2O_3 eventually approached the Lowenstein limit of 2. For the value 2 if the rule is valid there would be ordering of Al and Si, each occupying tetrahedral sites alternately. In zeolites rich in 5-rings (e.g. the mordenite and heulandite groups), alternation of Si and Al may not be possible and so if the Lowenstein rule were applicable the ratio SiO_2/Al_2O_3 could then not reach the lower limit of 2. Thus the question of the validity or otherwise of the rule can be important in indicating the possible most aluminous compositions. From Figure 6 of Chapter 4 when $SiO_2/Al_2O_3 = 2$ in the initial mixture this same ratio was preserved in the crystals of K-F formed from it.

Changes in the values of the ratios SiO_2/Al_2O_3 when parent hydrogels crystallized are illustrated in Fig. 1[10] for edingtonite type K-F, chabazites K-G, phillipsite-like K-M (which may in fact be merlinoite[11]) and the felspathoid kalsilite, K-N. The crystals of K-F and K-N had SiO_2/Al_2O_3 very near or at the Lowenstein limit of 2. Thus, starting with gel compositions having $SiO_2/Al_2O_3 = 1$, the crystals were *more* siliceous than the gel, and with gels having $SiO_2/Al_2O_3 > 2$ the crystals were *less* siliceous than the gel. The K-M crystals had values of SiO_2/Al_2O_3 grouped about the value 3, the spread being much less than in the parent gels. The chabazites were all less siliceous than the parent gels but showed a considerable spread in the ratios SiO_2/Al_2O_3. Thus Fig. 1 shows three kinds of behaviour which may be considered as typical. Even with zeolites as rich in silica as the ZSM-5 series the crystals in one study were somewhat less siliceous than the parent mixtures,[12] the behaviour being like that of the chabazites in Fig. 1. For analcimes grown at 100°C from alkaline sodium aluminosilicate solutions very low in alumina (Chapter 3, §2.5), the ratios SiO_2/Al_2O_3 in the solutions were from 40 to 320, with Na_2O/Al_2O_3 from 70 to 200.[13] The analcimes were much richer in Al_2O_3 than the reactant compositions, having SiO_2/Al_2O_3 between 3·24 and 4·39 and thus being grouped about the value 4. The increase in the proportion of Al_2O_3 after crystallization is similar to the behaviour of the chabazites K-G (Fig. 1), while the grouping for analcimes of values of SiO_2/Al_2O_3 around a particular value recalls the behaviour of K-M, where the preferred value was 3 (Fig. 1).

Zdhanov,[14] for a series of zeolites grown from alkaline Na- and K-hydrogels, illustrated the role of excess alkali upon the zeolites formed and upon the Si/Al ratios in the crystals. The abscissa was the ratio $(Na_2O - Al_2O_3)/SiO_2$ in the reaction mixture (or this ratio with Na_2O replaced by K_2O) and the correlation shown in Fig. 2 was then obtained.

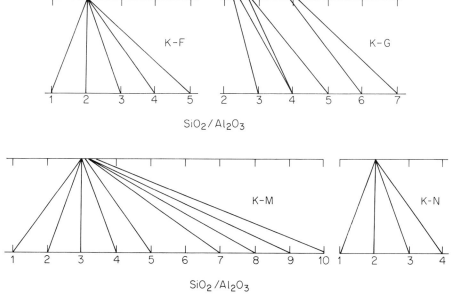

Fig. 1. Composition changes when alkaline aqueous potassium aluminosilicate gels were crystallized.[10] Horizontal scales give the ratios SiO_2/Al_2O_3 in parent gels (lower) and in products (upper).

K-F = edingtonite type zeolite   K-M = phillipsite type zeolite
K-G = chabazite type zeolite   K-N = felspathoid kalsilite

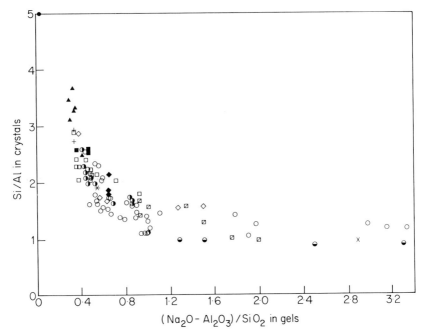

Fig. 2. Si/Al ratios in various zeolite crystals correlated with excess alkali, expressed as the ratio $(Na_2O–Al_2O_3)/SiO_2$ in the parent gels.[14]

● = mordenite	+ = zeolite L
○, ◖ = faujasite	◇ = Na-P
☐ = Na-chabazite	◆ = K-M
▨ = K-chabazite	◓ = Na-A
■ = K, Na-chabazite	* = analcime
△, ▲ = erionite	× = K-F

Zeolites with low values of Si/Al were favoured by large excesses of Na_2O (or K_2O) over Al_2O_3 (faujasite, zeolite A, edingtonite type K-F, chabazite type K-G), sometimes reaching the Lowenstein limit Si/Al = 1. As the ratio $[(K, Na)_2O – Al_2O_3]/SiO_2$ decreased, the zeolites became progressively more siliceous. This ratio decreases either when the SiO_2 content is made larger or when the excess of alkali over alumina is made smaller. Both these effects are therefore involved in the figure.

Robson[14a] has listed a number of zeolites extremely rich in silica and with the approximate composition ranges given in Table 1(b). Not only do the ratios SiO_2/Al_2O_3 often vary greatly, but also the values of these ratios and the conditions for their synthesis, involving specific organic template bases, were considered to place them in a different category from the zeolites made under extremely alkaline conditions, usually with inorganic bases, which (Table 1(a)) do not exceed $SiO_2/Al_2O_3 \simeq 10$, and can have these ratios as low as 2. Even the siliceous natural zeolites mordenite, dachiardite and clinoptilolite do not significantly exceed $SiO_2/Al_2O_3 \simeq 10$. For the zeolites of Table 1(b), on the other hand, it is difficult to obtain them with SiO_2/Al_2O_3 below ~ 25. One zeolite which appears able to bridge the gap is ferrierite for which the following ratios have been reported for the naturally occurring zeolite and its synthetic variants:[14a]

Type	Range in SiO_2/Al_2O_3
Natural ferrierite	7·6–14
Synthetic Sr-ferrierite	7·4–17
Synthetic Li-ferrierite	7·4–16
ZSM-38	12–31
ZSM-35	14–37
Synthetic Na-ferrierite	22–69

In addition, svetlozarite has $SiO_2/Al_2O_3 \simeq 12$,[14b] but its synthesis has not been reported and it is not known within what limits this ratio can be varied.

The extreme of the replacement of Al by Si is represented by crystalline silicas iso-structural with given zeolites. Crystalline silicas exist in considerable variety, as shown in Table 2. The densest crystals (stishovite and coesite) form under high-pressure conditions only. Keatite appears as an intermediate stage in the hydrothermal conversion under autogenous pressure of amorphous silica gel to quartz, as noted in Chapter 2, §8.3. Iler[22] gave the following approximate temperature and pressure ranges for stability of the three phases:

	$T(°C)$	P (kbar)
Keatite	400–500	0·8–1·3
Coesite	300–1700	15–40
Stishovite	1200–1400	160

These materials and quartz are too dense to have any zeolitic intracrystalline porosity. This property first appears in cristobalite and tridymite which were shown to sorb helium and neon at high pressures and at temperatures from 293 K upwards.[23]

Melanophlogite, a rare naturally occurring crystalline silica, has not yet been synthesized. It has the structure of clathrate hydrate of type I[18] and as such has appreciable intracrystalline porosity. There is no known zeolite with this structure, but such a zeolite is possible. The melanophlogite lattice is obtained by stacking dodecahedra with 12 pentagonal faces and tetradecahedra with 12 pentagonal and 2 hexagonal faces in the ratio two dodecahedra to six tetradecahedra. The stacking involves sharing

TABLE 2
Densities and polymorphs of some crystalline silicas

Crystals	Densities (g cm^{-3})	Iso-structural with
Stishovite[15] (tetragonal)	4·3	
Coesite[16] (monoclinic)	2·93	
Quartz (hexagonal)	2·66	
Keatite[17] (tetragonal)	2·50	
Cristobalite (cubic)	2·32	Ice Ic
Tridymite (hexagonal)	2·28	Ice I
Melanophlogite[18] (cubic)	1·99–2·10 1·9 (calc. for calcined anhydrous crystal)	Clathrate hydrate type I
Silicalite 1[19] (orthorhombic)	1·99 ± 0·05 1·79 (calc. for calcined compound, from unit cell)	Zeolite ZSM-5
Silicalite 2[20] (tetragonal)	1·78 (calc. for calcined compound, from unit cell)	Zeolite ZSM-11
Faujasite silica[21] (cubic)	1·33 (calc. for calcined compound, from unit cell)	Faujasite

pentagonal faces. The dodecahedra and tetradecahedra are shown in Fig. 3,[24] together with the stacking of the dodecahedra which automatically creates the tetradecahedral cavities. Melanophlogite appears to contain intracrystalline SO_3 or H_2SO_4 which can be driven off by heating, and the optical properties suggest that organic matter is present in films between the crystals. The cubic unit cell edge is 13·4 Å as compared with 12 Å for clathrate hydrate of type I. The clathrate hydrate can during its formation include large amounts of various gases[25,26] so that empty melanophlogite may have comparable potential. It is stable to heating up to 900°C when it converts to cristobalite. There are various clathrate hydrates of known structure[24,27] in which the host lattices of water molecules are analogous to aluminosilicate frameworks in zeolites, but which, except for melanophlogite and zeolite ZSM-39, *MTN* (Chapter 1 §3), have no counterparts among tectosilicates. The clathrate host lattices are, like tectosilicate lattices, bonded by shared tetrahedra, this time of the type where the dashed lines denote hydrogen bonding to O and the full lines are for H covalently linked to O. The clathrate hydrates are therefore models for possible, so far unknown, zeolites or porous crystalline silicas.

The silicalites represent examples of crystalline silicas which can be made directly, and which are isostructural with known zeolites, silicalite $1^{(19)}$ with zeolite ZSM-5 and silicalite $2^{(20)}$ with ZSM-11. As noted in Chapter 2 §8.3, silicalite 1 is made from aqueous silica gel with tetra-propylammonium hydroxide. It can in addition be made$^{(14a)}$ using diamines $H_2N(CH_2)_5NH_2$ or $NH_2(CH_2)_6NH_2$ and also $NH_4OH +$ C_4H_9OH or $NH_4OH + C_2H_5OH$. Silicalite 2 can be prepared from the gel with tetrabutylammonium hydroxide, or diamines $H_2N(CH_2)_n NH_2$ where $n = 7$ or 12. The structures are known and the channel patterns are given in Figure 4.$^{(28)}$ The channels have free diameters of ~6 Å and the intracrystalline pore volumes are ~0·33 cm^3 per cm^3 of crystal. The silicalites are non-polar sorbents, stable to heat and acids.

The faujasite silica of Table 2 is the most open of the crystalline silicas in that ~50% of each cm^3 of crystal is available as pore space for molecules able to traverse the 12-ring windows with free diameters of ~8 Å. It is not prepared directly but can be made from zeolite Na-Y. Thus Beyer and Belenykaja$^{(21)}$ passed a stream of SiCl$_4$ vapour in

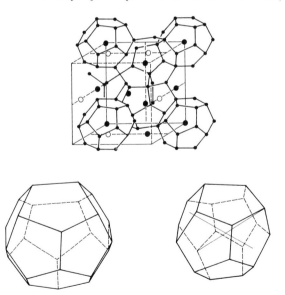

Fig. 3. The polyhedra (12- and 14-hedra) present in melanophlogite and in clathrate hydrate of Type 1, together with the mode of stacking of the 12-hedra by sharing pentagonal faces. This kind of stacking creates the 14-hedral cavities.

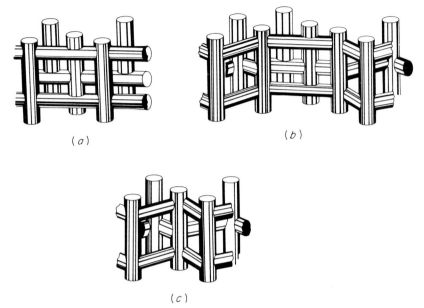

Fig. 4. Formal representation of channel patterns in pentasil zeolites.[28] (*a*) is for ZSM-11; (*c*) for ZSM-5; and (*b*) for one of the intermediate structures of which there may be many types.

nitrogen as carrier gas over outgassed Na-Y at sample temperatures rising at 4 K per minute to a final value in the range 730–830 K. The Al was removed from the framework and Si substituted:

$$Na_{52}[Al_{52}Si_{140}O_{384}] + 52SiCl_4 \rightarrow (SiO_2)_{192} + 52NaCl + 52AlCl_3$$

The $AlCl_3$ condensed outside the hot zone. Some $NaAlCl_4$ formed but it and the NaCl could be removed from the SiO_2 by washing in distilled water. The product was excellently crystalline and a subsequent study of another preparation using high-resolution magic angle sample spinning n.m.r. showed only Si combined through oxygen bridges with four other Si.[29] The product was thermally stable and is presumably a non-polar sorbent like the silicalites.

The above method was not successful with Na-X, zeolite L or mordenite,[21] and so may not be generally applicable. The zeolite must in any case be open enough to admit $SiCl_4$ and must not become choked with reaction products, a situation which could arise most easily with one-dimensional channel systems like those in mordenite and zeolite L. The failure of the aluminous form of faujasite, Na-X, to give the faujasite-silica may have been due to side-reactions involving sintering. In this connection, if Na-Y was first heated to temperatures over 750 K before exposing it to the $SiCl_4$ stream, a vigorous exothermal reaction took place which yielded an amorphous product.[21] It was suggested that Al removal occurred more rapidly than insertion of Si into the vacancies left by Al so that at high rates of de-alumination the vacancy concentration built up to values at which the lattice became unstable.

Other methods are available for removing Al from tectosilicates, often however without insertion of any Si in the vacancies, so that defect- and silica-rich products result. Murata[30] found that tectosilicates in acids behaved in a manner determined by the Si/Al ratios. For values of 1 up to nearly 2 the frameworks were dispersed. For ratios of 2 and upwards the frameworks were swollen but not dispersed. As the aluminosilicates became still richer in silica, the degree of swelling and breakdown became less and less marked. Barrer and Makki[31] found that clinoptilolite, where Si/Al $\sim$ 5, could be refluxed in strong hydrochloric acid with almost complete removal of Al to give a defect-containing silica framework which still gave an X-ray pattern not very different from that of clinoptilolite. It is also possible to remove much of the Al from mordenite or ferrierite by similar treatments with strong acid and likewise to enrich less siliceous zeolites like faujasite in silica by removal of Al using the weak ethylenediaminetetracetic acid (EDTA).[32] The initial process in these treatments is thought to be

$$\begin{bmatrix} O \quad \quad O \\ \quad Al^{\ominus} \\ O \quad \quad O \end{bmatrix} + M^+ + 4HCl \longrightarrow \begin{bmatrix} OH \quad HO \\ \\ OH \quad HO \end{bmatrix} + MCl + AlCl_3$$

Secondary reactions involving the "nests" of four hydroxyls may ensue. Especially on heating water may be lost:

$$\begin{bmatrix} Si-OH \quad HO-Si \\ Si-OH \quad HO-Si \end{bmatrix} \longrightarrow \begin{bmatrix} Si-O-Si \\ Si-OH \quad HO-Si \end{bmatrix} \longrightarrow \begin{bmatrix} Si-O-Si \\ Si-O-Si \end{bmatrix}$$
$$+ H_2O \qquad\qquad\qquad +2H_2O$$

If the proportion of Al atoms is low the resultant framework remains a similar if defective version of that of the parent structure. When such a de-aluminated cation-free product is heated sufficiently for it to lose all hydroxyl groups as water it becomes, like the silicalites, a non-polar hydrophobic sorbent.[33]

The formation of ultrastable forms[34] of H-Y by suitable heat-treatment in steam also liberates Al from the framework as oxy-aluminium cations or Al^{3+}, with shrinkage of the unit cell. Eventually a sharp, clear X-ray pattern of a faujasite lattice with $a \sim 24 \cdot 3$ Å is obtained, very much like that of the faujasite silica of Table 2, prepared by de-alumination of Na-Y by $SiCl_4$. It seems as though silica migrates, at the high temperature of the ultra-stabilization treatment in steam, to occupy vacancies in the framework left when the oxy-aluminium cations break away from it. The source of this silica is not clear: part of the structure could decompose to provide it as silica mobile in steam, or some detrital silica could already be present within the intracrystalline pore space. Some evidence of the first possibility is provided by a sorption study[35] in which micropore and secondary pore volumes were measured using several n-paraffins, benzene and nitrogen as sorbates, with US-Ex catalyst derived from Na-Y as sorbent. This material was 99% SiO_2, giving the X-ray powder pattern of faujasite. However, it contained secondary pores with radii in the range $1 \cdot 5 – 1 \cdot 9$ nm, while the micropore volume was reduced to $\sim 75\%$ of that in the parent Na-Y. The pore system was thus bi-disperse and this could be interpreted to mean that in the ultra-stabilization process about a quarter of the lattice had yielded a more open amorphous gel and supplied silica needed to heal lattice vacancies in the remainder of the framework which had formed during de-alumination.

The results described in this section and in §4 lead to groupings of 3-dimensional frameworks based on fully linked $(Al, Si)O_4$ tetrahedra according to their Si/Al ratios, following and extending the categories suggested by Robson:[14a]

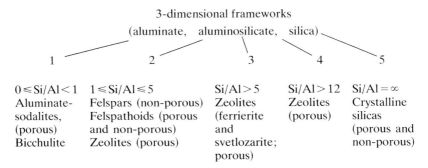

1	2	3	4	5
$0 \leqslant Si/Al < 1$	$1 \leqslant Si/Al \leqslant 5$	$Si/Al > 5$	$Si/Al > 12$	$Si/Al = \infty$
Aluminate-sodalites, (porous) Bicchulite	Felspars (non-porous) Felspathoids (porous and non-porous) Zeolites (porous)	Zeolites (ferrierite and svetlozarite; porous)	Zeolites (porous)	Crystalline silicas (porous and non-porous)

In these categories the terms "porous" and "non-porous" refer to scales of molecular dimensions, and to the framework only. These frameworks, as in the case of salt-bearing felspathoids (Chapter 7), can be so filled by the guest species as to make the crystals impermeable to other species.

Very open zeolite frameworks may be similarly blocked. The aluminate-sodalites in the first category are referred to in §4 in connection with Lowenstein's Ål–O–Ål avoidance rule. Zeolites are found in each of the categories 2, 3 and 4. In category 4, however, reference to Table 1(b) raises the question at what ratio Si/Al one should refer to the crystals as impure porous crystalline silicas rather than as zeolites. Such a distinction might be considered when there is less than one Al atom per unit cell. Thereafter, as the aluminium content becomes smaller and smaller, this aluminium can be increasingly regarded as an impurity in the silica end-member. Examination of the categories 1 to 5 given above indicates that Si/Al ratios, taking all frameworks into account, may range from 0 to ∞, but, as exemplified in Table 1(a), in any one kind of framework the range is restricted. This range and factors controlling it are considered in §3 in terms of a statistical thermodynamic model. There may tend to be a gap between Si/Al = 0 and Si/Al = 1 (category 1), although at least one intermediate composition is known. This is the mineral bicchulite of composition $Ca[Al_2SiO_6].Ca(OH)_2$, which has the sodalite structure with all the Al in the tectosilicate framework.[35a]

The silicalites and faujasite-silica of Table 2 provide non-polar molecular sieves which are virtually devoid of cations and hydroxyl groups. Provided no detrital silica is present within the intracrystalline channels, the sieve properties should be determined only by the framework geometries and so should be the same for all preparations. This aspect has not been investigated. Because of the absence of polar centres in the "clean" frameworks, electrostatic components of the sorption energies should be reduced to a minimum so that these porous silicas could have greater energetic homogeneity (i.e. smaller variation of energy of sorption with amount taken up) than their zeolite counterparts, rich in polar centres. Quantitative comparisons between a given porous silica and its zeolite counterpart (in terms of heats, entropies and free energies of sorption, selectivities and isotherm contours) have so far been very limited. Irrespective of such comparisons sorption studies on porous crystalline silicas are so far in no way comparable in depth or number with those made on a range of zeolites.[36]

3. A Statistical Thermodynamic Formulation of Isomorphous Replacement

On the basis of the examples given in §2 one may consider how far a statistical thermodynamic model can interpret and rationalize the observed ranges in Si/Al in different tectosilicates. Such a treatment has been developed[37] for replacements

$$M^+Al^{3+} \rightleftarrows Si^{4+}$$

where M^+ denotes an electrochemical equivalent of cations $A, B, \ldots, .$ In an aluminosilicate three kinds of framework bond are possible:

(i)   Si–O–Si

(ii)   Si–O–$\overset{\ominus}{Al}$

(iii) $\overset{\ominus}{Al}$–O–$\overset{\ominus}{Al}$

The reference state was chosen to be the Si–O–Si bond. If Si in (i) is replaced by Al to give (ii) the energy change was taken to be ϵ. Such a replacement involves in addition the uptake of a cation, say A^+, to maintain electrical neutrality, the binding energy of which is ϵ_A. If two Si atoms in (i) are replaced by two Al atoms to give (iii) in addition to the energy $2(\epsilon + \epsilon_A)$ an extra energy $2w/\nu$ per $\overset{\ominus}{Al}$–O–$\overset{\ominus}{Al}$ bond was assumed, where ν is the co-ordination number, 4, of a tetrahedral site, so that $2w/\nu = w/2$. This additional energy represents that for the re-arrangement reaction

$$2M^+[\overset{\ominus}{Al}\text{–O–Si}] \rightarrow M^+[\overset{\ominus}{Al}\text{–O–}\overset{\ominus}{Al}]M^+ + [\text{Si–O–Si}]$$

It was assumed that $w/2$ was pairwise additive for each $\overset{\ominus}{Al}$–O–$\overset{\ominus}{Al}$ bond formed. In systems obeying Lowenstein's $\overset{\ominus}{Al}$–O–$\overset{\ominus}{Al}$ avoidance rule, $w/2$ should be large and endothermal. However this is not a necessary assumption of the model which can allow for any values of $w/2$, positive or negative.

A further assumption of the model was that all tetrahedral sites were located on one sub-lattice and that all were occupied. The cations were likewise all located on one sub-lattice, but there could be more cation sites than were required for electrical neutrality, so that cation vacancies were allowed.

The Helmholtz free energy F of the crystal is given in terms of the partition function P by

$$F = -kT \ln P$$

and the partition function of the whole crystal is the product of those of the anionic framework, the intracrystalline cations, and the guest molecules, if any are present. For the framework the notation used was:

N_γ = total number of tetrahedral sites.

N_{Al}, N_{Si} = numbers of tetrahedral sites occupied by Al and Si respectively.

J_{Si} = partition function of an SiO_4 tetrahedron with all oxygens shared, i.e. of SiO_2.

J_{Al} = corresponding partition function of an AlO_4 tetrahedron with all oxygens shared, i.e. of $\overset{\ominus}{Al}O_2$, excluding extra terms due to $w/2$.

P_{Al}^{int} = contribution to the partition function arising from $w/2$.

The framework partition function can then be written as

$$P_f = \frac{N_\gamma!}{N_{Al}!\, N_{Si}!} \cdot J_{Si}^{N_{Si}} J_{Al}^{N_{Al}} P_{Al}^{int} \tag{2}$$

Likewise for the cations A, B, ... and with the notation:

N = total number of cation sites.

$N_A, N_B, \ldots$ = numbers of cation sites occupied respectively by cations A, B, ...

$J_A, J_B, \ldots$ = partition functions of cations A, B, ... in the crystal.

P_G = the partition function of the N_G guest molecules in the crystal, including configurational and bonding energy contributions.

The partition function $P_{c,G}$ for cations and guest molecules was then

$$P_{c,G} = \frac{N!}{(N - N_A - N_B - \cdots)!\, N_A!\, N_B! \ldots} J_A^{N_A} J_B^{N_B} \ldots P_G \tag{3}$$

The partition function $P = P_f P_{c,G}$; and the resulting expression for the Helmholtz free energy, when the only cations present are A, is then

$$\begin{aligned} F = -kT\{&N_\gamma \ln N_\gamma - N_{Al} \ln N_{Al} - N_{Si} \ln N_{Si} + N \ln N \\ &- (N - N_A) \ln (N - N_A) \\ &- N_A \ln N_A + N_{Si} \ln J_{Si} + N_{Al} \ln J_{Al} + N_A \ln J_A\} + F_G + F_{Al}^{int} \end{aligned} \tag{4}$$

where $F_G = -kT \ln P_G$ and F_{Al}^{int} is the contribution arising from the energy term $w/2$ through P_{Al}^{int}.

The contribution F_{Al}^{int} has an analogy in the adsorption of molecules upon adsorption sites when there is an extra pairwise additive interaction energy $w/2$ if two adsorbed molecules occupy adjacent sites. The empty sites correspond with tetrahedral sites occupied by Si; the Al atoms replacing them with energy ϵ correspond with the adsorbed molecules. This situation, first treated by Lacher,[38] gives in the present case

$$F_{Al}^{int} = 2kT \left[2N_{Al} \ln \frac{2X_{Si}}{D - 2X_{Al}} + N_\gamma \ln \frac{D - 2X_{Al}}{X_{Si}D} \right] \tag{5}$$

where

$$D = [1 - 4X_{Al}X_{Si}\alpha]^{1/2} + 1 \tag{6}$$

$$\alpha = 1 - \exp(-w/2kT) \tag{7}$$

The following relations arise partly from stoichiometry and the electrical neutrality conditions:

$$N_{Si}/N_\gamma = X_{Si}, \qquad Z_A N_A = N_{Al}$$

$$\left.\begin{array}{ll} N_{Al}/N_\gamma = X_{Al}, & \dfrac{N - N_A}{N_\gamma} = r_c - X_{Al}/Z_A \\[2mm] N/N_\gamma = r_c, & N_A/N_\gamma = X_{Al}/Z_A \end{array}\right\} \qquad (8)$$

where Z_A is the number of charges carried by each cation A.

If we now divide both sides of Eqn (4) by kTN_γ, and use the above relations, the result is

$$\frac{F}{kTN_\gamma} = X_{Al} \ln X_{Al} + X_{Si} \ln X_{Si} - r_c \ln r_c + \left(r_c - \frac{X_{Al}}{Z_A}\right) \ln \left(r_c - \frac{X_{Al}}{Z_A}\right)$$

$$+ \frac{X_{Al}}{Z_A} \ln \frac{X_{Al}}{Z_A} + X_{Al} \ln \frac{J_{Si}}{J_{Al} J_A^{1/Z_A}} - \ln J_{Si}$$

$$+ 2 \left[2 X_{Al} \ln \frac{2 X_{Si}}{D - 2 X_{Al}} + \ln \frac{D - 2 X_{Al}}{X_{Si} D} \right] + \frac{F_G}{kTN_\gamma} \qquad (9)$$

It is now assumed that the J, and F_G, are independent of X_{Al} (or if there are no guest molecules F_G is zero). Then, after introducing the constants t_1 and t_2, where

$$\left.\begin{array}{l} t_1 = \ln \dfrac{J_{Si}}{J_{Al} J_A^{1/Z_A}} \\[3mm] t_2 = \ln J_{Si} - F_G/kTN_\gamma \end{array}\right\} \qquad (10)$$

one obtains as a working equation

$$f(F) = F/kTN_\gamma + t_2 = X_{Al} \ln X_{Al} + X_{Si} \ln X_{Si} - r_c \ln r_c + \frac{X_{Al}}{Z_A} \ln \frac{X_{Al}}{Z_A}$$

$$+ \left(r_c - \frac{X_{Al}}{Z_A}\right) \ln \left(r_c - \frac{X_{Al}}{Z_A}\right) + X_{Al} t_1$$

$$+ 2 \left[2 X_{Al} \ln \frac{2 X_{Si}}{D - 2 X_{Al}} + \ln \frac{D - 2 X_{Al}}{X_{Si} D} \right] \qquad (11)$$

If N_γ denotes one Avogadro number of tetrahedral sites then $kTN_\gamma = RT$.

The behaviour of $f(F)$ plotted as a function of X depends on the values of the constants Z_A, r_c, t_1 and w/kT. The plots have the same slopes and positions of maxima or minima as plots of $F/N_\gamma kT$ against X_{Al}. $f(F)$ cannot be considered in the region in which $(r_c - X_{Al}/Z_A)$ is negative, because once all the cation sites are filled no further replacement of Si by Al is possible. $f(F)$ contains the term $X_{Al} t_1$ and involves, through the D, the term $\exp(-w/2kT)$. To obtain an estimate of the order of magnitude and sign of t_1 it was assumed that $J_{Si}/J_{Al} = \exp(-\epsilon/kT)$ and that cations behaved as isotropic three-dimensional harmonic oscillators, so that

$$J_A = \left[2 \sinh\left(\frac{h\omega}{2kT}\right)\right]^{-3} \exp(\epsilon_A/kT)$$

where ϵ_A is the binding energy of the cation A in the lattice. Then

$$t_1 = \frac{3}{Z^A} \ln\left[2 \sinh\left(\frac{h\omega}{2kT}\right)\right] - \frac{\epsilon_A}{Z_A kT} - \frac{\epsilon_{Al}}{kT} \tag{12}$$

For $\omega = 10^{13}\ s^{-1}$ and $T = 300$ K

$$t_1 = \frac{1 \cdot 7265}{Z_A} - 0 \cdot 401\left(\frac{\epsilon_A}{Z_A} + \epsilon_{Al}\right) \tag{13}$$

where ϵ_A and ϵ_{Al} are in kJ mol^{-1}. t_1 can thus have positive or negative values, according in the above example as $\epsilon_A + Z_A\epsilon_{Al} < 4 \cdot 306$ kJ mol^{-1} or is $> 4 \cdot 306$ kJ mol^{-1} respectively. One is now able to assign values of t_1, w/kT, Z_A and r_c in order to find how the contours of curves $f(F)$ plotted against X_{Al} behave.

Examples of these curves are given in Figures 5 and 6 and lead to the following generalizations:

(i) For negative t_1 and positive w/kT there is a minimum in $f(F)$ the position and depth of which depend on the magnitudes of t_1 and w/kT (Fig. 5(a) and (b)). For high values of w/kT this minimum becomes nearly V-shaped and occurs very close to the Lowenstein limit $X_{Al} = 0 \cdot 5$. Since chemical systems tend to produce resultants of the lowest free energy, tectosilicates with compositions at or close to the minima would be expected. The value $X_{Al} = 0 \cdot 5$ is usually observed with sodalites and cancrinites, and often with zeolite K-F (cf. Figs 1 and 2).

(ii) Changing Z_A from 1 to 2 did not alter the behaviour significantly (curves 1 and 2 of Fig. 5(b), while changing r_c from 1 to 5 increased the depth of the minimum considerably but not its position (curves 1 and 3 of Fig. 5(b)).

(iii) For positive t_1 and w/kT, the minima in curves of $f(F)$ against X_{Al} become shallower and move to lower values of X_{Al} (Fig. 5(c)). With $Z_A = 1$, $r_c = 1$ and $w/kT = 40$ the maxima occur as follows for several values of t_1:

$t_1 =$	2	4	6	8	10
$X_{Al}^{min} =$	0·18	0·10	0·043	0·017	0·0065

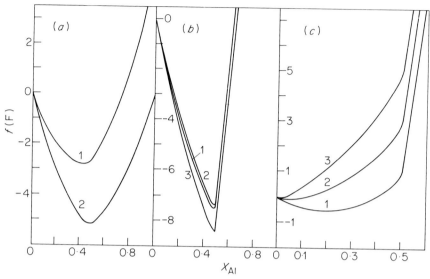

Fig. 5. Plots of $f(F)$ against $X_{al} = Al/(Al + Si)$.[37] (a) $Z_A = 1$; $r_c = 1$; $w/kT = 10$. For curve 1, $t_1 = -5$; for curve 2, $t_1 = -10$. (b) $t_1 = -15$; $w/kT = 40$. For curve 1, $Z_A = 1$ and $r_c = 1$; for curve 2, $Z_A = 2$ and $r_c = 1$; for curve 3, $Z_A = 1$ and $r_c = 5$. (c) $Z_A = 1$, $r_c = 1$, $w/kT = 40$. For curve 1, $t_1 = 2$; for curve 2, $t_1 = 6$; for curve 3, $t_1 = 10$.

Again, since the crystals formed will favour compositions close to the minima in $f(F)$ the model gives a possible criterion governing the formation of silica-rich zeolites (mordenite, ferrierite) right up to the pentasils ZSM-5 and -11 (Table 1(b)) and even in the limit their pure silica end members silicalites 1 and 2.

(iv) The behaviour with $r_c < 1$ is shown in Fig. 6(a) in which $Z_A = 1$, $t_1 = -15$ and $w/kT = 40$. Composition limits for $r_c = 0.167$, 0.25 and 0.5 occur respectively at $X_{Al} = 0.167$, 0.25 and 0.5, at which values all cation sites are filled and so the substitution $Si^{4+} \rightarrow Al^{3+}$, M^+ ceases. In these situations crystals of lowest free energy would have fixed compositions. This behaviour could represent that found with anorthite or celsian where $X_{Al} = 0.5$ and albite or orthoclase where $X = 0.25$.

(v) In Fig. 6(b) with $w/kT = 40$, $Z_A = 1$ and $r_c = 0.5$ (curve 1) or 1.0 (curve 2) the assignment of small negative values of t_1 gives curves with shallow flat minima. The same behaviour is found with small positive t_1 (Fig. 5(c), curve 1). These conditions in a given tectosilicate would help the crystals to form with a considerable spread in the

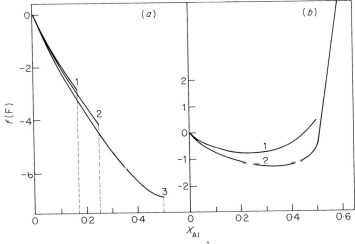

Fig. 6. Plots of $f(F)$ against X_{Al},[37] drawn for $Z_A = 1$ and $w/kT = 40$. (a) $t_1 = -15$. For curve 1, $r_c = 1/6$; for curve 2, $r_c = 1/4$; and for curve 3, $r_c = 1/2$. (b) For curve 1, $t_1 = -0.2$ and $r_c = 1/2$; for curve 2, $t_1 = -1.1$ and $r_c = 1$.

values of X_{Al} (or of the ratio Si/Al), as observed for example with ferrierites (§2), chabazites or faujasites (Table 1(a)).

It is therefore encouraging that one can model in a simple way a number of the situations arising in isomorphous substitutions of Al for Si in tectosilicates. More elaborate modelling might allow for more than one sub-lattice both for tetrahedral sites and for cation sites. One might also remove the assumption that F_G, the free energy contribution from guest molecules (if any are present), is independent of X_{Al} by introducing a specific partition function for the guests, allowing their bonding energy and number to vary as the crystals become richer in Al. However the above considerations would increase considerably the number of disposable parameters.

4. Lowenstein's $\overset{\ominus}{Al}$–O–$\overset{\ominus}{Al}$ Avoidance Rule and Al,Si Ordering

Aluminosilicate tectosilicates do not in general have ratios Si/Al < 1. Lowenstein[8] considered that this indicated that linking of two AlO_4 tetrahedra was energetically unfavourable. Such behaviour could be given a physical basis by assuming that the adjacent negative charges on the paired AlO_4 tetrahedra render the groupings

$$[\overset{\ominus}{Al}-O-\overset{\ominus}{Al}]+[Si-O-Si]$$

less stable than

$$2[\overset{\ominus}{A}-O-Si]$$

The limiting composition for tectosilicates with $Si/Al = 1$ would, if the rule were exactly valid, require strict alternation of Al and Si on tetrahedral sites throughout the framework, and hence a fully ordered framework. Thus Al, Si ordering on these sites can be linked with the Lowenstein rule. Also the rule could not be generally valid if ratios $Si/Al < 1$ were found in the framework of a tectosilicate. A ratio Si/Al somewhat less than unity was reported[9] for a preparation of Linde zeolite of type A, but as noted earlier it was considered that this was due to intracrystalline $NaAlO_2$. It is not unlikely that in the open structures of certain zeolites some detrital material (aluminate, silicate, base, or salts) may be included during synthesis. Indeed, with sodalites and cancrinites numerous salts can be entrained in large amounts[39,40] (Chapter 7).

One may therefore consider whether there are known instances where the $\overset{\ominus}{Al}-O-\overset{\ominus}{Al}$ avoidance rule is clearly invalid, based on chemical and structural evidence. One example is found in the case of the sodalite structure for which stable pure aluminate forms are known (cf. §2). The compound $Ca_6[Al_{12}O_{24}]2CaWO_4$ can be made, for example by sintering $CaCO_3$, $Al(OH)_3$ and WO_3 at 1350°C for 12 hours.[41] Also, the compound $Ca_6[Al_{12}O_{24}]2CaO$ can be obtained by heating calcium aluminate hydrate, $4CaO.3Al_2O_3.3H_2O$, at 650°C.[42] In both these compounds AlO_4 tetrahedra alone form the sodalite framework. Relevant crystallographic information is given below:

Sodalite, $Na_6[Al_6Si_6O_{24}]2NaCl$ cubic, $P\bar{4}3n$ $\qquad a = 8\cdot87\ \text{Å}$[43]
Aluminate sodalite,
$\quad Ca_6[Al_{12}O_{24}]2CaWO_4$ $\qquad$ tetragonal,† $P\bar{4}c2$ $\quad a = 26\cdot132\ \text{Å}; c = 18\cdot60\ \text{Å}$[41]
Aluminate sodalite,
$\quad Ca_6[Al_{12}O_{24}]2CaO$ $\qquad$ cubic, $I\bar{4}3m$ $\qquad a = 8\cdot85\ \text{Å}$[42]

Because of the high negative charge on the aluminate framework Ca^{2+} ions replace the Na^+ of ordinary sodalite with $Si/Al = 1$. Thus the sodalite aluminates compared with sodalite represent isomorphous replacements

$$Na, Si \rightarrow Ca, Al$$

together with isomorphous replacements of intercalated species:

$$NaCl \rightarrow CaWO_4$$

$$NaCl \rightarrow CaO$$

† In this instance the formula is not the unit cell content.

The mineral bicchulite, $Ca_8[Al_8Si_4O_{24}](OH)_8$, is another tectosilicate having the sodalite framework.[35a] The Al and Si atoms are distributed statistically on the tetrahedral sites, while the framework composition is such that Lowenstein's rule must be invalid. The aluminate-sodalites and bicchulite, as noted in §2, belong to the category 1 of 3-dimensional frameworks for which category $0 \leqslant Si/Al < 1$. They raise the question whether other synthetic phases of sodalite type with $0 \leqslant Si/Al < 1$ could be made, for example by sintering Na_2CO_3, $CaCO_3$, silica gel, $Al(OH)_3$ and WO_3 in appropriate molar proportions. If so, for all such phases the Lowenstein rule would not be valid.

Although not a three-dimensional network of AlO_4 tetrahedra the aluminate $Na_7Al_3O_8$ provides rather complex chains of tetrahedra[43a] in which rings of six tetrahedra alternate with rings of four tetrahedra to give infinite chains. Part of one such chain is shown below:

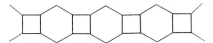

Also, the micaceous mineral xanthophyllite is reported to have an Si/Al ratio of about 1/3 on the tetrahedral sites,[43b] and like bicchulite has a statistical distribution of Si and Al on tetrahedral sites. Clearly, in certain structures which are three-dimensional networks, and in others which are not, the Ål–O–Ål avoidance rule is not maintained. It might be argued that the rule could become more applicable for aluminosilicates of category 2 of §2, i.e. those with $1 \leqslant Si/Al < 4$ to 5. Evidence on this point is already accumulating for this category, some of which is outlined below.

In Chapter 3 §2.2, reference was made to high-resolution n.m.r. spectroscopy of ^{29}Si in crystalline silicates, made possible by magic angle rapid spinning of the sample. The chemical shifts relative to $Si(CH_3)_4$ (Chapter 3, Table 4) should enable one to determine for a tectosilicate whether there are present groupings with Si linked through oxygen to 4, 3, 2, 1 or 0 other Si atoms (and hence to 0, 1, 2, 3 or 4Al). This is possible without ambiguity only where there is no overlap in the ^{29}Si chemical shifts for the above groupings. For a particular framework topology, such as that of faujasite, this appears to be the case. Thus in a particular study of faujasites with different Si/Al ratios[43f] the measurements gave the following results:

Second co-ord. sphere	4Al	1Si, 3Al	2Si, Al	3Si, 1Al	4Si
Range in ratios Si/Al	1·18–2·5	1·18–5·0	1·18–5·0	1·18–67	1·18–67
Range in chemical shifts	−83·8 to −85·3	−89·0 to −89·6	−94·0 to −95·9	−98·8 to −101·5	−103·1 to −107·8

On the other hand, when a range of different framework topologies is considered, as illustrated by Table 3, some overlap is observed (see also Table 4 of Chapter 3), so that allocations of n.m.r. peaks within the overlap regions may be uncertain, in particular if they are the only peaks. The n.m.r. spectra are particularly informative concerning the Lowenstein rule when the Si/Al ratio is unity because then, where the rule applies, the shift due to Si linked to 4Al is the only one possible. This is termed 4:0 ordering. On the other hand, again for a compound with Si/Al = 1, if the chemical shift indicates that each Si is linked to 3Al and 1Si (3:1 ordering), then to preserve the stoichiometry there must be numbers of Al atoms in the framework equal to the numbers of Si atoms showing the ordering 1:3 (Al joined to 1Al and 3Si) which breaks the Ȧl–O–Ȧl avoidance rule. This argument has been applied in the case of the following tectosilicates having Si/Al = 1 with the conclusions shown below:

Anorthite	Ordering	$4:0^{(43c)}$	Rule valid
Nepheline	Ordering minimal	$4:0$, with $3:1^{(43c)}$	Rule nearly valid
Eucryptite	Ordering	$4:0^{(43c)}$	Rule valid
Zeolite Li-ABW	Ordering	$4:0^{(43c)}$	Rule valid
Sodalite hydrate	Ordering	$3:1^{(43c)}$	Rule invalid
Sodalite	Ordering	$4:0^{(43g)}$	Rule valid
Cancrinite hydrate	Ordering (i)	$3:1^{(43g)}$	Rule invalid
	Ordering (ii)	$4:0^{(43g)}$	Rule valid
Zeolite Losod	Ordering	$3:1^{(43g)}$	Rule invalid
Zeolite Na-A	Ordering	$3:1^{(44,45,43h)}$	Rule invalid

Some of the above results are surprising. Thus in sodalite and cancrinite frameworks either 3:1 or 4:0 ordering was reported. Although a connection is not established, this behaviour may be related to differing synthesis conditions and hence synthesis mechanism. The shifts attributed to 3:1 ordering in sodalite and cancrinite are, however, very near or in the region of overlap of Q^4(4Al) and Q^4(3Al, 1Si) as seen in Table 3, and in Table 4 of Chapter 3. Ordering in the distribution of Al and Si is the rule in all the above compounds, but 3:1 ordering, where it occurs, is not in accord with Lowenstein's rule when Si/Al = 1. Neutron and electron diffraction have extended the above results on 3:1 ordering in zeolite A, to the proposal for the new structure seen in Fig. 7(b).[45,43h] This figure can be compared with Fig. 7(a) which shows the structure previously assumed, with 4:0 ordering. In the new structure with space group R$\bar{3}$ the rhombohedral unit cell has $a = 17·40 \pm 0·001$ Å and $\alpha = 59·53 \pm 0·01°$ when Si/Al = 1. In it there are two kinds of 6-ring differing in Si, Al sequence, and two kinds of 4-ring. The pseudo-cubic unit cell to correspond with the new structure is 24·6 Å, as with the old. Alternatively to

Fig. 7(*b*) a cubic structure with unit cell edge 12·3 Å has been pro-posed[44] in which Al–O–Al bonds are bridging between two sodalite cages. The cubic unit then has two Si–O–Si and two Al–O–Al as the bridging bonds. The two Al–O–Al form diagonally opposite edges of the cube. There is however a difficulty in reconciling a unit cell of 12·3 Å with X-ray data.

The deduction that Lowenstein's rule is invalid for zeolite A with Si/Al = 1 rests primarily upon the assumption that the single peak at about −89 p.p.m. as compared with Si(CH₃)₄ is due to Si(3Al, 1Si). However, a further study of zeolites A with Si/Al = 1·13 and 1·40 has made this assumption doubtful and supports Lowenstein's rule.[45a] The ²⁹Si n.m.r. absorptions, their assignments and relative populations were:

Si/Al = 1·40	Si/Al = 1·13
−89·1; Si(0Si, 4Al); 0·317	−89·1; Si(0Si, 4Al); 0·642
−93·9; Si(1Si, 3Al); 0·375	−93·9; Si(1Si, 3Al); 0·246
−99·5; Si(2Si, 2Al); 0·168	−99·5; Si(2Si, 2Al); 0·036
−106·1; Si(3Si, 1Al); 0·103	−106·1; Si(3Si, 1Al); 0·026
−110·7; Si(4Si, 0Al); 0·037	−110·5; Si(4Si, 0Al); 0·051

Provided Lowenstein's rule is valid these results yield compositions of 1·41 and 1·18 for the two Si/Al ratios, in satisfactory agreement with the elemental analyses. The authors conclude that zeolites Na-A having Si/Al ratios 1, 1·13, and 1·40 "exhibit a consistent and systematic variation constrained by Lowenstein's rule ... *provided* it is assumed that the single ²⁹Si n.m.r. absorption in Na-A"—with Si/Al = 1—"represents a Si(0Si, 4Al) environment". These observations, confirmed by Thomas *et al*[45b] make it desirable to look again at other phases in which, on the basis of the single ²⁹Si n.m.r. absorptions when Si/Al = 1, Lowenstein's rule has been questioned (sodalite, cancrinite and losod).

An attempt was made to relate 4:0 and 3:1 ordering when Si/Al = 1 with framework open-ness, in that denser frameworks might favour 4:0 and more open ones 3:1 ordering. However additional investiga-tions[46,46a,43e,46b] all agree that faujasites, which provide the most porous frameworks, show only 4:0 ordering, in the limit when Si/Al = 1. The faujasites examined had Si/Al ratios 1·18 and 1·4,[46] between 1·19 and 2·45,[46a] and between 1·26 and 2·66.[46b] The ratio 1·18 is obtained by symmetrical replacement of one Al by one Si in each sodalite cage, starting with a faujasite having Si/Al = 1 and 4:0 ordering in which all 6-rings are meta substituted with respect to both Al(●) and Si (○), to give

Through the substitution giving Si/Al = 1·18 two 6-rings

TABLE 3

Some ^{29}Si isotropic Q^4 chemical shifts in tectosilicates (p.p.m. difference from $Si(CH_3)_4$)

Compound	Si/Al	Second co-ordination sphere of Si†					Refs.
		4Al	1Si, 3Al	2Si, 2Al	3Si, 1Al	4Si	
Low quartz	8					−107·4	(43c)
Low cristobalite	8					−109·9	(43c)
Na-Y silica	8					−107·4	(29)
Clinoptilolite	5					−106·8, −112·8	(43d)
Heulandite	3·5			−95·0	−100·6	−108	(43d)
Harmotome	3			−95·0	−105·3, −99·0	−108	(43d)
Sanidine	3			−95·7	−102·6, −98·6		(43c)
Adularia	3			−95	−100·9, −96·8		(43c)
Orthoclase	3			−95	−100·5, −98·2		(43c)
Albite	3			−92·5	−101, −98, −104·2, −96·7		(43c)
Zeolite Na-Y	2·66		−90·1	−95·2	−100·8	−106·4	(43e)
Stilbite	2·6			−98	−103·6, −101·5	−108	(43d)
Zeolite Na-Y	2·5	−83·8	−89·2	−94·5	−100	−105·5	(43d)
Zeolite Na-Y	2·0	−84·8	−89·5	−94·8	−100	−104·5	(43f)
Zeolite Na-Y	—		−89·7	−94·6	−100·6	−105·5	(29)
Gmelinite	2		−92·0	−97·2	−102·5		(43d)
Chabazite	2		−94·0	−99·4	−104·8	−110	(43d)

Analcime	2		−92		−101·3	−108	(43d)
Natrolite	1·5		−87·7	−95·4			(43c)
Zeolite Na-X	1·5	−85·3	−89·5	−94·9	−99·5	−103·6	(43f)
Zeolite Na-X	1·26	−86·5	−90·5	−95·2	−100	−103·9	(43e)
Zeolite Na-X	1·18	−84·6	−89·0	−94·2	−98·8	−103·1	(43d)
Anorthite	1	−83·1					(43c)
Nepheline	1	−84·8					(43c)
Thomsonite	1	−83·5					(43d)
Ga-Thomsonite	1	−82·9					(43g)
Zeolite Na-A	1	−85·0	−89·6	−94·5			(43d)
Zeolite Na-A	1		−88·3				(43h)
Zeolite Na-A	1		−89				(45)
Zeolite Na-A	1		−88·9				(43g)
Sodalite (Cl⁻)	1	−84·8					(43d)
Sodalite hydrate	1	−83·5					(43d)
Sodalite hydrate	1	−84·0					(43g)
Sodalite hydrate	1		−86·4				(43g)
Sodalite hydrate	1		−86·3				(43g)
Cancrinite I	~1	−82·0					(43g)
Cancrinite II	1		−87·2				(43g)
Zeolite Losod	1		−88·9				(43g)
Zeolite Li-ABW	1	−80·1					(43g)
Eucryptite	1	−81·2					(43g)

† The first co-ordination sphere consists of four O atoms.

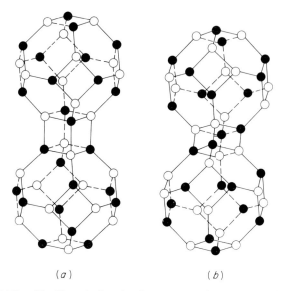

(a)   *(b)*

Fig. 7. (*a*) The Si, Al ordering in the accepted structure of zeolite A. (*b*) The Si, Al ordering proposed in Refs 45 and 43h but not acceptable according to Refs 45a and b.

are obtained in a sodalite unit each with two Al in *meta* positions. For the resultant structure the neighbours of the Si atoms and the proportions of each distribution are as follows:

$$\mathrm{Si(4Si, 0Al) : Si(3Si, 1Al) : Si(2Si, 2Al) : Si(1Si, 3Al) : Si(0Si, 4Al)}$$

$$1 \quad : \quad 0 \quad : \quad 0 \quad : \quad 4 \quad : \quad 8$$

These proportions agreed with the intensity distributions of the corresponding signals in the ^{29}Si n.m.r. spectrum.[47] For faujasite (Na-X) with Si/Al = 1·4 (14Si and 10Al per sodalite unit) a second Al is replaced by Si in each sodalite unit and several possibilities of ordering arise. The intensity distribution in the ^{29}Si spectrum gave:

$$\mathrm{Si(4Si, 0Al) : Si(3Si, 1Al) : Si(2Si, 2Al) : Si(1Si, 3Al) : Si(0Si, 4Al)}$$

$$0 \quad : \quad 2 \quad : \quad 3 \quad : \quad 4 \quad : \quad 5$$

which accords with one non-centrosymmetric Al, Si distribution of the several possibilities. For both the above faujasites the Al, Si ordering satisfies Lowenstein's rule. At the ratio Si/Al = 1·40 *para* substitution of

Si begins in 6-rings to give ⬡ The various possible configurations were also found for the additional Si/Al ratios 1·67, 2·00, 2·42, 3·00, 3·8 and 5·00,[46a] but as these ratios increase it becomes more difficult to use magic angle high-resolution ^{29}Si n.m.r. to distinguish betwen alternative ordered structures having the same Si/Al ratio. The n.m.r. method is at its best when Si/Al is near or equal to unity. Si, Al orderings sometimes different from those suggested in Refs (46), (46a) and (47) were proposed by Melchior *et al.*[43e,46b] These latter authors introduced as an extension of Lowenstein's rule the principle of maximum possible avoidance of the $\overset{\ominus}{Al}$–O–Si–O–$\overset{\ominus}{Al}$ grouping. For example, with faujasite having Si/Al = 1·40, the two representations of the ordering give the following weightings of the distributions:

	4Al	3Al, 1Si	2Al, 2Si	1Al, 3Si	4Si
Refs (46) and (46a):	10	8	6	4	0
Refs (43e), (46b):	8	10	8	2	0

Melchior *et al.* reported relative areas for the corresponding peaks of their n.m.r. spectrum as 7·8 : 10·1 : 7·8 : 2·3 : 0 in good accord with their representation of the Si, Al distribution. Currently, therefore, there are differences as well as a measure of agreement in views on the distribution of tetrahedral atoms in faujasites of different composition.

The cation distribution may be an important factor in modifying the energy change in the rearrangement

$$2M^+[\overset{\ominus}{Al}\text{–O–Si}] \rightarrow Si\text{–O–Si} + M^+[\overset{\ominus}{Al}\text{–O–}\overset{\ominus}{Al}]M^+$$

The associated energy change $w/2$ is partly electrostatic and partly short-range and covalent in origin. The electrostatic component could vary according to the distribution of M^+. For example, it could differ for the two arrangements

$$\begin{array}{cc} M^+ & \\ \overset{\ominus}{Al}\text{–O–}\overset{\ominus}{Al} & \text{and} \quad \overset{\ominus}{Al}\text{–O–}\overset{\ominus}{Al} \\ M^+ & M^+ \quad M^+ \end{array}$$

The covalent component of $w/2$ would not be expected to vary in sign. It might represent the endothermal part of $w/2$ favouring the Lowenstein rule.

According to the treatment[37] in §3 there are two ways in which fixed or nearly fixed Si/Al ratios can arise in aluminosilicates. Firstly, as noted in §3, negative values of $t_1 = \ln[J_{Si}/J_{Al}J_A^{1/Z_A}]$ combined with values of r_c (the number of cation sites per tetrahedral framework site) of 0·5 or less, as in Fig. 6(a), lead to free energies of the crystal which are least when $X_{Al} = r_c$. Secondly, when r_c is greater than 0·5, negative t_1 combined with large positive values of $w/2$ result in deep, sharp minima in the free energy of the crystal very near the Lowenstein limit Si/Al = 1 (Fig. 5(b)). In the felspars the first of these possibilities would govern the fixed stoichiometry. However, this does not necessarily mean that, where Lowenstein's rule is valid and where Si/Al is greater than unity, there need be ordering of Si and Al on tetrahedral sites. Indeed, disordered felspars such as albite are well known, and Si, Al disorder in zeolites is also possible, despite the apparent ordering in faujasite and zeolite A of particular compositions. In tectosilicates the detailed statistical thermodynamic treatment of Si, Al order and disorder would be expected to differ according to the framework topology. Hitherto attention has been directed mainly to order and disorder in felspars and accordingly a recent treatment[48] of the problem for albite will be outlined. This treatment permits in principle the occurrence of $\overset{\ominus}{\text{Al}}$–O–$\overset{\ominus}{\text{Al}}$ bonds in the structure.

The tetrahedral framework of albite was represented formally by a two-dimensional square lattice (Fig. 8(a)). The corners of the squares correspond with the four types of tetrahedral site (T_1o, T_1m, T_2o and T_2m). In this picture oxygens are not shown and the connecting rods are rings of tetrahedra. However, the square lattice represents correctly the nearer co-ordination of each tetrahedral site. Because the Al concentration on T_1m, T_2o and T_2m is the same during Si, Al ordering in albite these sites may be considered as energetically equivalent. This situation is represented by a second lattice (Fig. 8(b)) where the net made by the α (i.e. T_1o) and β (i.e. $T_1m = T_2o = T_2m$) sites serves as a model for the albite framework for purposes of the calculation.

In each group of four linked sites there are 16 possible distributions of Si (denoted by "S") and Al (denoted by "A"), to each of which a specific configurational energy u_m ($m = 0$ to 15) was assigned (Table 4). The energy u_0 of the ASSS group, corresponding with the ordered state, was the reference energy level and all other group energies were then found in terms of the energy of interchange, ϵ_1, between α and β sites according to the equation

$$A_\alpha + S_\beta \rightarrow S_\alpha + A_\beta \tag{14}$$

To allow for the energy of interaction between nearest neighbours, the energy change ϵ_2 was assumed for the re-arrangement

$$2AS \rightarrow AA + SS \tag{15}$$

In Fig. 8(b) each α and β site is shared by two groups of four. Therefore a single group has $\epsilon_1/2$ of the interchange energy. If interchange occurs within a single group, as for

$$ASSS \rightarrow SSAS$$

TABLE 4

Assignment of group energies and interchange energies for the model of
Fig. 8(b)[48]

Type number m	Occupancy of sites $\alpha\beta_1\beta_2\beta_3$	Group energy	Transformation	Transformation energy
0	ASSS	u_0	ASSS → SASS	$(\frac{1}{2})\epsilon_1$
1	SSSS	$u_0+(\frac{1}{4})\epsilon_1$	ASSS → SSAS	$(\frac{1}{2})\epsilon_1$
2	SASS	$u_0+(\frac{1}{2})\epsilon_1$	ASSS → SSSA	$(\frac{1}{2})\epsilon_1$
3	SSAS	$u_0+(\frac{1}{2})\epsilon_1$	2ASSS → AASS+SSSS	$(\frac{1}{2})\epsilon_1+\epsilon_2$
4	SSSA	$u_0+(\frac{1}{2})\epsilon_1$	2ASSS → ASAS+SSSS	$(\frac{1}{2})\epsilon_1$
5	AASS	$u_0+(\frac{1}{4})\epsilon_1+\epsilon_2$	2ASSS → ASSA+SSSS	$(\frac{1}{2})\epsilon_1+\epsilon_2$
6	ASAS	$u_0+(\frac{1}{4})\epsilon_1$	2ASSS → SAAS+SSSS	$\epsilon_1+\epsilon_2$
7	ASSA	$u_0+(\frac{1}{4})\epsilon_1+\epsilon_2$	2ASSS → SASA+SSSS	ϵ_1
8	SAAS	$u_0+(\frac{3}{4})\epsilon_1+\epsilon_2$	2ASSS → SSAA+SSSS	$\epsilon_1+\epsilon_2$
9	SASA	$u_0+(\frac{3}{4})\epsilon_1$	3ASSS → AAAS+2SSSS	$\epsilon_1+2\epsilon_2$
10	SSAA	$u_0+(\frac{3}{4})\epsilon_1+\epsilon_2$	3ASSS → AASA+2SSSS	$\epsilon_1+2\epsilon_2$
11	AAAS	$u_0+(\frac{1}{2})\epsilon_1+2\epsilon_2$	3ASSS → ASAA+2SSSS	$\epsilon_1+2\epsilon_2$
12	AASA	$u_0+(\frac{1}{2})\epsilon_1+2\epsilon_2$	3ASSS → SAAA+2SSSS	$(\frac{3}{2})\epsilon_1+2\epsilon_2$
13	ASAA	$u_0+(\frac{1}{2})\epsilon_1+2\epsilon_2$	4ASSS → AAAA+3SSSS	$(\frac{3}{2})\epsilon_1+2\epsilon_2$
14	SAAA	$u_0+\epsilon_1+2\epsilon_2$		
15	AAAA	$u_0+(\frac{3}{4})\epsilon_1+4\epsilon_2$		

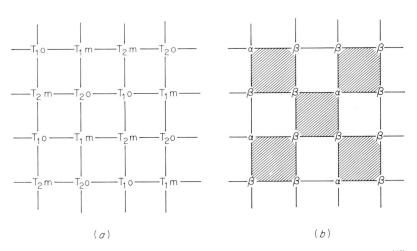

(a)   (b)

Fig. 8. The two-dimensional model of the framework of Na-felspar.[48]
(a) The relation between T_1o, T_1m, T_2o and T_2m on a planar net. (b)
The model used for the calculations. Sites T_1m, T_2o and T_2m are
equivalent.

the energy increase is $\epsilon_1/2$. The same energy increase must occur in Si, Al interchanges betwen different groups such as

$$2ASSS \rightarrow ASAS + SSSS$$

Thus, the group energy of SSAS is $u_0 + \frac{1}{2}\epsilon_1$ and the group energy of ASAS + SSSS is $2u_0 + \frac{1}{2}\epsilon_1$. It was assumed that during the Al, Si redistribution the energy was equally divided between ASAS and SSSS so that, for example,

$$u_1 = u_0 + \tfrac{1}{4}\epsilon_1$$

On this basis the group energies were assigned as in Table 4, and also the transformation energies associated with different re-arrangements are given in that table.

If N is the number of Al atoms and α sites, and r denotes the fraction of α sites occupied by Al, the partition function $P(r, T)$ takes the form

$$\ln P(r, T) = \ln P(r, \infty) + \int_{\infty}^{T} \frac{\bar{E}(r, T)}{kT^2} \, dT \tag{16}$$

When the temperature $T \rightarrow \infty$, $P(r, T)$ becomes the maximum number of ways of arranging the Al and Si atoms on the α and β sites, retaining the specified numbers of "right" and "wrong" atoms on the sites:

$$P(r, \infty) = \frac{N!}{[rN]! \, [(1-r)N]!} \cdot \frac{3N!}{[(1-r)N]! \, [(2+r)N]!} \tag{17}$$

$\bar{E}(r, T)$ is the average internal energy for the given r and T. It is determined by the number of groups of each of the types in Table 3 corresponding with the distribution r multiplied by the energy of each group:

$$\bar{E}(r, T) = \sum_m \bar{M}_m u_m \tag{18}$$

In Eqn (18), $\bar{M}_m$ is the average number of groups of type m, Table 4, and u_m is the energy of a group of type m. The $\bar{M}_m$ for different values of m are inter-dependent according to the quasi-chemical equilibria illustrated in column 4 of Table 4. The free energy of the crystal is

$$F(r, T) = -NkT \ln P(r, T)$$

and the equilibrium condition, for which $r = \bar{r}$, is then given by

$$\left[\frac{\partial F(r, T)}{\partial r} \right]_{r=\bar{r}} = 0 \tag{19}$$

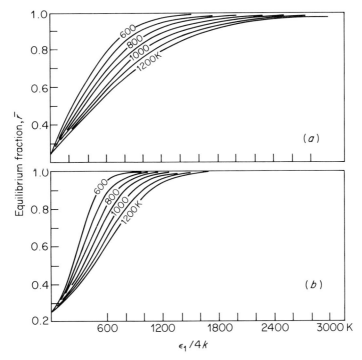

Fig. 9. The equilibrium fraction, $\bar{r}$, of α-sites occupied in the net of Fig. 8(b), in relation to ϵ_1 and T for two different values of ϵ_2.[48] In (a), $\epsilon_2/k = 0$ and in (b) $\epsilon_2/k = 40\,000K$.

From the equilibrium condition $\bar{r}$ could be calculated as a function of T, ϵ_1 and ϵ_2. Figures 9(a) and (b)[48] illustrate such calculations for $\epsilon_2/k = 0$ and 40 000 K respectively. The temperature-dependent continuous change of $\bar{r}$ with $\epsilon_1/4k$ is shown in the temperature range 600–1200 K. For a given $\epsilon_1/4k$ order diminishes with rising temperature, but order increases with $\epsilon_1/4k$. The influence of ϵ_2 is seen primarily in the shape of the curves of $\bar{r}$ vs $\epsilon_1/4k$.

It is concluded from the foregoing account that Lowenstein's Ål–O–Ål avoidance rule is usually valid. It is grossly violated only in the aluminate sodalites and also in bicchulite. High-resolution magic angle ²⁹Si n.m.r. can give required information on ordering among Si, Al distributions by identifying and quantifying the numbers of groupings with Si joined to each of n Si nearest neighbours ($n = 0, 1, 2, 3$ or 4). This information might enable one to say whether Ål–O–Ål bonds are or are not present when Si/Al = 1, but results must be interpreted with caution.

[27]Al high-resolution n.m.r. may in the future give more information, especially for systems where Si/Al > 1.

5. Ga and Ge Substitution

Ga and Ge are next to Al and Si in their respective groups of the periodic table and could therefore be expected to replace these two elements in aluminosilicates. Indeed, as illustrated in Chapter 2, §§8.1, 8.2 and 8.4, Ga can proxy for Al and Ge for Si in smectites and micas and in felspars analogous to albites, orthoclases and anorthites.

Among zeolites an early example was the synthesis by Goldsmith[49] in 1952 of a Ga-bearing thomsonite. It was grown hydrothermally at 245°C from a glass of composition $Ca_4Ga_3Al_5Si_8O_{32}$. In 1959 Barrer et al.[10] reported a number of Ga- and Ge-bearing zeolites made from alkaline aqueous mixtures of the required compositions (aluminate + GeO_2; gallate or Ga_2O_3 + silicate; gallate + germanate). They reported a Ga-thomsonite; faujasite, zeolite A and a harmotome type zeolite each containing Ge; and faujasite, zeolite A and thomsonite containing both Ga and Ge. Subsequent microprobe analyses of their products showed that the Ge content of their zeolites was reduced considerably by Si, presumably due to the presence in the mixtures of silica derived from the glass reaction vessels.[50] Little or no alumina came from this source, however, and so the Ga-content of the crystals was not diluted by Al. In 1961 Selbin and Mason[51] recorded the synthesis of Ga-faujasite and Ga-sodalite as well as an unidentified gallosilicate, $Na_2O.Ga_2O_3.2\cdot8SiO_2$. Their reaction mixtures were also alkaline aqueous hydrogels under reflux at about 100°C. In addition a number of mixed (Al, Ga)-faujasites were prepared. All the faujasites had the good sorption capacities of normal zeolite Na-X. The next synthesis reported was in 1974 when Lerot et al.[52] prepared from aqueous alkaline mixtures of $NaAlO_2$ and GeO_2 a pure Ge-faujasite having the oxide composition at the Lowenstein limit, $Na_2O.Al_2O_3.2GeO_2.xH_2O$. The cubic unit cell of edge $a = 25\cdot59$ Å shows the lattice increase expected when Si is replaced by the larger Ge atom. The Ge-faujasite was less stable thermally than zeolite X, with severe structural damage at 400°C. The NH_4-, Ni- and Cu-forms were even less stable than the Na-form.

Results in the system $K_2O–Al_2O_3–GeO_2–H_2O$ were described in 1975.[53] The only zeolite obtained was Ge-phillipsite; in addition a germanate $KH_3Ge_2O_6$, an aluminogermanate $KAlGeO_4$ and an unidentified phase were prepared. Crystallization fields were plotted at 90, 150 and 225°C. Although 250 runs were made the appearance of one zeolite only c█████ts with the variety of potassium zeolites grown from alkaline aqueou█ █████inosilicate gels (Chapter 5). It seems likely that further investi██ ████would add to the number of Ge-bearing potassium zeolites. Since ████ several papers have described the catalytic and sorptive properties of the aluminogermanate form of faujasite with Ge/Al = 1.[54,55,56,57]

The examples given indicate that under mild zeolitization conditions it is comparatively easy to replace Al by Ga and Si by Ge in zeolite frameworks. This area of exploration could be extended to Li, Ca, Sr and Ba zeolites bearing Ga and/or Ge; and to such zeolites in which the cation is derived from an organic base. Silica-rich gallium-bearing zeolites for example could be of considerable interest in view of the importance as catalysts of silica-rich zeolites.

6. Substitution by Be and B

Beryllium, like aluminium, has an amphoteric oxide which might enable it to replace aluminium in tectosilicates. Examples of natural minerals demonstrating this replacement are helvine, $Mn_6[Be_6Si_6O_{24}]2MnS$, and tugtupite, $Na_8[Al_2Be_2Si_8O_{24}]2Na(Cl, S_{1/2})$, both of which have the framework structure of sodalite, $Na_6[Al_6Si_6O_{24}]2NaCl$. The compositions of helvine and sodalite demonstrate the isomorphous replacement

$$Na^+Al^{3+} \rightarrow Mn^{2+}Be^{2+}$$

while those of tugtupite and sodalite correspond with

$$2Al^{3+} \rightarrow Be^{2+}Si^{4+}$$

Synthetic forms of helvine and tugtupite have not been prepared. However, a beryllo-analcime has been made,[58] in which, compared with analcime, the substitution

$$2Al^{3+} \rightarrow 3Be^{2+}$$

was assumed. Ideal beryllo-analcime would then be $NaBe_{1.5}Si_2O_6.H_2O$† by analogy with analcime, $NaAlSi_2O_6.H_2O$. The above ideal composition was used in the parent mixtures and also the composition $Na_3Be_{1.5}Si_{4.5}O_{12}.NaCl$. The mixtures were made from appropriate amounts of silica sol, powdered BeO and 1 M NaOH and NaCl. The stirred mixtures were evaporated to dryness and ground to a powder.

---

† However, to preserve the number of tetrahedral sites per unit cell, this would have to be $NaBe_{0.5}[BeSi_2O_6]H_2O$.

Subsequent hydrothermal crystallization over 1–3 weeks at ████ 225°C resulted, using the ideal composition, in the analcime together ████ some beryllia and unknown phases. The cubic unit cell had $a = 1$██ compared with 13·7 Å for analcime. The composition containing NaCl also yielded beryllo-analcime at 170–225°C, this time associated with chaklovite, $Na_2BeSi_2O_6$.

Stabenow et al.[59] have reported syntheses of Be-bearing faujasites. Solutions of NaOH, Na-aluminate and Na-beryllate were made with stirring, and appropriate amounts of silica sol were added and well mixed. After ageing for 16–24 hours at room temperature the mixtures were crystallized with stirring at 90–100°C for 48–80 hours. The ratios BeO/Al_2O_3 in three preparations of faujasites were given as 0·395, 0·66 and 0·35 respectively and were surprisingly almost the same as in the parent mixtures. If the Be was all in the framework this would mean a distribution coefficient of Al and Be between reaction mixture and crystals of unity. The ratio $Na_2O/(Al_2O_3 + BeO)$ in the products was unity, but since $Na_2O/BeO = 1$ in sodium beryllate this does not mean that the Be is all in the anionic framework; it could be present as intercalated Na_2BeO_2. It was established that mild acid treatments removed the beryllium from the crystals much more rapidly than the aluminium which would certainly be expected if Na_2BeO_2 was intercalated. Further work seems needed to establish the nature of the faujasites of Stabenow et al.

In 1951 Roedder[60] in a study of the system $K_2O–MgO–SiO_2$ reported four new ternary compounds. One of these, $K_2O.MgO.5SiO_2$, was considered to be iso-structural with leucite, hence with analcime, and another, $K_2O.MgO.3SiO_2$, to be iso-structural with the felspathoid kalsilite. This would require Mg^{2+} in tetrahedral framework sites and further exemplifies isomorphous substitutions of the type

$$2Al^{3+} \rightarrow Mg^{2+}Si^{4+}$$

However Roedder's compounds were prepared under anhydrous high-temperature conditions. The hydrothermal formation of zeolites with Mg in the tetrahedral sites seems unlikely. On the other hand, ZnO and PbO are both amphoteric oxides like BeO, and the possibility that Zn and Pb could occupy tetrahedral sites should not be overlooked. Evidence that Zn may be present in tetrahedral sites in a zinc mica was obtained by Barrer and Sieber[61] (Chapter 2, §8.2).

Borosilicates are not uncommon,[62] but boron often exists in three-fold planar co-ordination with oxygen as units which are not part of a tectosilicate framework. In a few borosilicates, however, the boron is in tetrahedral co-ordination with oxgen as part of a tectosilicate framework.

Compounds of this type include reedmergnerite,[63] $NaBSi_3O_8$, which is the boron analogue of the felspar albite, $NaAlSi_3O_8$. It is also of interest that the borate $Zn_4O(BO_2)_6$ has been reported to have the sodalite structure.[64] Borosilicate zeolites analogous to their aluminosilicate counterparts might be possible and attempts to synthesize them have been made. The system $Na_2O–B_2O_3–SiO_2–H_2O$ under a water pressure of 2000 bar in the range 300–500°C yielded reedmergnerite and searlsite,[65] but the temperatures were not very favourable for zeolite synthesis. Examination of the system $Na_2O–B_2O_3–Al_2O_3–SiO_2–H_2O$ was subsequently made to see whether under mild alkaline conditions boron could replace aluminium in zeolites.[66] Analcime was made at 200°C in steel autoclaves, and the gismondine type zeolite Na-P1, zeolite A, faujasite (Na-X) and sodalite hydrate were made at 80°C in polypropylene containers.

Analcime has too compact a structure to intercalate borate during synthesis, so that any boron in its composition would necessarily be a framework substituent. The zeolite was made from hydrogels $1·5Na_2O.Al_2O_3.4SiO_2.24OH_2O + n(0·5Na_2O + B_2O_3)$ where $n = 0$, $0·2$, $0·4$ and $2·0$. Large cubic crystals about 50 μm in diameter were formed which, whatever the boron content of the parent mixture, showed no variation in unit cell dimension. After washing, the boron content of the crystals was not significantly above the blank of the analytical method, indicating that if any replacement of Al by B had occurred it was less than 1 atom of B per 250 of Al.

The Na-P1 was made from hydrogels $6Na_2O.Al_2O.6SiO_2.250H_2O + n(0·5Na_2O + B_2O_3)$ where $n = 0$, $0·05$, 1, 2, 4, 6, 8, 10 and 12, and zeolite A from hydrogels $4Na_2O.Al_2O_3.2SiO_2.250H_2O + n(0·5Na_2O + B_2O_3)$ where $n = 0$, $0·5$, 1, 2, 4 and 8. Both zeolites contained boron which, however, was reduced by extended washing to very low levels and thus was primarily adsorbed or intercalated borate. The reduced boron content for Na-P1, estimated as B_2O_3, was normally less than $0·01\%$ of the dry weight, while for Na-A it was $0·65\%$ or less. In Na-A any borate taken into sodalite cages during synthesis would be permanently trapped and since about 2% B_2O_3 would represent one BO_2^- or $B(OH)_4^-$ inside each sodalite cage, the $0·65\%$ maximum observed B_2O_3 content could mean that one in every three sodalite cages had intercalated one such boron anion.

Zeolites Na-X were made from aqueous compositions $6Na_2O.Al_2O_3.6SiO_2.250H_2O + n(0·5Na_2O + B_2O_3)$ with $n = 0$, $0·5$, 1, 2, 4 and 6, and all contained boron even after thorough washing. This residual B_2O_3 content is plotted against the concentration of B_2O_3 in the parent mixture in Fig. 10(a). The maximum uptake is well below the

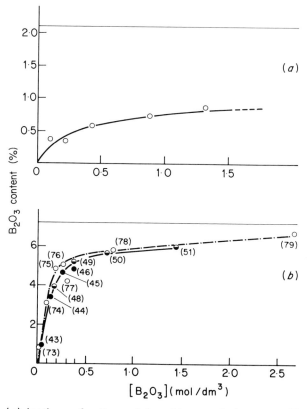

Fig. 10. (*a*) Isotherm for the uptake of boron during crystallization of zeolite Na-X. The horizontal line represents one boron atom per sodalite cage.[66] (*b*) Isotherms for the trapping of boron during crystallization of sodalite hydrate. The horizontal line represents one boron per sodalite cage. The numbers correspond with the numbers of the experiments in Table 5.[66]

amount of B_2O_3 required to give one B atom per sodalite cavity, and for reasons given below when considering the sodalites, was thought to represent borate trapped inside sodalite cages. The sodalites were prepared from the several types of initial mixture given in Table 5. The B_2O_3 content of well washed crystals is shown in Fig. 10(*b*) as a function of the B_2O_3 concentration in the parent mixtures. Like that in Fig. 10(*a*) the isotherm has the form of those measured when $NaClO_4$ or $NaClO_3$ are intercalated in sodalite during its formation[40] (Chapter 7, Fig. 3). The maximum uptake of boron corresponds with 93% of the value expected for one B per sodalite cage (as BO_2^- or $B(OH)_4^-$), and the shape of the

TABLE 5
Synthesis of sodalite hydrate with borate at 80°c[66]

Expt No.	Reactants	Initial oxide composition	Reaction time (days)	Products
42–46	$Na_2SiO_3.5H_2O$ Na-aluminate $Na_2B_4O_7.10H_2O$	$5Na_2O.2SiO_2.$ $(1-x)Al_2O_3.xB_2O_3.130H_2O$ ($x = 0, 0·1, 0·3, 0·6$ and $0·9$)	6	44–46: sodalite 42–43: sodalite and traces unidentified
47–51	$Na_2SiO_3.5H_2O$ Na-aluminate $Na_2B_4O_7.10H_2O$ NaOH	$5Na_2O.2SiO_2.Al_2O_3.$ $x(Na_2O.B_2O_3).150H_2O$ ($x = 0, 0·25, 0·5, 1·0$ and $2·0$)	5	47, 48, 50: sodalite 49, 51: sodalite + traces unidentified
52–55	metakaolin $Na_2B_4O_7.10H_2O$ NaOH	$10Na_2O.2SiO_2.Al_2O_3.$ $x(Na_2O.B_2O_3).180H_2O$ ($x = 0, 1, 2$ and 4)	4	54, 55: sodalite 52, 53: sodalite + traces unidentified
72–79	$Na_2SiO_3.5H_2O$ Na-aluminate $Na_2B_4O_7.10H_2O$ NaOH	$8Na_2O.2SiO_2.xAl_2O_3.$ $(1-x)B_2O_3.y(Na_2O.B_2O_3).$ $180H_2O$ (72 to 77: $y = 0$; $x = 0$, $0·1, 0·6, 0·8$ and $0·95$) (78 and 79: $y = 1$ or 4; $x = 0·6$)	6	sodalite

isotherm suggests that as with other salts ($NaClO_4$, $NaClO_3$, NaCl, etc.) one such borate anion only can be accommodated per cage. The cubic unit cell dimensions increased with B_2O_3 content whereas framework substitution of Al by B should reduce the cell size as found when B-felspar (reedmergnerite) was compared with albite and B-phlogopite mica with phlogopite.[66] Moreover, when the yield of sodalite crystals was plotted against the ratio $B_2O_3/(B_2O_3 + Al_2O_3)$ this yield decreased linearly to zero as the ratio rose from 0 to 1 (Fig. 11). Whereas Al was essential, the B apparently played no part, beyond being intercalated in sodalite cages.

These several lines of evidence indicate little if any substitution of Al by B in the anionic frameworks under the mild conditions used, but show that intercalation of borate during crystal growth occurs in some zeolites open enough to allow this.

Taramasso et al.[67] made high silica zeolites or their silica end members in presence of B_2O_3. They considered the products to contain framework boron and called them "boralites". The syntheses effected were in the range 140–175°C for times of 5–15 days. $B(OC_2H_5)_3$ or H_3BO_3 were the

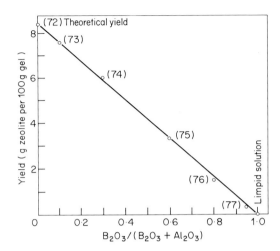

Fig. 11. The yield of sodalite in relation to the ratio $B_2O_3/(B_2O_3+Al_2O_3)$ in the reaction mixture. The numbers are those of experiments in Table 5.[66]

sources of B_2O_3 and $Si(OC_2H_5)_4$ was the normal source of silica. The usual compositions crystallized were characterized as below:

SiO_2/B_2O_3	1–5
OH^-/SiO_2	0·2–0·5
H_2O/SiO_2	25–40
M^+/SiO_2	0–0·1
R^+/SiO_2	0·3–0·9

Here M^+ denotes an alkali metal ion and R^+ an organic ion. However, the inorganic ions were dispensed with in many preparations, including those of which the boron contents were investigated. The boron contents were studied only for the products analogous to zeolite ZSM-5, although synthesis of silica-rich zeolites analogous to Nu-1, Beta and ZSM-11 was also claimed. The chemical analyses of the ZSM-5 type products yielded inconsistent results ascribed to the presence of glassy phase with the crystals. It was then assumed that all organic material in the precursors to the ZSM-5 (subsequently obtained by burning out the organic matter in air at 500°C) was present as the cations, R^+, neutralizing the framework charge which would develop if B replaced Si. Thus the organic content would serve to give the framework boron content if this assumption were correct. It led to the values of $B/(B+Si)$ in Table 6. The assumption used omits the possibility that organic base or the borate of the organic base as well as charge-balancing organic cations are present in the precursor to ZSM-5. Thus the boron contents of the frameworks given in Table 6 can be upper limits only. In this connection it is seen that the apparent boron content is a function not only of $B/(B+Si)$ in the parent mixture but also depends on the base used in synthesis for a fixed $B/(B+Si)$ in the parent mixtures. The best evidence of framework substitution of Si by B came from the determination of unit cell dimensions, which decreased as the boron contents (determined by the above questionable assumption and given in Table 6) increased.

Partial explorations were also made of the systems $CaO-B_2O_3-SiO_2-H_2O$ and $BaO-B_2O_3-SiO_2-H_2O$ at 250 and 415°C.[68] No zeolites were prepared but there was a rich harvest of silicates and borates and a number of borosilicates. The calcic system yielded the borosilicates dato-lite, $2CaO.B_2O_3.2SiO_2.H_2O$, and danburite, $CaO.B_2O_3.2SiO_2$. The former has a structure which consists of sheets of interlinked SiO_4 and BO_3OH tetrahedra.[69] Danburite has a three-dimensional tetrahedral framework of Si_2O_7 and B_2O_7 pairs.[70] The appearance of B_2O_7 units means that the Lowenstein Al–O–Al avoidance rule cannot be extended to B–O–B in danburite. Six Ba-borosilicates were prepared in the baric system, one of which (Ba-BS) could be identified with a known synthetic product,[71] and the remaining five appeared to be new phases. Compositional data are given in Table 7. For Ba-BS, Ba-α8 and Ba-α10 the ratios Ba/Al_2O_3 are ~1·0 and the ratios $O/(B+Si)$ are ~2. They may therefore be tectosilicates, but if so they do not, from the values of the ratios B/Si, follow a B–O–B avoidance rule.

In the above calcic and baric systems, in order to have crystallization in a reasonable time, it was necessary to employ higher temperatures. However, 250°C is within the range at which Ca, Sr and Ba zeolites

TABLE 6

Upper limits to boron content of frameworks of ZSM-5 type products[67]

Base used in synthesis	$B/(B+Si)$	
	in reaction mixture	in product
$N(C_3H_7)_4OH$	0	0
$N(C_5H_{11})_4OH$	0·5	0·022
$N(C_2H_5)_4OH$	0·5	0·049
$NH_2(CH_2)_2NH_2$	0·5	0·102
$N(C_3H_7)_4OH$	0·5	0·041
$N(C_3H_7)_4OH$	0·074	0·016
$N(C_3H_7)_4OH$	0·020	0·007

TABLE 7
Compositions of some Ba-borosilicates[68]

Compound designation	BaO/B_2O_3	B/Si	$O/(B+Si)$	OH present
Ba-BS	1·0	3	2	No
Ba-α4	2·0	0·5	$2·1_7$	Yes
Ba-α7	2·0	0·31	$2·4_0$	Yes
Ba-α8	$0·9_2$	$2·8_3$	$1·9_7$	Yes
Ba-α10	$0·9_3$	$1·2_2$	$1·9_8$	Little
Ba-α14	2·0	0·5	$2·1_7$	No

formed from aluminosilicate hydrogels (Chapter 5) so that the absence of zeolites from borosilicate compositions suggests that boron zeolites with full replacements of Al by B are not readily prepared. Equally, when B competes with Al in zeolite synthesis the boron content of the frameworks of the zeolites seems to be very low indeed. Boron in these respects has a hydrothermal chemistry which differs considerably from that of aluminium, whereas gallium demonstrates very marked similarities to aluminium (§5). In the system $CaO–B_2O_3–GeO_2–H_2O$ studied at 415°C the borogermanate analogue of datolite was among the products.[72] This again emphasizes the similarities between Si and Ge in hydrothermal systems already seen from the results in §5.

7. Substitution by Fe, Cr and Other Metallic Elements

That iron can replace other elements such as aluminium in tectosilicates has been demonstrated under high-temperature conditions for iron fels-par, $KFeSi_3O_8$,[73,74] and for iron leucite, $KFeSi_2O_6$,[75] which has the framework topology of analcime. These compounds were formed in high-temperature conditions in absence of water. Leucite type ferrisili-cates $KFeSi_2O_6$, $RbFeSi_2O_6$ and $CsFeSi_2O_6$ and aluminosilicates $KAlSi_2O_6$ and $RbAlSi_2O_6$ were prepared and compared by Hirao et al.[76] They were made from pure K, Rb or Cs carbonates and high purity SiO_2 and Al_2O_3 or Fe_2O_3. After thorough mixing they were melted at 1300 to 1650°C in platinum crucibles. The melts were then crystallized. The $RbAlSi_2O_6$ was also made hydrothermally by the procedures used by Barrer and McCallum.[77] Leucite is tetragonal at room temperature but becomes cubic at 620°C in the way shown in Fig. 12.[76] The similar behaviour of $RbAlSi_2O_6$ results in its becoming cubic at 360°C. This

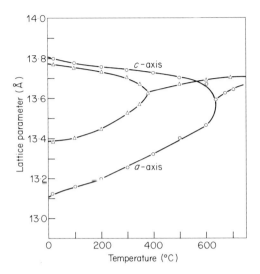

Fig. 12. Thermal expansion curves along *a* and *c* axes for (○) KAl-Si$_2$O$_6$ and (△) RbAlSi$_2$O$_6$, both having the analcime topology.[76]

property also characterized RbFeSi$_2$O$_6$, KFeSi$_2$O$_6$ and mixed phases Rb$_x$Cs$_{(1-x)}$FeSi$_2$O$_6$. However, as the cation increases in size the temperature at which tetragonal symmetry is replaced by cubic diminishes and CsAlSi$_2$O$_6$ (pollucite) is cubic even at room temperature.

In another series of high-temperature preparations Eitel *et al.*[78] introduced Y, La and Nd in place of Al in synthetic nepheline type phases. The similarities are brought out by comparing the *a* and *c* dimensions of the hexagonal unit cells:

	a (Å)	c (Å)
NaAlSiO$_4$	9·98	8·44
KAlSiO$_4$	10·41	8·70
NaYSiO$_4$	10·79	8·80
Ca$_{0.5}$YSiO$_4$	10·79	8·80
NaLaSiO$_4$	11·01	8·96
KLaSiO$_4$	11·01	8·96
NaNdSiO$_4$	10·89	8·85
KNdSiO$_4$	10·89	8·85
Ca$_{0.5}$NdSiO$_4$	10·89	8·85

However, all these replacements refer to syntheses at high temperatures in absence of water. Zeolite formation at low temperatures under alkaline aqueous conditions represents a very different situation, so that it is of interest to examine claims to have made zeolites bearing such metallic elements as Fe and Cr in the tectosilicate frameworks of zeolites.

Two patents exemplify such claims,[78,80] in both of which zeolites having the ZSM-5 topology were produced, and in which iron and chromium were considered to occupy some of the tetrahedral framework sites. The iron-bearing ZSM-5 was prepared from reaction mixtures containing sources of silica, ferric oxide and caustic soda together with hexamethylene diamine. Crystallization was effected in the temperature range 140–160°C in times of 2–4 days. The chromium-bearing ZSM-5 was made in similar conditions, but with a source of Cr_2O_3 replacing that of Fe_2O_3. For two particular preparations the analyses of the ZSM-5 zeolites, after heating to 550°C to eliminate the entrained organic base, gave the values in weight %:

(1)	(2)
$SiO_2 = 88 \cdot 6$	$SiO_2 = 93 \cdot 4$
$Al_2O_3 = 0 \cdot 64$	$Al_2O_3 = 0 \cdot 40$
$Fe_2O_3 = 2 \cdot 9$	$Cr_2O_3 = 0 \cdot 85$
$Na_2O = 0 \cdot 42$	$Na_2O = 0 \cdot 4$
Total $= 92 \cdot 56$	Total $= 95 \cdot 05$

The differences required to give totals adding to 100% were attributed to the zeolitic water contents, which were not determined. After the heating the iron and chromium could be present as oxides external to the crystals or intercalated; a little could be present as charge-balancing cations; or some Fe^{III} and Cr^{III} could occupy tetrahedral sites in the framework. It was not proven how the iron and chromium were distributed among these possibilities. Zeolite ZSM-5 can readily be made as its pure silica end-member, silicalite 1, so that even in absence of all Al there is no chemical requirement that Fe and Cr should be located on tetrahedral framework positions.

In another study Barrer et al.[39] synthesized pale brown iron-bearing cancrinite from kaolinite ($Al_2O_3.SiO_2.2H_2O$), 12 M NaOH and sodium ferrate maintained at 80°C for 5 days. The Mössbauer spectrum showed that some Fe^{III} was present, with some covalent bonding, for example to oxygen. However the Si/Al ratio was 1·06 so that no significant amount of this iron need be in the tetrahedral sites of the framework. The weight loss below 500°C was 7·7% and could represent zeolitic water. Between 500°C and 1000°C a further weight loss of 7·3% occurred, most rapidly at ~750°C. The products at 1000°C were nepheline and ferric oxide. These observations are consistent with cancrinite in which there is included ferrate, water, and ferric oxide or hydroxide.

Along with the iron-bearing cancrinite a second minor phase, termed Species H, appeared as tetrahedral crystals of about 0·1 mm edge. They were coloured pale green to pale red-brown and decomposed with some effervescence in dilute hydrochloric acid. This suggests the presence of ferrate in the structure. The compound could also be made in absence of alumina from ferrate, 10 M NaOH and precipitated silica, but it could not be made in absence of silica. It was not possible to obtain a pure sample for analysis, but it had a cubic unit cell with a 9·5 Å edge. Since the $Fe^{III}-O$ bond is 1·86 Å and the $Al^{III}-O$ is 1·75 Å, the results are compatible with a sodalite containing intercalated ferrate and with replacement of Al by Fe in the tectosilicate framework, although final proof of this was not obtained.

Physical methods have been employed to try to locate iron impurity in mordenite and faujasite.[81,82] X-band electron spin resonance (e.s.r.) of the fully hydrated states of the zeolites gave a signal with $g = 4·3$. This result was interpreted in terms of rhombic distortion of the co-ordination around the Fe^{III}, which can be achieved either through distorted octahedral or through distorted tetrahedral co-ordination.[83] The latter possibility was supported by phosphorescence spectroscopy which gave an emission band around 5000 Å rather than the 7000 Å expected for octahedral co-ordination. It was therefore considered that Fe^{III} was present in the framework.

However, this evidence was re-considered by Wichterlova and Jiru[84] who pointed out that the signal with $g = 4·3$ was compatible with charge-balancing Fe^{3+}Td complexes in cavities in (H, Na)-Y rather than in the framework. These authors found that the signal was enhanced by dehydrating the zeolite. Its maximum intensity was attained after heat treatment with oxygen; treatment *in vacuo* or in hydrogen weakened the signal, a behaviour ascribed to reduction of Fe^{3+}.

Titano- and zirconosilicates have been prepared under low-temperature aqueous conditions and were considered to exhibit zeolitic properties.[85] Peroxo-titanate and -zirconate provided soluble Ti and Zr constituents of the reaction mixtures. The anhydrous oxide compositions of the products were given as

$$x M_{2/n} O . XO_2 . y SiO_2$$

where X is Ti or Zr and the cation M^{n+} was NH_4^+, H_3O^+ or a mono- or divalent metal ion. In the titanosilicates $1·5 < x < 3$ and $1·0 < y < 3·5$, while in the zirconosilicates $1·5 < x < 4$ and $4·5 < y < 8·5$. The above compositions do not correspond with those expected for tectosilicates. A true tectosilicate framework in which Zr^{IV} or Ti^{IV} replaces Si^{IV} would be uncharged like the crystalline silicas. The X-ray powder lines were limited in number and crystallographic information was not given.

8. Substitution by P and N

Phosphorus and nitrogen are more electronegative than any of the elements so far considered, and nitrogen is a special case which will be discussed separately from phosphorus. Oxygen compounds of phosphorus are often based on tetrahedral units PO_4^{3-} and, neglecting differences in electronegativity, might be expected to replace SiO_4^{4-} in certain silicates. The crystalline silicas are indeed imitated by $AlPO_4$:

SiO_2	$AlPO_4$
Quartz	Berlinite
$\downarrow\uparrow$ 867°C	$\downarrow\uparrow$ 815 ± 4°C
Tridymite	Tridymite type
$\downarrow\uparrow$ 1470°C	$\downarrow\uparrow$ 1025 ± 50°C
Cristobalite	Cristobalite type

$AlPO_4$ (cristobalite type) melts near or above 1600°C and cristobalite melts at 1713°C. Crystalline BPO_4 also forms a distorted high-cristobalite structure and, in the anhydrous system B_2O_3–P_2O_5–SiO_2, is the stable form over a wide range of compositions. BPO_4 and $AlPO_4$ can both precipitate from an aqueous mixture. Despite isomorphism between $AlPO_4$, BPO_4 and also $FePO_4$ and certain of the crystalline silicas, there is little evidence of silica-bearing $AlPO_4$, BPO_4 or $FePO_4$ in which Si substitutes for the other elements, e.g.

$$2Si^{4+} \rightleftarrows P^{5+} + Al^{3+}$$

$$Si^{4+} \rightleftarrows P^{5+} + OH^-$$

Nevertheless two naturally occurring phosphatic minerals, viseite and kehoite, are reported to have the analcime structure.[86,87] McConnell formally represented the structures of such minerals in terms of the units AlO_2^-, PO_2^+, $H_3O_2^-$ and SiO_2:

Viseite, $Na_2Ca_{10}[(AlO_2)_{20}(PO_2)_{10}(SiO_2)_6(H_3O_2)_{12}]16H_2O$

Kehoite, $Zn_{5.5}Ca_{2.5}[(AlO_2)_{16}(PO_2)_{16}(H_3O_2)_{16}]32H_2O$

Analcime, $Na_{16}[(AlO_2)_{16}(SiO_2)_{32}]16H_2O$

A further compound in which PO_4^{3-} tetrahedra replace some SiO_4^{4-} is the phosphate garnet, griphite.[88] These compounds have not been made synthetically but they suggest the possibility of aluminophosphosilicates having frameworks

$$x\,AlO_2.y\,PO_2.z\,SiO_2$$

Where $x > y$ the framework would carry a nett negative charge $(x - y)$ and if $x < y$ it would carry a nett positive charge. In the latter case this charge would be neutralized by intracrystalline anions such as OH^-, Cl^-, F^-, and provided the framework was open like those in zeolites a crystalline anion exchanger would be possible. Therefore it is of interest to consider the system $(M_2^I, M^{II})O–Al_2O_3–P_2O_5–SiO_2–H_2O$ under aqueous alkaline zeolitization conditions. M^I and M^{II} are respectively mono- and divalent cations.

In one investigation of this system[89] phosphate was supplied as phosphoric acid, aluminium phosphate or sodium phosphate; aluminium as aluminate or aluminium hydroxide; and silica as a colloidal dispersion; the bases were NaOH or CaO. Temperatures were from around 100°C to 450°C. The products included phosphates (apatite, crandallite, angelite and scorzalite), aluminosilicates (montmorillonite, paragonite, analcime, albite and cancrinite) together with quartz, cristobalite and some unidentified phases. However, phosphates and aluminosilicates co-precipitated without providing evidence of the substitutions $2SiO_4^{4-} \rightleftarrows AlO_4^{5-} + PO_4^{3-}$ or $SiO_4^{4-} \rightleftarrows PO_4^{3-} + OH^-$. In a second investigation[90] the reactions were studied of kaolin and Na-faujasite with aqueous alkalis plus phosphates at temperatures up to 350°C. Again aluminophosphates were obtained along with aluminosilicates (potassium felspar, albite, kalsilite, kaliophilite, muscovite mica, phosphatic and basic cancrinite, sodalite hydrate, analcime, the gismondine type Na-P and the gmelinite type Na-S). Although the cancrinite contained intercalated phosphate, this phosphate is not part of the anionic framework and there was no evidence among compounds identified of aluminophosphosilicates.

Kuhl[91,92,4] also studied zeolite synthesis in the presence of phosphate, and found that the phosphate, by complexing with some of the aluminium as a diphosphato-aluminate, helped the formation of silica-enriched forms of faujasite, chabazite, phillipsite and zeolite A. The complexing equilibrium was considered to be

$$Al(OH)_4^- + 2PO_4^{3-} \rightleftarrows Al(PO_4)_2^{3-} + 4OH^-$$

For all the above zeolites save zeolite A, Na^+ and K^+ were the only cations employed; but for zeolite A, Na^+ and $N(CH_3)_4^+$ were the cations. In the case of zeolite A, two variants were produced termed ZK-21 and ZK-22 respectively. ZK-21 crystallized in the Na-form with $Na/Al = 1$; ZK-22 contained non-exchangeable charge-balancing $N(CH_3)_4^+$ ions trapped in the structure so that the ratio Na/Al was less than unity. Each form contained trapped phosphate in amounts not more than one phosphorus atom per unit pseudo cell. Some analytical results are given in Table 8. The ZK-21 variant which does not contain $N(CH_3)_4^+$ as charge-balancing cations is on the whole less siliceous than ZK-22 which does. Thus ZK-21 shows primarily the effect of complexing between Al^{3+} and the PO_4^{3-} on

TABLE 8
Analytical results[a] for some variants of zeolite A[(4)]

$\dfrac{SiO_2}{Al_2O_3}$	$\dfrac{Na_2O}{Al_2O_3}$	P atoms per unit pseudo cell	$\dfrac{SiO_2}{Al_2O_3}$	$\dfrac{Na_2O}{Al_2O_3}$	P atoms per unit pseudo cell
2·06	1·00	1·00	3·94	0·96	0·58
3·09	1·03	0·71	5·64	0·67	0·08
3·26	1·01	0·74	4·06	0·90	0·50
3·89	0·98	0·49	6·16	0·51	0·17
2·79	1·04	0·66	6·15	0·67	0·21
3·16	1·08	0·70	5·66	0·65	0·13
2·93	1·05	0·76	5·34	0·78	0·38
3·34	1·12	0·76	2·38	0·91	0·29
3·45	1·18	0·74	2·71	0·93	0·29
3·47	1·03	0·66	3·05	0·88	0·28
3·23	1·02	0·58	4·86	0·69	0·25
3·82	1·06	0·68	3·68	0·71	0·95
3·04	1·06	0·58	5·70	0·54	0·13
3·73	1·02	0·54	5·70	0·62	0·16
			6·24	0·63	0·12
			6·82	0·53	0·10

[a] On left, Z-21; on right, Z-22.

the Si/Al ratio while ZK-22 shows the effect of this complexing plus that of $N(CH_3)_4^+$ within the structure upon the Si/Al ratio. From the present point of view the phosphorus content is of particular interest. Per unit pseudo cell there is one 26-hedral and one 14-hedral sodalite cage, and, as noted previously, the sodalite cages are good traps for anions. Thus the phosphorus contents in Table 8 are compatible with intercalation of phosphate anions in the sodalite cages. The phosphorus contents of ZK-22 are on average below those of ZK-21, which would be expected if some of the charge-balancing $N(CH_3)_4^+$ cations in ZK-22 are also in the sodalite cages, to the exclusion of phosphate anions from such cages.

In contrast with these low phosphorus contents, Flanigen and Grose[(93)] reported the synthesis of zeolites with phosphorus contents in the range 5–25%, estimated by analysis as weight % of P_2O_5 and determined by microprobe analyser on the fine crystalline powders. These were analcime, phillipsite, chabazite, zeolite A, zeolite L and the gismondine type Na-Pl. Other zeolites were also prepared which contained less than 5% of P_2O_5 (gmelinite, mordenite and faujasite). In the more open zeolites intercalation of a few per cent by weight of phosphate is possible, but in no case would intercalated phosphate be expected to reach 25%. Also in analcime the framework is too compact for intercalation of phosphate to occur, so that the high phosphorus content must mean either that there is phosphate external to the crystals included in the analysis or that phosphorus in considerable amounts substitutes for Si on the tetrahedral framework sites as $2Si^{IV} \rightleftarrows Al^{III} + P^{V}$. Flanigen and Grose were of the latter opinion. The wet chemical analysis and electron probe gave respectively 22 and 19·5 weight % of P_2O_5 associated with the analcime. The molar proportions reported from the chemical analyses were those in Table 9. They are such as to suggest extremely large replacements of Si by P and Al as seen in the last two columns.

Because of the remarkable phosphate contents reported by Flanigen and Grose which contrast with the quantitative results of Kuhl (Table 8), and in view of the co-precipitations observed by Barrer and Marshall, a further study was made by Barrer and Liquornik[94] for aqueous alkaline mixtures in the range 80–150°C. As sources of Al, P and Si $AlPO_4$ and silica gels were prepared from metakaolinite and phosphoric acid:

$$Al_2O_3.2SiO_2 + 2H_3PO_4 \rightarrow 2AlPO_4 + 2SiO_2 + 3H_2O$$

The bases used were LiOH, NaOH and KOH, and zeolites Li-*ABW*, edingtonite type K-F, zeolite A, gismondine type Na-Pl and sodalite hydrate were prepared, the last four in the pure state. After thorough washing samples of each pure product were decomposed by acid and examined for their phosphorus contents, first by the molybdenum blue test and secondly by the more sensitive quantitative analytical procedure of Kirkbright *et al.*[95] The molybdenum blue test was too insensitive to detect any phosphate while the quantitative procedure gave the results in Table 10. The extremely low phosphorus contents are in complete contrast with those in Table 9. Even so, they are upper limits only to the amounts of phosphorus substituting in the frameworks because some or all of the phosphorus could be PO_4^{3-} intercalated in cavities in the

TABLE 9
Chemical analyses of products of Flanigen and Grose[93]

Zeolite type	Na_2O	K_2O	Al_2O_3	SiO_2	P_2O_5	H_2O	P atoms per unit cell	Tetrahedral sites per unit cell
Analcime	0·65	–	1·00	1·11	0·49	2·07	6·7	48
Phillipsite	–	0·54	1·00	1·73	0·37	2·98	10·7	64
Chabazite	–	0·54	1·00	1·54	0·36	3·80	2·0	12
Chabazite	0·84	–	1·00	1·92	0·35	4·13	2·0	12
Zeolite A	1·00	–	1·00	1·71	0·24	4·32	2·8	24
Zeolite L	–	0·69	1·00	1·59	0·38	2·53	12·6	72
Zeolite Pl	0·58	–	1·00	0·99	0·52	3·12	3·8	16

TABLE 10
Phosphorus contents of some zeolites[94]

Zeolite	P content $(\mu g\,g^{-1})$	P atoms per Al†	P atoms per unit cell
Zeolite A	95 ± 8	$0 \cdot 0005_6$	$0 \cdot 0067$
Zeolite A	482 ± 26	$0 \cdot 0028_4$	$0 \cdot 034$
Zeolite A	1100 ± 72	$0 \cdot 006_5$	$0 \cdot 078$
Zeolite A	1100 ± 70	$0 \cdot 006_5$	$0 \cdot 078$
Sodalite hydrate	2230 ± 50	$0 \cdot 011_7$	$0 \cdot 070$
Zeolite Na-P	110 ± 7	$0 \cdot 0006_3$	$0 \cdot 0050$
Zeolite Na-P	190 ± 10	$0 \cdot 0011$	$0 \cdot 0066$
Zeolite K-F	175 ± 10	$0 \cdot 0011$	$0 \cdot 0055$

† Assuming $SiO_2/Al_2O_3 = 2$ for each zeolite. This may not be exact for Na-P and K-F but is a very good approximation for sodalite hydrate and zeolite A.

frameworks. This is especially likely for zeolite A and sodalite where sodalite cages occur. These are known, as mentioned earlier, to be good traps for anions.[39,40] If the results of Barrer and Liquornik for zeolite A are compared with those of Kuhl (Table 8) it is seen that the method of preparation can, within limits, influence the amount of phosphorus associated with the structure. However, the results of Flanigen and Grose represent an extreme situation. They varied the ways of making up the reaction mixtures. For example, the analcime was prepared by adding aqueous sodium metasilicate and phosphoric acid to aqueous $AlCl_3$ and NaOH was then added. Crystallization was effected in the range 175–210°C. Phillipsite was made by titrating an aqueous mixture of $AlCl_3$ and phosphoric acid with concentrated KOH to a pH of 7·5 and, for crystallising, this product was heated with silica sol and KOH solution. Nevertheless, however made all the products were characterized by the extremely high phosphorus contents of Table 9. The location of this phosphorus (on framework sites, intercalated or external to the crystals) may repay further study.

In the beginning of this section reference was made to forms of $AlPO_4$ isostructural with quartz, tridymite and cristobalite, representing complete replacements $Al^{III}P^V \rightleftarrows 2Si^{IV}$. This aspect of aluminophosphate chemistry has been extended by hydrothermal crystallizations in the temperature range 100–250°C and in the presence of template amines or quaternary ammonium salts (all designated R). About 20 novel structures were reported[95a] of composition $AlPO_4$, xR, yH_2O where x and y denote numbers of template and water molecules required per $AlPO_4$ to fill intracrystalline voids. Six were layer structures which collapsed when

TABLE 11
Properties of some aluminophosphate molecular sieves[95a]

AlPO$_4$-n	Structure	Approx free, diameter of windows (Å)	Pore volumes (cm g^{-1})	
			O$_2$($-183°$C)	H$_2$O (room temp.)
AlPO$_4$-5	Determined and novel	~8	0·18	0·3
AlPO$_4$-11	Unknown	6·1	0·11	0·16
AlPO$_4$-14	Unknown	4·1	0·19	0·28
AlPO$_4$-16	Unknown	~3	0	0·3
AlPO$_4$-17	Erionite/ offretite type	4·6	0·20	0·28
AlPO$_4$-18	Unknown	4·6	0·27	0·35
AlPO$_4$-20	Sodalite type	~3	0	0·24
AlPO$_4$-31	Unknown	~8	0·09	0·17
AlPO$_4$-33	Unknown	4·1	0·23	0·23

interlayer templates were burnt out by heating in air at 400–600°C. Fourteen were three-dimensional networks, many of which were stable to heating and yielded the narrow, intermediate and wide-pore molecular sieves of Table 11 on burning out the template. As apparent examples of complete replacements $Al^{III}P^V \rightleftarrows 2Si^{IV}$ these compounds suggest that

TABLE 12
Some N-containing silicon and aluminium compounds

Structural type	Examples
β-Si$_3$N$_4$	Si$_3$N$_4$
β'-sialons	$\square_{x/12}$Si$_{(6-3x/4)}$Al$_{2x/3}$O$_x$N$_{8-x}$ $\quad (0 < x < 6)$
	Mg$_{x/4}$Si$_{(6-3x/4)}$Al$_{x/2}$O$_x$N$_{8-x}$ $\quad (0 < x < 6)$
	Be$_{x/4}$Si$_{(6-3x/4)}$Al$_{x/2}$O$_x$N$_{8-x}$ $\quad (0 < x < 6)$
	Li$_{x/8}$Si$_{(6-3x/4)}$Al$_{5x/8}$O$_x$N$_{8-x}$ $\quad (0 < x < {\sim}5)$
α-Si$_3$(O, N)$_4$	Si$_{11\cdot4}$N$_{15}$O$_{0\cdot3}$ to Si$_{11\cdot5}$N$_{15}$O$_{0\cdot5}$
	Li$_x$(Si, Al)$_3$(O, N)$_4$
Oxynitride	Si$_2$N$_2$O
	LiAlSiN$_2$O
Mullite	AlSiNO$_2$
Cristobalite	Li$_2$AlSiO$_3$N
Wurtzite	AlN
	MgSiN$_2$
	MnSiN$_2$
	LiSi$_2$N$_3$
Melilite	Y$_2$Si$_3$O$_3$N$_4$
Eucryptite	LiAlSiO$_2$N$_{4/3}$

more novel, porous, crystalline silicas may remain to be synthesized, and also porous networks based on BPO_4 and $GaPO_4$.

Nitrogen is more electronegative and acidigenic than phosphorus and the question arises whether it could replace framework oxygen to give the nitrogen equivalents of tectosilicates. In ceramic systems this replacement has been established through high-temperature reaction for a series of compounds, [96–101] examples of which are given in Table 12. The structural unit in β-Si_3N_4 is the SiN_4 tetrahedron. Because each nitrogen is trivalent it is shared by three tetrahedra to give the formula Si_3N_4 in place of SiO_2. A section of the framework is indicated in Fig. 13 in which open circles are N and filled circles are Si.[97]

Silicon nitride and alumina or gallia react at high temperatures to give new compositions, still with the β-Si_3N_4 structure, termed β'-sialons. The reaction represents isomorphous replacements of the type

$$Al^{3+} + O^{2-} \rightleftarrows Si^{4+} + N^{3-}$$

The substitution produces atom vacancies, denoted by □ in Table 12, and determined in extent by the amount x of N in the Si_3N_4 replaced by O in the sialon. Alumina with magnesia, beryllia or lithia also react with

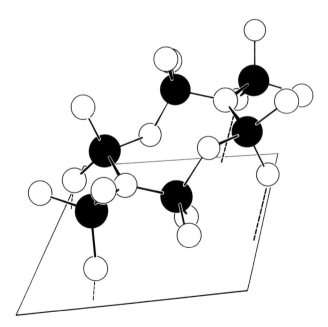

Fig. 13. The crystal structure of β-Si_3N_4.[97] Open circles denote N and filled circles denote Si.

β-Si_3N_4 so that the vacancies are filled, yielding the compositions in the table, still as β'-sialons. Another form of silicon nitride is known, termed α-Si_3N_4. This, however, always contains a small amount of oxygen, as shown in Table 12. It reacts with alumina and lithia to give α'-sialons as in the table.

Structures differing from either of the above are also shown in Table 10, one of these being isomorphous with mullite[97] and another with cristobalite.[97] There is also a group of phases with the wurtizite structure.[98] Aluminium nitride is based on the AlN_4 tetrahedral unit analogous to the SiN_4 unit in the sialons. Yttria reacts with Si_3N_4 to give a phase (Table 12) having the melilite structure typified by akermanite, $Ca_2Mg[Si_2O_7]$ and gehlenite, $Ca_2Al[SiAlO_7]$ with which it forms a complete series of solid solutions.[99] The yttrium-bearing structure consists of layers of $Si(O, N)_4$ tetrahedra cross-linked through yttrium. A final example in Table 12 is a Li-"sialon" having the structure[101] of eucryptite, $LiAlSiO_4$.

Eucryptite can be synthesized hydrothermally at relatively low temperatures,[102] and this raises the question whether synthesis of nitrogen-bearing aluminosilicates including zeolites could be made ammonothermally. Reaction mixtures would comprise ammonia or hydrazine as solvent in place of water, bases like $NaNH_2$ or KNH_2 as the equivalents of NaOH and KOH, and AlN, $NaAl(NH_2)_4$ or Al_2O_3 as sources of aluminium, with dry silica gel or Si_3N_4 as sources of silicon.

9. Concluding Remark

The evidence presented in this chapter shows that Ge and Ga can readily replace Si and Al in zeolites, and that a beryllium analcime has also been made. Under high-temperature conditions, often well outside the range of zeolite formation, iron, certain rare earth elements and boron can replace Si or Al, but the structures are not zeolites. The evidence of significant framework substitution by B^{III}, Fe^{III}, Cr^{III}, Zr^{IV} or Ti^{IV}, under the low-temperature aqueous alkaline conditions normally used in zeolitization, is incomplete and ambiguous. The evidence of substitution P^V in zeolite frameworks is contradictory in that one study has claimed remarkably large substitutions which have not so far been substantiated by other investigators. It appears also that a number of porous aluminophosphates, $AlPO_4$, can be prepared which imitate the behaviour of zeolite molecular sieves. In some ceramic compounds with structures analogous to particular silicates it has been found that N^{III} and O^{II} can replace each other in the frameworks. Whether such a replacement is possible under zeolitizing conditions remains unknown. A first requirement would be a change from hydrothermal to ammonothermal systems. Finally, there seems so far no clear exception to Lowenstein's rule of $\overset{\ominus}{Al}$–O–$\overset{\ominus}{Al}$ avoidance for aluminosilicates in which Si/Al ⩾ 1.

References

1. R.M. Barrer and J.W. Baynham, *J. Chem. Soc.* (1956) 2882.
2. A.J. Gude III and R.A. Sheppard, *Amer. Mineral.* (1966) **51,** 909.
3. H. Kacirek and H. Lechert, *J. Phys. Chem.* (1976) **80,** 1291.
4. G.H. Kuhl, *Inorg. Chem.* (1971) **10,** 2488.
5. R.M. Barrer and D.E. Mainwaring, *J. Chem. Soc. Dalton* (1972) 1259.
6. D.W. Breck and E.M. Flanigen, *in* "Molecular Sieves", p. 47. Society of Chemical Industry, London, 1968.
7. C. Baerlocher and W.M. Meier, *Helv. Chim. Acta* (1969) **52,** 1853.
8. W. Lowenstein, *Amer. Mineral.* (1954) **39,** 92.
9. R.M. Barrer and W.M. Meier, *Trans. Faraday Soc.* (1959) **55,** 130.
10. R.M. Barrer, J.W. Baynham, F.W. Bultitude and W.M. Meier, *J. Chem. Soc.* (1959) 195.
11. E. Passaglia, D. Pongiluppi and R. Rinaldi, *Neues Jb. Mineralog. Mh.* (1977), **8,** 355.
12. K.-J. Chao, T.C. Tasi, M.-S. Chen and I. Wang, *J. Chem. Soc. Faraday I* (1981) **77,** 547.
13. S. Ueda and M. Koizumi, *Amer. Mineral.* (1979) **64,** 172.
14. S.P. Zdhanov, *in* "Molecular Sieves", p. 70. Society of Chemical Industry, London, 1968.
14a. H.E. Robson, Pittsburgh Catalysis Soc. Symp., 20th May 1981.
14b. M.N. Maleyev, *Internat. Geol. Review* (1977) **19,** 993.
15. S.M. Stishov and S.V. Popova, *Geokhimiya* (1961) **10,** 837.
16. L. Coes, Jr., to Norton Co., 1959, U.S.P. 2,876,072.
17. P.P. Keat, *Science* (1954) **120,** 328.
18. B. Kamb, *Science* (1965) **148,** 232.
19. E.M. Flanigen, J.M. Bennett, R.W. Grose, J.P. Cohen, R.L. Patton, R.M. Kirchner and J.V. Smith, *Nature* (1978) **271,** 512.
20. D.M. Bibby, N.B. Milestone and L.P. Aldridge, *Nature* (1979) **280,** 664.
21. H.K. Beyer and I. Belenykaja, *in* "Catalysis by Zeolites" (Ed. B. Imelik *et al.*) p. 203. Elsevier, Amsterdam, 1980.
22. R.K. Iler, "The Chemistry of Silica", p. 16. Wiley-Interscience, New York, 1979.
23. R.M. Barrer and D.E.W. Vaughan, *Trans. Faraday Soc.* (1967) **63,** 2275.
24. R.M. Barrer, *in* "Non-stoichiometric Compounds" (Ed. L. Mandelcorn), pp. 314, 315. Academic Press, London and New York, 1964.
25. R.M. Barrer and D.J. Ruzicka, *Trans. Faraday Soc.* (1962) **58,** 2239.
26. R.M. Barrer and A.V.J. Edge, *Proc. Roy. Soc.* (1967) **A300,** 1.
27. R. McMullan and G.A. Jeffrey, *J. Chem. Phys.* (1959) **31,** 1231.
28. G.T. Kokotailo and W.M. Meier, *in* "The Properties and Applications of Zeolites" (Ed. R.P. Townsend), p. 133. Chemical Society Special Publication No. 33, 1979.

29. J. Klinowski, J.M. Thomas, M. Audier, S. Vasudevan, C.A. Fyfe and J.S. Hartman, *Chem. Comm.* (1981) 570.
30. K.J. Murata, *Amer. Mineral.* (1943) **28,** 545.
31. R.M. Barrer and M.B. Makki, *Canad. J. Chem.* (1964) **42,** 1481.
32. G.T. Kerr, *J. Phys. Chem.* (1968) **72,** 2594.
33. N.Y. Chen, *J. Phys. Chem.* (1976) **80,** 60.
34. R.M. Barrer, "Zeolites and Clay Minerals as Sorbents and Molecular Sieves", p. 350. Academic Press, London and New York, 1978.
35. von U. Lohse, H. Stach, H. Thamm, W. Schirmer, A.A. Isirikjan, N.I. Regent and M.M. Dubinin, *Zeit. Anorg. Allg. Chem.* (1980) **460,** 179.
35a. K. Sahl, *Zeit. Krist.* (1980) **152,** 13.
36 R M Barrer, "Zeolites and Clay Minerals as Sorbents and Molecular Sieves", Ch. 3, 4, 5. Academic Press, London and New York, 1978.
37. R.M. Barrer and J. Klinowski, *Phil. Trans.* (1977) **A285,** 637.
38. J.R. Lacher, *Proc. Camb. Phil. Soc.* (1937) **33,** 518.
39. R.M. Barrer, J.F. Cole and H. Sticher, *J. Chem. Soc.* A (1968) 2475.
40. R.M. Barrer and J.F. Cole, *J. Chem. Soc.* A (1970) 1516.
41. W. Depmeier, *J. Appl. Cryst.* (1979) **12,** 623.
42. V.I. Ponomarev, D.M. Kheiker and N.V. Belov, *Soviet Physics—Crystallography* (1971) **15,** 799 (English translation).
43. L. Pauling, *Zeit. Krist.* (1930) **74,** 213.
43a. M.G. Barker, P.G. Gadd and M.J. Begley, *Chem. Comm.* (1981) 379.
43b. K. Takeuchi and R. Sadanaga, *Mineral. J. Jap.* (1966) **4,** 424.
43c. E. Lippmaa, M. Mägi, A. Samoson, G. Engelhardt and A.H.R. Grimmer, *J. Amer. Chem. Soc.* (1980) **102,** 4889.
43d. E. Lippmaa, M. Mägi, A. Samoson and G. Engelhardt, *J. Amer. Chem. Soc.* (1981) **103,** 4993.
43e. M.T. Melchior, D.E.W. Vaughan and A.J. Jacobson, *J. Amer. Chem. Soc.*, in press.
43f. G. Engelhardt, U. Lohse, E. Lippmaa, M. Tarmak and M. Mägi, *Zeit. Anorg. Allg. Chemie,* (1981), **482,** 49.
43g. J. Klinowski, J.M. Thomas, C.A. Fyfe and J.S. Hartman, *J. Phys. Chem.,* 1981, **85,** 2590.
43h. L.A. Bursill, E.A. Lodge, J.M. Thomas and A.K. Cheetham, *J. Phys. Chem.* (1981) **85,** 2409.
44. G. Engelhardt, D. Zeigan, E. Lippmaa and M. Magi, *Zeit. Anorg. Allg. Chemie* (1980) **468,** 35.
45. J.M. Thomas, L.A. Bursill, E.A. Lodge, A.K. Cheetham and C.A. Fyfe, *Chem. Comm.* (1981) 276.
45a. M.T. Melchior, D.E.W. Vaughan, R.H. Jarman and A.J. Jacobson, private communication, and to appear in *Nature*, 1982.
45b. J.M. Thomas, C.A. Fyfe, S. Ramdas, J. Klinowski and G.C. Gobbi, private communication, and to appear in *Nature*, 1982.
46. G. Engelhardt, E. Lippmaa and M. Mägi, *Chem. Comm.* (1981) 712.
46a. S. Ramdas, J.M. Thomas, J. Klinowski, C.A. Fyfe and J.S. Hartman, *Nature* (1981) **292,** 228.
46b. M.T. Melchior, D.E.W. Vaughan and A.J. Jackson, private communication.
47. E. Lippmaa, M. Mägi, A. Samoson and G. Engelhardt, *J. Amer. Chem. Soc.* (1981) **103,** 4993.
48. E.E. Senderov, *Phys. Chem. Mineral* (1980) **6,** 251.
49. J.R. Goldsmith, *Min. Mag.* (1952) **29,** 952.

50. J. Klinowski, private communication.
51. J. Selbin and R.B. Mason, *J. Inorg. Nucl. Chem.* (1961) **20**, 222.
52. L. Lerot, G. Poncelet and J.J. Fripiat, *Mat. Res. Bull.* (1974) **9**, 979.
53. G. Poncelet and M. Lauriers, *Mat. Res. Bull.* (1975) **10**, 1205.
54. L. Lerot, G. Poncelet, M.K. Dubru and J.J. Fripiat, *J. Catal.* (1975) **37**, 396.
55. L. Lerot, G. Poncelet and J.J. Fripiat, *J. Solid State Chem.* (1975) **12**, 283.
56. G. Poncelet, M.L. Dubru and T. Lux, *Mat. Res. Bull.* (1976) **11**, 813.
57. G. Poncelet and M.L. Dubru, *J. Catal.* (1978) **52**, 321.
58. S. Ueda and M. Koizumi, *in* "Molecular Sieve Zeolites-1" (Ed. R.F. Gould), p. 135. American Chemical Society Advances in Chemistry Series, No. 101. 1971.
59. J. Stabenow, L. Marosi and M. Schwarzmann, to BASF, 1976, Ger. Pat. 2,429,182.
60. E.W. Roedder, *Amer. J. Sci.* (1951) **249**, 81 and 224.
61. R.M. Barrer and W. Sieber, *Chem. Comm.* (1977) 905.
62. E.L. Muetterties, "The Chemistry of Boron and its Compounds". Wiley, New York, 1967.
63. D.E. Appleman and J.R. Clark, *Amer. Mineral.* (1965) **50**, 1827.
64. P. Smith, S. Garcia Blanco and L. Rivoir, *Zeit. Krist.* (1961) **115**, 460.
65. H.P. Eugster and N.L. McIver, *Bull. Amer. Geol. Soc.* (1959) **70**, 1598.
66. R.M. Barrer and E.F. Freund, *J. Chem. Soc., Dalton* (1974) 1049.
67. M. Taramasso, G. Perego and B. Notari, *in* "Proceedings of the 5th International Conference on Zeolites" (Ed. L.V.C. Rees) p. 40. Heyden, London, 1980.
68. R.M. Barrer and E.F. Freund, *J. Chem. Soc., Dalton* (1974) 20 and 2060.
69. T. Ito and H. Mori, *Acta Cryst.* (1953) **6**, 24.
70. C. Dunbar and F. Machatschki, *Zeit. Krist.* (1930) **70**, 133.
71. E.M. Levin and G.M. Urgrinic, *J. Res. Nat. Bur. Stds.* (1953) **51**, 37.
72. R.M. Barrer and E.F. Freund, *J. Chem. Soc. Dalton* (1974) 2123.
73. P. Hautefeuille, *Compt. Rend.* (1880) **90**, 303 and 378.
74. A. Perrey, *Compt. Rend.* (1888) **107**, 1150.
75. A. Duboin, *Compt. Rend.* (1927) **185**, 416.
76. K. Hirao, N. Soga and M. Kunugi, *J. Phys. Chem.* (1976) **80**, 1612.
77. R.M. Barrer and N. McCallum, *J. Chem. Soc.* (1953) 4029.
78. W. Eitel, E. Herlinger and G. Tromel, *Naturwiss.* (1930) **18**, 469.
79. L. Marosi, J. Stabenow and M. Schwarzmann, 1980, Ger. Pat. 2,831,611.
80. L. Marosi, J. Stabenow and M. Schwarzmann, 1980, Ger. Pat. 2,831,630.
81. B.D. McNicol and G.T. Pott, *Chem. Comm.* (1970) 438.
82. B.D. McNicol and G.T. Pott, *J. Catal.* (1972) **25**, 223.
83. J.S. Griffith, *Mol. Phys.* (1964) **8**, 213.
84. B. Wichterlova and P. Jiru, *React. Kinet. Lett.* (1980) **13**, 197.
85. D.A. Young, 1967, U.S.P. 3,329,480 and U.S.P. 3,329,481.
86. D. McConnell, *Min. Mag.* (1964) **33**, 799.
87. D. McConnell, *Amer. Mineral.* (1952) **37**, 609.
88. D. McConnell and F.H. Verhoek, *J. Chem. Educ.* (1963) **40**, 512.
89. R.M. Barrer and D.J. Marshall, *J. Chem. Soc.* (1965) 6616.
90. R.M. Barrer and D.J. Marshall, *J. Chem. Soc.* (1965) 6621.
91. G.H. Kuhl, to Mobil Co., 1967, U.S.P. 3,355,246.
92. G.H. Kuhl, *in* "Molecular Sieves", p. 85. Society of Chemical Industry, London, 1968.

93. E.M. Flanigen and R.W. Grose, *in* "Molecular Sieve Zeolites-1" (Ed. R.F. Gould), p. 76. American Chemical Society Advances in Chemistry Series, No. 101, 1971.
94. R.M. Barrer and M. Liquornik, *J. Chem. Soc. Dalton* (1974) 2126.
95. G.F. Kirkbright, A.M. Smith and T.S. West, *Analyst* (1967) **92,** 411.
95a. S.T. Wilson, B.M. Lok, C.A. Messina, T.R. Cannan and E.M. Flanigen, *J. Amer. Chem. Soc.*, (1982) **104,** 1146.
96. K.H. Jack and W.I. Wilson, *Nature* (1972) **238,** 28.
97. K.H. Jack, *Trans. and J. Brit. Ceram. Soc.* (1973) **72,** 376.
98. A. Hendry, D.S. Perera, D.P. Thompson and K. H. Jack, *in* "Special Ceramics, 6", p. 321. British Ceramic Research Association, 1975.
99. A.W.I.M. Rae, D.P. Thompson, N.J. Pipkin and K.H. Jack, *in* "Special Ceramics 6" p. 347. British Ceramic Research Association, 1975.
100. A. Hendry and K.H. Jack, *in* "Special Ceramics 6", p. 199. British Ceramic Research Association, 1975.
101. S.A.B. Jama, D.P. Thompson and K.H. Jack, *in* "Special Ceramics 6", p. 299. British Ceramic Research Association 1975.
102. R.M. Barrer and E.A.D. White, *J. Chem. Soc.* (1951) 1267.

Chapter 7

Synthesis and Some Properties of Salt-bearing Tectosilicates

1. Introduction

Porous tectosilicates may form in the presence either of water or of salts. The water or salt then functions as lattice filler and stabilizer, as considered in thermodynamic terms in Chapter 2 §4. The replacement of water by salts or of one salt by another is an isomorphous replacement differing from ion exchange of charge-balancing cations, or from the replacements of elements in the anionic frameworks considered in the previous chapter. Because salts are relatively non-volatile and high melting, dry way syntheses of salt-bearing tectosilicates can be expected from salt melts and sources of silica, alumina and alkali. Salt-bearing tectosilicates can also form in the presence both of salts and water. A number of these, either occurring naturally or synthesized with the salts incorporated during formation, are given in Table 1. The diversity of salts which can be incorporated in the sodalite and cancrinite groups is noteworthy (see §2.3). It is of interest that structurally these two groups can be related *inter se* and with the chabazite family of zeolites, in terms of sequences of layers, either of six-membered rings of TO_4 tetrahedra (T = Si or Al) or of double 6-rings (hexagonal prisms). There are three ways of stacking such layers on the layer below, denoted by a, b and c for layers composed of single 6-rings or A, B and C for layers composed of hexagonal prisms. Then the structural relations referred to above are summarized in Table 2.[1a] While instances of the chabazite family of zeolites incorporating salts during synthesis have not yet been reported this may be because the relevant experiments have not been made. However, for structures composed only of layers of single 6-rings in various stacking sequences, Table

TABLE 1
Instances of salt-bearing tectosilicates

Mineral type and examples	Examples of salts included in natural or synthetic products
Scapolites Marialite Wernerite Meionite	NaCl, Na_2SO_4, Na_2CO_3
Sodalite-nosean group[a] Sodalite Nosean Hauyne Ultramarine Helvite Danalite	NaCl, Na_2SO_4, $CaSO_4$, Na_2CO_3
Cancrinite group[a] Cancrinite Davynite Vischnevite Microsommite	NaCl, Na_2SO_4, $CaCO_3$
Cancrinite-related Liottite Afghanite Franzinite	$(Na, K, Ca)_{1-2}(CO_3, SO_4, Cl, OH)_{1-2}$
Zeolite types Zeolite ZK-5 Zeolite K-F (edingtonite type) Phase A[b] (merlinoite type[1])	$BaCl_2$, $BaBr_2$, H_2O KCl, KBr, KI, H_2O $BaCl_2 + H_2O$

[a] See §2.2 for additional examples of intercalated salts.
[b] Unit cell composition $Ba_6[Al_{12}Si_{20}O_{64}]4BaCl_2.2H_2O$.

1 shows examples in which there is salt uptake. The table also makes it clear that other tectosilicate frameworks in zeolite and felspathoid groups can during formation show a propensity for salt uptake analogous to that of the sodalite and cancrinite minerals.

2. Syntheses of Salt-bearing Tectosilicates

Syntheses of salt-bearing tectosilicates may be made, as noted above, either in the dry way from salts and sources of Al_2O_3, SiO_2 and alkali or in the presence of both water and salts. Examples of both methods are given in §§2.1 and 2.2.

TABLE 2

Layer sequences in sodalite–cancrinite–chabazite minerals

No. of layers in repeat unit	Layer sequence	Space group	a (Å)	c (Å)	Name
2	ab	P6̄	12·72	5·19	Cancrinite
3	abc	P4̄3m	8·87	–	Sodalite
4	abac	P6₃/mmc	12·91	10·54	Losod
6	ababac	P6̄m2	12·85	16·10	Liottite
8	ababacac	P6₃mc	12·77	21·35	Afghanite
10	abcabcbacb	P3̄m1	12·88	26·76	Franzinite
2	AB	P6₃/mmc	13·75	10·05	Gmelinite
3	ABC	R3̄m	13·78	15·06	Chabazite
2	Ab	P6̄m2	13·29	7·58	Offretite
4	AbAc	P6₃/mmc	13·26	15·12	Erionite
6	AbCaBc	R3̄m	13·34	23·01	Levynite

2.1. *Dry Way Syntheses*

The sodalite framework is readily made under pyrolytic and sintering conditions. Because of its excellence as a blue pigment, the synthesis and properties of ultramarine, which has this framework, has been the subject of study since 1828 when Guimet made the first synthesis and thereby won a prize offered by Napoleon.[2] Large-scale syntheses have been based on heating fine-grade specially treated china clay mixed with sulphur and anhydrous sodium sulphate or carbonate. Carbon, pitch or rosin are further components of the reaction mixture and serve with the sulphur to reduce the sodium salts to give the intercalated polysulphides, Na_2S_x, needed to produce the eventual vivid blue colour. The mixture is fired at temperatures brought to 850–900°C over a period of 4–5 hours, and held at these values for 6–12 hours. After completion of the reduction the furnaces are cooled over a period of 2–5 days, according to the scale of the operation.

The cooled, reduced ultramarine is at this stage usually green in colour and is ground in preparation for final conversion to the blue pigment. The powder is mixed with 7–10% more sulphur and heated to a bright red heat in muffle furnaces for 3–6 days. Small amounts of air enter the furnaces oxidizing sulphur to SO_2 which in turn converts the green powder to the blue pigment. Alternatively, air may be admitted in controlled amounts during the cooling stage. The ultramarine has the composition $Na_6[Al_6Si_6O_{24}]Na_2S_x$, where $x > 1$, and the higher the sulphur content the better tends to be the quality of the pigment.

A number of variants of the above process have also been successful.

One of these[3] employed initial mixtures of Na_2CO_3, Al_2O_3, SiO_2 and Na_2S in the molar proportions $3:3:6:1$. The mixtures were heated in nitrogen to 900°C for 15 hours. The green product was then heated in H_2S for 2 hours followed by ignition in air, to yield a deep blue ultramarine. When Na_2S in the above mixtures was replaced by $2NaCl$ the product of heating was sodalite, while if the sodalite was heated in H_2 plus Se vapour and then in air a red-brown ultramarine was obtained in which Se replaced the S in ultramarine blue. Jaeger[4] reported that blood-red and yellow ultramarines were obtained when Se and Te respectively replaced S.

In other examples[5] the reaction mixtures were sodium aluminate, silica (as diatomaceous earth), anhydrous sodium acetate and sulphur; or zeolite, sulphur, anhydrous sodium acetate and a little soap binding agent. In each case the initial reduction stage yielded the ultramarine structure and the secondary calcination gave the improved blue pigment. The beautiful blue colour seems to be the result of three simultaneously operating factors: the alkali; the high sulphur content; and the sodalite type framework.[6] The sulphur content can be reduced in a controlled way by treatment with aniline hydrochloride, which also removes Na^+ ions. However some of the sulphur remains firmly bound within the sodalite framework. Reduction of ultramarine blue with molten sodium formate removes only alkali, but by heating the product in air at 550°C the blue colour is restored. Heating in Cl_2 gas at 400°C changes ultramarine blue to a substance containing $\sim 1\cdot 5\%$ sodium, and sulphur, but still having the sodalite framework. When the white product is heated in molten sodium nitrate the blue colour re-appears, presumably through restoration of the sodium content. If ultramarine blue is heated in a current of nitrogen above 950°C sulphur is driven off. When heated in hydrogen gas hydrogen is absorbed within the sodalite cavities, probably by formation of H_2S which remains encapsulated. However, at 850–875°C the H_2S is evolved and in this way the sulphur is eliminated. Above 500°C the sulphur in ultramarine blue reacts with oxygen and above 750°C the product is similar to nosean, with trapped Na_2SO_4 in the sodalite cages.[7] The sodium ions in ultramarine can readily be replaced by ions such as Ag^+, Li^+, K^+ and Pb^{2+} often with striking colour changes.[4,8,9] For this purpose salt melts or aqueous solutions may be used, but in pure water there is evidence that intracrystalline sodium can in part be replaced by hydrogen.[10] Ultramarine evidently has a varied chemistry involving both anhydrous and aqueous media.[24]

Yamaguchi and Kubo[10,11,12] made sodalite structures containing Na_2S and Na_2S_2 from a starting composition $NaAlSiO_4$. This composition was made by heating kaolinite and Na_2CO_3 to about 700°C. It was then mixed with different amounts of Na_2S and heated in N_2 or H_2 at temperatures of 850–950°C for an hour. Starting with equimolar proportions ($NaAlSiO_4/Na_2S = 1$) the sodalite type products had per unit cell $\sim$1·5 atoms of sulphur at 850°C and $\sim$1·8 atoms of sulphur at 950°C. The sulphur content could be increased by heating the parent $NaAlSiO_4 +$ Na_2S mixtures in nitrogen atmospheres with controlled vapour pressures of sulphur, again in the range 850–950°C. As the sulphur content increased, the blue colour associated with ultramarine began to develop, but was much less intense than in ultramarine blue. The maximum sulphur content corresponded approximately with Na_2S_2 in the sodalite cages, and the product was called leuco-ultramarine.

Eitel[13] reported melt equilibria in the reciprocal system (Na_2^{2+}, Ca^{2+})– (CO_3^{2-}, $Al_2Si_2O_8^{2-}$) under a carbon dioxide pressure of about 112 kg cm^{-2}. Carbonate-bearing cancrinites formed readily, and other phases were $CaCO_3$, Na_2CO_3, (Na_2, $Ca)CO_3$ and the non-porous silicates anorthite, nepheline and carnegieite. The appearance of another porous tectosilicate, meionite, a member of the scapolite group of Table 1, was also indicated. In another pyrolytic process Lambertson[14] observed that glass melts containing sodium sulphate reacted in a reducing atmosphere with aluminosilicates of the furnace linings to give blue nosean together with nepheline ($NaAlSiO_4$). Also sodalites have been prepared by fusing a large excess of $NaCl$ with nepheline,[15] with kaolinite and sodium carbonate[16] and with mixtures of Al_2O_3, SiO_2 and $NaOH$.[17] Morozewicz[18] cooled substantial amounts of a multicomponent aluminosilicate melt at 10 hour intervals from 1600°C to 800°C. When $CaSO_4$, Na_2SO_4 and $NaCl$ were added to the mixture hauyne and sodalite were reported among the products. Such preparations are thus not uncommon, as these examples show.

2.2. *Hydrothermal Syntheses of Sodalites and Cancrinites*

Exploration in hydrothermal systems of reactions yielding salt-bearing sodalites and cancrinites began nearly a hundred years ago, the work of Lemberg,[19] Thugutt[20] and the Friedels[21] being of note. Thugutt's work was especially detailed and showed that many sodium salts could be "bound" by sodalite. From later work (Table 4 and Ref. (22)) one may add $NaMnO_4$, Na_2FeO_4, NaN_3 and $NaClO_2$ to the salts in Table 3 which can be incorporated into either cancrinite or sodalite. In hydrothermal preparations the intracrystalline cavities and channels are shared by the salt, water and to a lesser extent sodium hydroxide so that the products are intermediate between zeolites and felspathoids. Examples of the compositions obtained which demonstrate this are given in Table 3.

Thugutt did not normally determine the caustic soda content, but the results tend to show that the greater the salt content the smaller the amount of zeolitic water.

Lemberg[19] made the noseans of Table 3 from natrolite, orthoclase, albite and labradorite, digested with strong caustic soda and Na_2SO_4 at temperatures and times ranging from 100°C and $6\frac{1}{2}$ months to about 190°C for much shorter times. In other experiments NaCl, Na_2CO_3, $NaNO_3$ and Na_2S were incorporated. Thugutt made the salt-bearing felspathoids from kaolinite and aqueous caustic soda containing the relevant dissolved salts at temperatures ranging from about 185°C to 220°C. All were termed sodalites, but recently, using kaolinite and 4 M NaOH plus salt at 80°C,

TABLE 3
Earlier syntheses of salt-bearing felspathoids

Date and reference	Product	Example of analytically found salt and water content per unit cell
1883[19]	Nosean	$0\cdot90Na_2SO_4, 2\cdot8H_2O$
	Nosean	$0\cdot96Na_2SO_4, 2\cdot9H_2O$
	Nosean	$0\cdot68Na_2SO_4, 3\cdot4H_2O$
1892[20]	Cl^--sodalite	$2NaCl, H_2O$
	Br^--sodalite	$1\cdot5NaBr, H_2O$
	ClO_3^--sodalite	$2NaClO_3, 0\cdot5H_2O$
	BrO_3^--sodalite	$0\cdot5(NaBrO_3, 2NaOH), H_2O$
	IO_3^--sodalite	$1\cdot2NaIO_3, 3H_2O$
	ClO_4^--sodalite	$2NaClO_4, H_2O$
	CO_3^{2-}-sodalite	$Na_2CO_3, 3H_2O$
	SO_3^{2-}-sodalite	Na_2SO_3, H_2O
	SO_4^{2-}-sodalite	$Na_2SO_4, 3H_2O$
	CrO_4^{2-}-sodalite	$0\cdot75Na_2CrO_4, 3\cdot8H_2O$
	SeO_4^{2-}-sodalite	$0\cdot75Na_2SeO_4, 3H_2O$
	MoO_4^{2-}-sodalite	$0\cdot75Na_2MoO_4, 5H_2O$
	WO_4^{2-}-sodalite	$0\cdot5Na_2WO_4, 5H_2O$
	PO_4^{3-}-sodalite	$0\cdot5Na_3PO_4, 4\cdot5H_2O$
	VO_4^{3-}-sodalite	$0\cdot5Na_3VO_4, 3\cdot5H_2O$
	AsO_4^{3-}-sodalite	$0\cdot5Na_3AsO_4, 5H_2O$
	$S_2O_3^{2-}$-sodalite	$0\cdot75Na_2S_2O_3, 2\cdot3H_2O$
	Formate-sodalite	$2HCOONa, H_2O$
	Acetate-sodalite	$2CH_3COONa$
	Oxalate-sodalite	$0\cdot75Na_2C_2O_4, 4\cdot5H_2O$
1945[22]	CO_3^{2-}-sodalite	$0\cdot42Na_2CO_3, 7\cdot3H_2O$
	Cl^--sodalite	$1\cdot6NaCl, 3\cdot3H_2O$
	Br^--sodalite	$1\cdot5_1NaBr, 2\cdot2H_2O$
	NO_3^--sodalite	$1\cdot5_7NaNO_3, 1\cdot9H_2O$
	SO_4^{2-}-cancrinite	$0\cdot86Na_2SO_4, 4H_2O$

TABLE 4

Sodalite and cancrinite formation from kaolinite and alkaline salt
solutions under prescribed conditions[24,25]a

Sodalite with:	Cancrinite with:
Alkali solution alone	Na_2SO_4
NaCl	Na_2SeO_4
NaBr	Na_2CrO_4d
NaI	Na_2MoO_4
NaFb	2 g Na_2FeO_4 + Fe^{III} + 12 M $NaOH$e
$NaClO_3$	Na_3VO_4
$NaClO_4$	$NaMnO_4$ (large excess)
Na_2SO_3	$NaNO_3$
Na_2Sc	$Cu(NH_3)_4SO_4$ + excess NH_3f
Na_2WO_4	
Na_3PO_4	
HCOONa	
CH_3COONa	
$(COONa)_2$	

a The prescribed conditions were: 2 g kaolinite reacted for 5 days at 80°C with 200 ml of 4 M NaOH containing 10 g of dissolved salt, or the amounts of salt specified in the table.
b No evidence of intercalated NaF.
c Pale blue crystals.
d Pale yellow crystals.
e Pale brown crystals mixed with species H (Chapter 6 §7).
f Pale blue crystals.

Barrer *et al.*[24,25] found that the salts employed by Thugutt sometimes yielded sodalites and sometimes cancrinites (Table 4). Also Lemberg[26] found that kaolin and prehnite with aqueous Na_2CO_3 yielded cancrinite rather than sodalite. The Friedels[21] made nosean from micas with aqueous $NaOH + Na_2SO_4$ at 500°C, and sodalites with aqueous $NaOH + NaCl$ at the same temperature. With aqueous $NaOH + Na_2CO_3$ the product was cancrinite. Evidently various aluminosilicates can with alkaline salt solutions yield salt-bearing felspathoids.

Sodalites and cancrinites can also be made very easily in absence of any salts. They then contain only zeolitic water and a limited amount of caustic soda. According to at least one analysis[27] there may also be a little detrital silica. As the hydrates they are no longer felspathoids but zeolites. In Chapter 5 §§3, 4, 5, 6 and 8 there are numerous examples of the synthesis of sodalite and cancrinite hydrates from gels, tetramethyl-ammonium aluminosilicate solutions, kaolinite, metakaolinite, mont-morillonite, analcime and volcanic and synthetic glasses, at temperatures ranging from 80 to 450°C. In the ideal cancrinite structure (Fig. 1) there are wide parallel channels circumscribed by 12-rings. However, so far synthetic cancrinite hydrates when outgassed have behaved as very narrow-pore structures.[28] This behaviour may arise in part because of stacking faults, in which the ab... stacking of layers (Table 2) is interrupted by, for example, abc... stacking, i.e. by sodalite layers. Further obstruction may be due to NaOH and other detrital material in the wide channels. Sodalite hydrate (Fig. 2) contains only 14-hedral cages of Type 1,[29] access to which occurs via 6-ring windows of free diameter ~2·2 Å. Despite this, outgassed sodalite hydrates at high pressures and temperatures can sorb Kr and Ar with crystallographic diameters of $3·9_4$ Å and $3·8_3$ Å respectively, up to a limit corresponding with about one rare gas molecule per sodalite cage.[28] Because of the disparity between the diameters of the gases sorbed and the free diameter of the windows, if the zeolite charged with Ar or Kr is rapidly cooled to room temperature and the pressure then released, the crystals retain the trapped gases more or less indefinitely. Cancrinite hydrate after outgassing did not under corresponding conditions of temperature and pressure sorb as much as sodalite hydrate, though the blocked channels could on quenching also retain whatever had been sorbed at the high temperature.

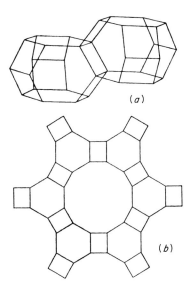

Fig. 1. Structural elements of the cancrinite framework: (a) shows the 11-hedral cages and (b) shows a view down the 6_3 axis indicating a cross-section of the wide channel circumscribed by puckered 12-membered rings. Cations are not shown. Al or Si is centred at each corner and oxygen near the mid-point of each edge.

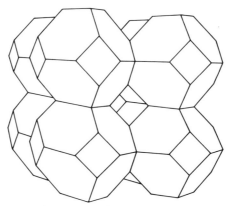

Fig. 2. The framework pattern of sodalite made of 14-hedral cages. Cations are not shown, Al or Si is centred at each corner and oxygen near the mid-point of each edge.

Because of the presence of entrained NaOH as well as the water, sodalite and cancrinite hydrates have been alternatively termed hydroxy- or basic sodalite and cancrinite. Barrer and White[30] made sodalites in presence and absence of sodium chloride, noseans in presence and absence of sodium sulphate and cancrinites in presence and absence of sodium carbonate. Unit cell dimensions then showed appreciable differences among the various specimens of each type:

Sodalites: cubic cells with $a = 8\cdot77\text{–}8\cdot90$ Å
Noseans: cubic cells with $a = 9\cdot03\text{–}9\cdot10$ Å
Cancrinites: hexagonal cells with $a = 12\cdot47\text{–}12\cdot71$ Å
$c = 5\cdot07\text{–}5\cdot20$ Å

Thus cell dimensions are within limits, functions of the contents of the intracrystalline channels (water, NaOH and salts) and of the relative proportions of these guest species. The refractive indices were also somewhat variable, no doubt for the same reason, as illustrated for some cancrinites in Table 5. Specimens of the outgassed sodalites and cancrinites were exposed to dry hydrogen chloride gas at elevated temperatures, and the results compared with the behaviour of silicate glass and analcime. The amounts of HCl irreversibly taken up are given in Table 6, and can be interpreted in terms of exchange reactions and lattice interactions between HCl and the silicates:

$$HCl\,(o) + NaOH\,(i) \longrightarrow H_2O\,(i) + NaCl\,(o)$$
$$HCl\,(o) + Na^+\,(i) \longrightarrow H^+\,(i) + NaCl\,(o)$$

TABLE 5

Some properties of synthetic cancrinites grown hydrothermally from alkaline aqueous gels[30]

Aqueous medium	Refractive indices ϵ	ω	Elongation	Cell compared with a natural cancrinite
NaOH aq	1·504	1·501	+	Slightly expanded
Na$_2$CO$_3$+NaOH aq	1·498	1·502	−	Considerably collapsed
NaOH aq	1·497	1·500	−	Identical
NaOH aq	1·498	1·494	+	Slightly collapsed
NaOH aq	1·504	1·498	+	Slightly collapsed
Na$_2$CO$_3$ aq	1·498	1·501	−	Slightly collapsed

where (o) denotes outside and (i) denotes inside the silicates. The presence of crystalline NaCl outside the silicates was confirmed by the X-ray
powder patterns. In all the samples of Table 6, mobility of protons and of
Na$^+$ ions within the silicates is expected but migration of molecules of
HCl into the glass will not occur, and into analcime or the felspathoids
will take place only with difficulty and, at the pressures, times and
temperatures involved, in small amounts.

The competitive uptakes of water, NaOH and salts during synthesis
have been investigated quantitatively in sodalites and cancrinites.[23,25,31]
In Chapter 6, Fig. 10, the isotherms for uptake of borate in sodalite
hydrate and zeolite Na-X were seen to be of Type 1 in Brunauer's
classification.[31] This is the usual contour of other salt intercalation
isotherms in which salts are incorporated in sodalite during crystal formation, as illustrated in Fig. 3[25] for NaClO$_4$, NaClO$_3$, NaBr and NaCl. The

TABLE 6

Exchange sorption of dry HCl on some silicates[30]

Silicate	Max. Temp. reached (°C)	HCl sorbed (cm^3 at s.t.p./g)	% total Na$^+$ exchanged	Refractive indices Before exchange	After exchange
Silicate glass	200	25·7	18·2	−	−
Analcime	360	28·5	~28	1·487	1·455
Nosean (made in presence SO$_4^{2-}$)	250	30·8	17·1	1·486	~1·46
Nosean (made in absence SO$_4^{2-}$)	250	50·1	26·5	1·494	−
Sodalite (made in absence Cl$^-$)	250	99·1	52·5	1·488	~1·47
Cancrinite (made in absence CO$_3^{2-}$)	250	125	66·3	1·50 (mean)	1·46$_0$ (mean)

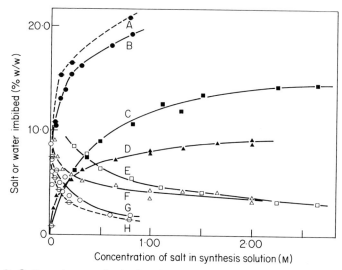

Fig. 3. Salt and water inclusion isotherms in sodalite. Sodalites were made at 80°C from kaolinite, 4 M NaOH, and systematically varied salt contents.[25]

$$A = NaClO_4; \quad\quad B = NaClO_3; \quad\quad C = NaBr;$$
$$D = NaCl; \quad\quad\quad E = H_2O(NaBr); \quad F = H_2O(NaCl);$$
$$G = H_2O(NaClO_3); \quad H = H_2O(NaClO_4)$$

sodalites were made at 80°C from 2 g of kaolinite in 200 ml of 4 M NaOH to which increasing amounts of each salt were added. The diagram shows also the diminution in the zeolitic water content corresponding with increased uptake of each of the salts. Figure $4(a)^{[23]}$ shows isotherms in sodalites for NaCl, for NaCl in presence of $NaClO_2$ and for $NaClO_2$ in presence of the NaCl. The figure also shows the decrease in zeolitic water as the combined uptake of $NaCl + NaClO_2$ increases. Figure $4(b)^{[23]}$ gives the isotherm for sodium azide. The sigmoid shape is associated with a change-over from sodalite to cancrinite formation as the azide concentration increases. In these systems there may also be a limited uptake of NaOH which was not determined. The maximum uptakes of the anions ClO_4^-, ClO_3^-, ClO_2^-, Br^-, Cl^-, SO_4^{2-}, etc., cannot for reasons of space be more than one per sodalite cavity, but the cavity has room to accommodate two OH^- ions (as NaOH) or four H_2O molecules. The intercalated salt molecules cannot leave or enter the sodalite cavities under the mild synthesis conditions and so their incorporation must occur in each cage before that cage is completed and thereby sealed off. The inference is that the salt molecules act as templates around which the sodalite cages grow, either in solution or on the exposed surfaces of the sodalite crystals. The first of these possibilities yields aluminosilicate precursor species in solution incorporating a salt molecule (or water and NaOH). Some OH groups at the outer periphery of the precursor species undergo condensation-polymerization with terminal OH groups at the crystal/solution interface and so extend the growing crystal. According to the second possibility, once a viable nucleus develops, crystal growth could proceed by additions of 6-ring units in layers in the abc. . . sequence of sodalite (Table 2). This traps salt molecules adsorbed in incomplete sodalite cages by capping and completing them. If the 6-ring units are added in layers in the ab... sequence of cancrinite the adsorbed salt molecules are similarly trapped in the 11-hedral cancrinite cages and in the wide channels as growth proceeds. These mechanisms, while rational, are of course speculative.

Under the conditions of synthesis giving the results of Table 4 as indicated in the footnote to the table, it is seen that the sodium hydroxide solution in absence of salts favoured crystallization of sodalite hydrate

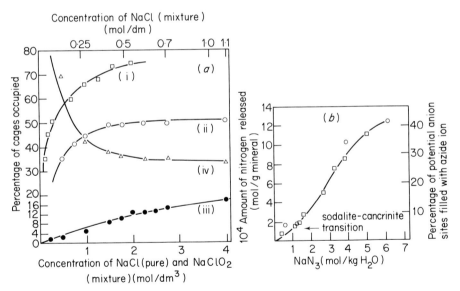

Fig. 4. (a) Isotherms in sodalite of (i) NaCl, (ii) NaCl in presence of NaClO$_2$ (iii) NaClO$_2$ in presence of NaCl, (iv) Water content for increasing uptakes of NaCl+NaClO$_2$.[23] Synthesis conditions similar to Fig. 3. (b) Isotherm for uptake of NaN$_3$ in sodalite (for low azide concentration) and in cancrinite (for higher azide concentration). ○ = first synthesis series; □ = second synthesis series.[23] Synthesis conditions as in Fig. 4(a).

rather than cancrinite hydrate. However, at high temperatures with different excesses of NaOH and with gels as sources of alumina and silica, Barrer and White could make either sodalite hydrate or cancrinite hydrate.[30] Likewise at high temperatures (170–350°C) kaolinite and Na_3PO_4 solution with no added NaOH yielded phosphatic cancrinites exclusively.[32] In these and other preparations the structure-directing role of the salts can vary according to the conditions. Thus the results in Table 4 are to be regarded as appropriate only for the conditions described in the footnote to the table. The correct weightings to be attached to temperature, alkali concentration, salt concentration and type, and sources of alumina and silica remain undetermined.

Since one series of salts yielded sodalites and another series yielded cancrinites, the possibility was considered of using pairs of salts to try to produce phases intermediate between sodalites and cancrinites, for example with layer sequences ababc. . . . However, although the possibility was examined for many salt pairs, no such intermediate species were identified by X-ray powder photography.[25] Instead, one salt dominated to give either sodalite or cancrinite, or else sodalite and cancrinite co-crystallized. Some results are given in Table 7, for equimolar salt pairs, each 0·1 M. The reaction conditions were otherwise like those for Table 4. It must be remembered that for the reaction conditions of Table 7, NaOH as well as at least one of the chosen salts according to Table 4 favours the formation of sodalites. Nevertheless, strongly cancrinite-directing anions such as nitrate and chromate can dominate the other anion and the hydroxyl ion together, and yield cancrinites as the table illustrates.

TABLE 7

Crystallization with salt pairs, each 0·1 M[25] (2 g kaolinite, 200 ml 4 M NaOH, 80°C agitated in plastic bottles[a])

Salt anions	Products	Salt anions	Products
Cl^-/CrO_4^{2-}	$S^\circ + C^\times$	WO_4^{2-}/NO_3^-	C
Cl^-/NO_3^-	$C^\circ + S^\times$	WO_4^{2-}/MoO_4^{2-}	S
Cl^-/MoO_4^{2-}	$S^\circ + C^\times$	CO_3^{2-}/CrO_4^{2-}	S
Br^-/CrO_4^{2-}	$S^\circ + C^\times$	CO_3^{2-}/NO_3^-	S
Br^-/NO_3^-	$S^\circ + C^\times$	CO_3^{2-}/MoO_4	$S^\circ + C^\times$
Br^-/MoO_4^{2-}	$S^\circ + C^\times$	ClO_3^-/CrO_4^{2-}	S
ClO_4^-/CrO_4^{2-}	S	ClO_3^-/NO_3^-	$S^\circ + C^\times$
ClO_4^-/NO_3^-	$S^\circ + C^\times$	ClO_3^-/MoO_4^{2-}	S
ClO_4^-/MoO_4^{2-}	$S^\circ + C^\times$	NO_3^-/CrO_4^{2-}	C
WO_4^{2-}/CrO_4^{2-}	C	NO_3^-/MoO_4^{2-}	C

[a] S = sodalite; C = cancrinite. Superscripts ○ and × denote major and minor yields respectively.

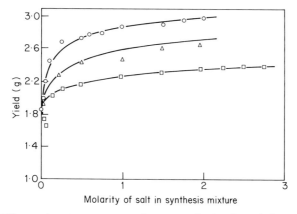

Fig. 5. Effect of salt concentration on yield of sodalite–salt solid solution for synthesis conditions of Fig. **4.**[23] $\bigcirc = NaClO_4$; $\triangle = NaClO_3$; $\square = NaClO_2$.

The yields of crystals from a given amount of kaolinite are a function of the concentrations of alkali and salt in the mineralizing solutions. Thus, in absence of salts, the yield of sodalite hydrate with 4 M NaOH in one instance was ~53% of theoretical and with 1 M NaOH this yield was ~75%.[25] The remainder was retained in solution. In the presence of salts, for constant concentration of NaOH, the yields can be increased substantially. This is illustrated in Fig. 5 for syntheses at 80°C from 2 g kaolinite with 4 M NaOH (200 ml) and various amounts of $NaClO_4$, $NaClO_3$ or $NaClO_2$.[23] The maximum yields and corresponding compositions (neglecting intercalated NaOH) were estimated as:

$$Na_6[Al_6Si_6O_{24}].8H_2O \qquad\qquad 62\%$$
$$Na_6[Al_6Si_6O_{24}].1{\cdot}8_3NaClO_3, 0{\cdot}6_8H_2O \qquad 85\%$$
$$Na_6[Al_6Si_6O_{24}].1{\cdot}7NaClO_4, 1{\cdot}0_3H_2O \qquad 92\%$$

Parallel with the increase in yield of crystals the amounts of dissolved species remaining in solution must be decreased.

2.3. Hydrothermal Syntheses of Other Salt-bearing Species

From Table 1 it is seen that various kinds of tectosilicate framework containing intercalated salts have formed naturally or have been synthesized. Sodalites and cancrinites form from sodic media, but zeolite ZK-5 (as Species P and Q) grew only from baric[33,34] and the edingtonite type zeolite K-F from potassic[33,34] media. Examples of the relevant syntheses are given in Chapter 5 §5, Table 15 and §8, Table 21. Table 1 of this

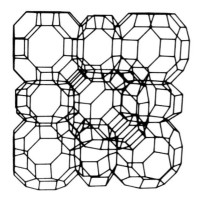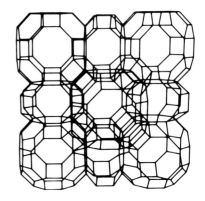

Fig. 6. The framework of Species P and Q (the ZK-5 structure) shown as a stereoscopic pair.[36] Al or Si is at each corner; oxygen is near the mid-point of each edge.

chapter, which is not intended as a complete inventory of salt-bearing tectosilicates, includes eight different framework topologies (those of scapolite, sodalite, cancrinite, liottite, afghanite, franzinite, zeolite ZK-5 and edingtonite type K-F). Some time ago Species P (salt-bearing ZK-5) was subjected after its formation to hydrothermal extraction above 200°C of $BaCl_2$ (which was progressively replaced by water). It was then found to be a good molecular sieve sorbent.[35] It can separate n-paraffins from iso-, neo- and cyclo-paraffins and from naphthenes and aromatics in much the same way as zeolite Ca-A or chabazite, in accordance with its framework topology shown in Fig. 6.[36] Its molecular sieving behaviour is determined by the 8-ring windows, seen in the figure.

The hydrothermal extraction of $BaCl_2$ is however inconvenient and is inclined to bring about some decomposition of the crystal lattice. Accordingly Barrer and Marcilly[34] re-investigated the extraction of the salt under mild conditions. It was thought that, because extraction of $BaCl_2$ with distilled water involved removal of three ions simultaneously ($Ba^{2+} + 2Cl^-$), extraction with aqueous salt solutions containing $NaNO_3$ for example would be more readily effected. Sodium ions could exchange with barium and subsequent extraction of NaCl would involve simultaneous removal of a pair of ions only ($Na^+ + Cl^-$). This should be more readily effected, not only because the ions are to be removed in pairs than in groups of three, but also because Ba^{2+} is probably more heavily hydrated than Na^+ so that Na^+ should pass more easily through the 8-ring windows. Moreover, the quantitative study of salt inclusion from aqueous solutions (§5) leads to the expectation that, for modest concentrations of salts in solution, the amounts of intercalated salts at equilibrium with the solution would be small.[37] The extraction of barium halide from the crystals was studied at 20, 80, 100 and 140°C in stirred solutions of several concentrations of $NaNO_3$, $NaNO_2$, $LiNO_3$, KNO_3, NH_4NO_3 and $Ca(NO_3)_2$ as well as with distilled water. At 140°C in the presence of sodium salts there was a strong tendency of Species P and Q to recrystallize to analcime, but at 100°C or below P and Q did not recrystallize. Four progressive extractions of a 2 g sample of Species P, each of 70 hours duration using 125 ml of 8 M $NaNO_3$ at 80°C, yielded a sodium form of P with about 90% removal of Ba^{2+} and of Cl^-. This represents a big improvement over extraction with distilled water. The same conditions removed only about 15% of the Br^- and total Ba^{2+}. Thus much of the resistance to removal of the barium salt may arise from the anions. Br^- being larger than Cl^- probably migrates through 8-ring windows of the ZK-5 structure more slowly than Cl^-.

Analyses of some of the preparations of Barrer and Marcilly[34] give the molar proportions in Table 8. The compounds P' and Q' are the aluminous forms of salt-bearing ZK-5 made from zeolite Na-X. They have larger unit cells (Chapter 5 §8) and recrystallize more easily during extraction of barium halide than the siliceous forms P and Q. The table shows the expected increase in the water content of the extracted form of Species P. In the cubic unit cell ($a \sim 18.7$ Å) there are 192 oxygens so that the unit cell loadings of halide, given in the penultimate column of Table 8, are very considerable. It is not known whether and if so what critical content of barium halide is required in the reaction mixture to enable conversion of other framework topologies to that of zeolite ZK-5.

TABLE 8
Analyses of compounds containing $BaCl_2$ and $BaBr_2$[34]

Compound (and origin)	SiO_2	Al_2O_3	BaO	Na_2O	$BaCl_2$ or $BaBr_2$	H_2O	Molecules per unit cell: Halide	H_2O
					Molar ratios:			
P (from analcime)	4·36	1	1·11	0	0·83	2·3	12·5	35
P (from zeolite Y)	4·91	1	1·00	0	0·90	2·5	12·5	35
P' (from zeolite X)	2·51	1	1·17	0	0·67	0·9	14·3	19
P (from analcime, $BaCl_2$ extracted)[a]	4·32	1	0	1	0·11	4·7$_5$	1·7	73
Q (from analcime)	4·30	1	1·09	0	0·78	2·0	11·9	31
Q (from zeolite Y)	5·10	1	1·01	0	0·74	2·3	10·0	31
Q' (from zeolite X)	2·46	1	1·14	0	0·66	1·6	14·2	35

[a] With $NaNO_3$ aq, which also exchanged Ba^{2+} by Na^+.

The salt-bearing forms of zeolite K-F when extracted hydrothermally with distilled water decomposed badly.[33] Analyses showed that some preparations of Barrer and Marcilly[34] had molar ratios $SiO_2/Al_2O_3 \sim$ 2·5. Because of the relatively aluminous character of the zeolite, and since edingtonite, unlike ZK-5, is not a good sorbent, further experiments on salt extraction were not made.

3. Dry-way Intercalation in Pre-formed Zeolites

The pore space within outgassed zeolite frameworks may be filled from the vapour phase, from salt melts or by heating zeolite with salt powders. The compounds so taken up are not equally ionic. They represent various gradations between the extremes of covalent and fully ionized. Representative examples of each are considered in §§3.1 and 3.2.

3.1. *Uptake of Less Ionic Compounds*

Comparatively non-volatile species which penetrate zeolite lattices, normally from the vapour phase and often at temperatures upwards of 100°C, include Hg_2Cl_2, HgS, $AlCl_3$, $TiCl_4$, S, P, Hg, I_2 and Br_2. Sorption of the last four elements has been described in a companion volume[38] and will not be considered here. Grandjean[39] in 1910 was one of the first to study the sorption of Hg_2Cl_2, HgS, Hg, I_2 and Br_2 in a number of zeolites, including chabazite, gmelinite, levynite, harmotome, gismondine, analcime, heulandite and stilbite. Little uptake was observed in the last three of these zeolites, in measurements normally made under atmospheric pressures often without special precautions to exclude air. One of the most characteristic processes described was the assumed uptake of mercury, which was reported to occur freely in chabazite, levynite and harmotome. However Barrer and Woodhead[40] and Barrer and Whiteman[41] showed that little mercury was sorbed in (Na, Ca)-zeolites. Copious uptake in (Na, Ca)-zeolites like chabazite did, however, take place at elevated temperatures, often in the range 200–300°C, provided oxygen was present.[40] Mercury and oxygen formed a mercury–oxygen complex within the chabazite. The zeolite became yellow, deepening to bronze as uptake increased. When exposed to H_2O, H_2S and NH_3 the Hg–O–chabazite complex darkened in each case. The lustrous black product obtained with water when heated became orange, yellow and eventually white once more.

Grandjean[39] found that even at 525°C chabazite sorbed calomel to give a 24·3% weight increase. On removing the source of calomel vapour desorption at 520°C was almost complete within 6 minutes. The crystals powdered during sorption and the calomel–chabazite complex showed much reduced capacity to sorb water. The vapour of mercury sulphide, when carried over chabazite in a stream of hot carbon dioxide, was also readily sorbed by the zeolite. Barrer and Wasilewski[42] found copious sorption of anhydrous $AlCl_3$ vapour in outgassed zeolite Na-X even at ~300°C. A little gas was evolved which was considered to be HCl formed by reaction between residual zeolite water and the $AlCl_3$ or between lattice OH groups and the chloride. Komarov et al.[43] passed a stream of dry air saturated with $TiCl_1$ vapour for about 25 hours through outgassed zeolites Y and L, mordenite and erionite and the NH_4- and H-forms. The zeolites became to varying degrees impregnated with $TiCl_4$. The titanium chloride in the zeolite was next hydrolysed in air or with ammonia and then, to remove volatile products formed by the hydrolysis, was heat treated at 450–550°C for 5–6 hours. The most open zeolites (Y, L and H-mordenite) sorbed $TiCl_4$ more copiously than erionite or Na-mordenite. The maximum TiO_2 contents of the zeolites were:

NH_4-Y	15·50 wt % TiO_2
H-L	19·63
(K, Na)-erionite	5·08
H-mordenite	10·30

The treatment, including the final heat treatment, resulted in some lattice damage which was more severe for the H-zeolites than for the (Na, K)-forms and least for the most silica-rich zeolite, mordenite. The presence of intercalated TiO_2 reduced the subsequent water uptake, and also that of benzene vapour. By reducing $TiCl_4$ sorbed in zeolites with hydrogen or metallic sodium, zeolites bearing up to 6·2% of Ti have been prepared that are active oxidation and polymerization catalysts.[44]

Other operations of the same kind as the above illustrative examples, $AlCl_3$ and $TiCl_4$, are possible. Some potential guest species are given below, primarily chosen from chlorides. Comparable behaviour with bromides and iodides could be expected in zeolites selected to have appropriately wide windows giving access to the channels and cavities:

BCl_3	$SbCl_5, SbCl_3$	UCl_4	ICl	$EuCl_3$
$GaCl_3$	$BiCl_3$	$ReCl_4$	ICl_4	$HgCl_2$
$InCl_3$	$TaCl_3$	$RuCl_3$	YCl_3	$SiCl_4$
$TlCl_3$	$FeCl_3$	$RhCl_3$	$SmCl_3$	$GeCl_4$
$ZrCl_4$	$CrCl_3$	$CoCl_3$	$GdCl_3$	$SnCl_4$
$HfCl_4$	$MoCl_5$	$PtCl_4$	$YbCl_3$	$TeCl_4$
PCl_5, PCl_3	WCl_6	$PdCl_4$	$DyCl_3$	$SO_2Cl_2, SOCl_2$

The examples given above range from covalent to partially ionic. When the zeolite is loaded with the halide it should be possible in many instances to hydrolyse the guest species to give oxide or hydroxide and to form structures heterogeneous on the molecular scale with oxide threads and clusters having the pattern of the channel and cavity systems and supported by the aluminosilicate framework of the zeolite. At the moment such structures remain only as chemical novelties. If hydrolysis is considered, the selected zeolite framework should be siliceous in order to minimize damage to the framework by the liberated acid. In a number of cases the intercalated halide can be readily reduced to the metal, as in the case of transition metal halides.

The borohydrides represent another group of compounds in which bonding can vary between largely ionic and largely covalent.[45] Thus $LiBH_4$, $NaBH_4$ and $LiAlH_4$ are saline but $Al(BH_4)_3$ is a liquid (b.pt ~ 44·5°C). The latter should be readily imbibed by appropriately selected zeolites. One interest in such compounds is the large amount of hydrogen liberated by reaction with water, as seen from the following figures:

$$Be(BH_4)_2 \rightarrow 4·7 \text{ (dm}^3 \text{ at stp of } H_2 \text{ per g of hydride)}$$

$$LiBH_4 \rightarrow 4·1$$

$$Al(BH_4)_3 \rightarrow 3·7$$

The reaction of the entrained borohydride with water could serve as means of introducing two elements (e.g. Al and B) simultaneously into the zeolite, as the hydroxides or oxides.

3.2. *Uptake of Salts*

Clarke and Steiger[46,47] first used salt melts or vapours as means of ion-exchanging zeolites. One may for this purpose select rather low-melting salts such as $NaNO_3$ (310°C), $AgNO_3$ (209°C), $LiNO_3$ (261°C) and KSCN (179°C). In addition volatile salts such as NH_4Cl or NH_4Br, or alkylammonium halides may serve to give ammonium or alkylammonium ion exchange via the vapour phase. Salt melts have been used in exchanges involving ultramarine,[9] chabazite[48] and zeolite A.[49,50] When the tectosilicate lattices are sufficiently open, as is the case for many zeolites, not only is there ion exchange but also salt molecules can penetrate and fill the intracrystalline pore space. Some examples of this behaviour will be summarized.

When dry ammonium chloride is vapourized, the vapour consists of a 1:1 mixture of HCl and NH_3. When exposed to zeolites Na-mordenite, H-mordenite, faujasites (Na-Y, H-Y), Na-A, or K-L it was found that the 1:1 mixture was imbibed much more strongly than either pure HCl or pure NH_3, to give reversible isotherms even at elevated temperatures, as

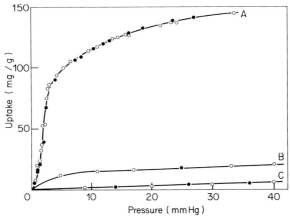

Fig. 7. Isotherms for uptakes in zeolite K-*L* at 245°C of:
A, vaporized NH_4Cl (i.e. $NH_3 + HCl$ in
1 : 1 proportions).
B, HCl alone.
C, NH_3 alone.[51]
○, sorption; ●, desorption

shown by those in Fig. 7.[51] There is a strong interaction within the zeolite of the co-sorbed HCl and NH_3 to give an intracrystalline complex which may be re-formed NH_4Cl. Imbibition of NH_4Cl in chabazite was also found, and in open structures there can also be uptake of the vapours of simple alkylammonium halides as well as ion exchange of inorganic by alkylammonium ions. However, by choosing zeolite and alkylammonium halide suitably, the alkylammonium can be made too large to enter the crystals. Thus, in chabazite, methylammonium chloride can both exchange and be sorbed, but trimethylammonium chloride can, through the sieve effect, only exchange. The relevant reactions are

$$CH_3NH_3Cl + Na\text{-chabazite} \rightarrow CH_3NH_3\text{-chabazite} + NaCl$$

$$nCH_3NH_3Cl + CH_3NH_3\text{-chabazite} \rightarrow CH_3NH_3\text{-chabazite}, nCH_3NH_3Cl$$

and

$$(CH_3)_3NHCl + Na\text{-chabazite} \rightarrow H\text{-chabazite} + (CH_3)_3N + NaCl$$

Some additional examples in which salts and zeolites have been heated together are summarized in Table 9. Most of these have been involved with zeolites A, X and Y and halides or nitrates. In zeolite A or the faujasites X and Y the penetrating salts may be found in the large cavities (26-hedra of types 1 and 2[29]) or in the sodalite cages. Entry into sodalite cages is through 6-ring windows which have free diameters of ~2·2 Å.

TABLE 9

Some preparations of salt-bearing zeolites

Date and reference	Zeolites	Salts	Conditions
1958[52]	Ag-A Na-A	$AgNO_3$ $NaNO_3$	Zeolite–nitrate mixture slowly raised to temperature a little above M.pt of nitrate. Heated for 8 hours. Cooled slowly to room temperature.
1968,[49] 1970[50]	Li-A Na-A	$LiNO_3$ $NaNO_3$ with dissolved Li, K, Cs, Ag, Tl, Ca and Sr nitrates	Heated at 330°C (350°C for exchanges with K).
1969[53]	Na-A	$Co(NO_3)_2$	Heated zeolite in nitrate melt in dry N_2.
1971,[54] 1981[55]	Na-A	Alkali halides, sulphates and nitrates	Zeolite heated in salt melts at 20°C above M.pt of salt. Some salts dissolved in $Na_2SO_4 + Li_2SO_4$ eutectic or in nitrate.
1974[56]	Na-A Cancrinite hydrate	$Cd(NO_3)_2$ $NaNO_3$	$Cd(NO_3)_2$ dissolved in molten $NaNO_3$.
1975[57]	Na-chabazite Mordenite Na-A Na-X and -Y	$NaNO_3$ $NaClO_2$	Zeolite powders heated with salt melts somewhat above M.pt of salts.
1975[58]	Na-Y Ca-Y Zn-Y Na-Y K-Y Ba-Y	$NaCl$, $NaBr$, NaI $CaCl_2$, KF $ZnCl_2$ $NiCl_2$, $LaCl_3$, KF KF KF	Strong aqueous salt solutions applied to anhydrous zeolite. Water solvent evaporated in vacuo. Product then heated at temperatures from 220 to 700°C for several hours and then extracted with boiling water until wash-water free of anion.
1980[59]	Na-Y Na-A Li-A Na-X Zn-X	$NaNO_3$, $AgNO_3$, $NaClO_3$ $NaNO_3$ $LiNO_3$ $AgNO_3$ $KSCN$	Zeolite discs heated with salt melts near the M.pt for 12 hours
1981[60]	Na-A, clinoptilolite Ag-A, Ag-clinoptilolite K-A, K-clinoptilolite	(Li, K)Cl eutectic $AgNO_3$ $KF.2H_2O$	Zeolite heated with melt at ~470°C for 8 hours. Zeolite heated with melt at 233°C for 8 hours. Zeolite heated with melt at 60°C for 8 hours.

Nevertheless, polar molecules such as water (diameter $\sim 2\cdot 8$ Å) can diffuse into outgassed sodalite hydrate which is composed only of sodalite cages, and at high temperature and pressure so do Ar (diameter $\sim 3\cdot 83$ Å) and Kr ($\sim 3\cdot 94$ Å). The 6-rings are lined on their inner peripheries by anionic oxygens and the negative charge of the anions and their large size could make entry into the sodalite cages difficult. For the halogens one has: F^-, $2\cdot 66$ Å; Cl^-, $3\cdot 62$ Å, Br^-, $3\cdot 92$ Å; and I^-, $4\cdot 40$ Å. The NO_3^- ion is triangular in shape and is also large. Thus it is of interest to consider evidence of penetration and the conditions for it.

The work of Rabo and Kasai[58] and of Skeels[61] showed that up to 330°C Cl^- showed negligible penetration into sodalite cages of Na-Y. Penetration was still slow at 440°C and substantially faster at 550°C. The method of preparation outlined in Table 9 ensured that all salt had been removed from the large cages before the analysis to determine the amount of NaCl trapped in the sodalite cages. The amount of NaCl which has entered the sodalite cages is shown as a function of time and temperature in Fig. 8.[52] With NaBr, even at 550°C penetration into sodalite cages was very slow, but at 650°C was considerably faster, giving 70% occupation of the cages in 24 hours. With NaI no penetration occurred at 500°C and penetration was still very slow at 645°C. Thus the halide ions show very well an ion sieve effect which should be fully adequate to separate Cl^-, Br^- and I^-.

For the nitrate ion, Barrer and Meier[54] with $AgNO_3$ melt near its melting point (209°C) found no evidence of penetration into the sodalite cages of Ag-A. They prepared a complex of unit cell composition $Ag_{12}[Al_{12}Si_{12}O_{48}]AgAlO_2, 9AgNO_3$. The failure to penetrate into the sodalite cages could be due to prior occupation by $AgAlO_2$. However, with Na-A + $NaNO_3$ heated to 330°C, Barrer and Villiger[55] found that nitrate had entered the sodalite cages, and at still higher temperatures Rabo[62] also found the uptakes recorded in Table 10. All the products were excellently crystalline as shown by their powder patterns. A comment can, however, be made regarding the reported trapping of $NaClO_3$. In sodalite synthesized with intercalation of $NaClO_3$, this salt decomposed within the structure, evolving oxygen gas, at temperatures below 650°C.[23] Thus the product of Table 10 obtained on heating Na-Y with $NaClO_3$ at 650°C may be one with Cl^- not ClO_3^- in the sodalite cages.

The mechanism of passage of the large anions and of Ar and Kr through 6-ring windows is of interest. It is possible that the migration involves momentary cleavage of an Al–O–Si bond in the ring, thus increasing the aperture. In this event the kinetics could be related to the rate of isotopic exchange between gaseous $^{18}O_2$ and lattice oxygens. This exchange was of first order with respect to the gaseous oxygen and

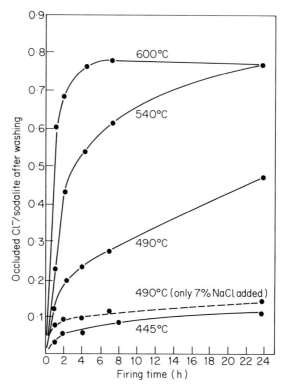

Fig. 8. Uptake of NaCl in sodalite cages of Na-Y as a function of temperature and time.[58]

occurred at a significant rate only above 600°C with an apparent activation energy for Na-Y of 190 kJ mol^{-1}.[63] However, salt trapping and uptake of Ar and Kr are observed below 600°C and one may consider a view less extreme than that of bond cleavage. The oxygens lining the inner periphery of the 6-ring and the anions or gas molecules are not hard

TABLE 10
Trapping of anions in sodalite cages of zeolite Y[62]

Zeolite-salt	T (°C)	Heating time (h)	Anions per sodalite cage
Na-Y + NaNO$_3$	550	64	0·59
Na-Y + NaNO$_3$	650	24	1·32
Na-Y + AgNO$_3$	550	48	0·37
Na-Y + NaClO$_3$	650	16	0·9

spheres and can be deformed. Also there are breathing frequencies for the 6-ring, which may stretch without such a large deformation as would be required for genuine bond dissociation. On the whole these factors seem more likely to be rate determining than does bond breaking.

Movement of salts through the 8-ring openings of zeolite A (free diameter 4·2 Å) occurs much more readily and so at lower temperatures than through 6-rings, but less readily than through the 12-ring openings in faujasite. Consequently, when zeolite A and the faujasites X or Y are each impregnated with salts from salt melts, it is possible on careful washing to retain most of the salt in the large cavities of zeolite A but it is much more difficult not to elute salt from the large cavities of X or Y. Araya and Dyer[60] studied the kinetics of elution of the chloride, nitrate and fluoride in the complexes (Table 9) made with zeolites A and clinoptilolite. The elution of Cl^- after different times of washing is shown for zeolite A in Fig. 9. If extrapolated to zero time the curves appear to give intercepts on the ordinate. This behaviour was first reported by Barrer and Meier[52] for elution of $AgNO_3$ from an Ag-A, $AgNO_3$ complex and was ascribed to fast solution of surface salt. After allowing for these intercepts, diffusivities for salt elution were estimated by Araya and Dyer and from the temperature coefficients of these the following activation energies in kJ mol^{-1} were derived:

	(Li, Na)Cl	AgNO$_3$	KF
Zeolite A	107	66	34
Clinoptilolite	86	87	61

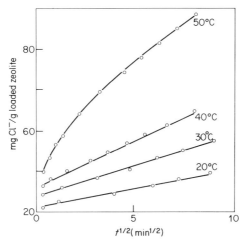

Fig. 9. The elution of (Li, Na)Cl from salt-loaded (Li, Na)-A as a function of time and temperature.[60]

Clinoptilolite has a layer structure, in the sense that the T–O–Si (T = Al or Si) bond density is greater in two directions than in the third, and salt penetration involves two-dimensional diffusion between layers. Each interlayer region provides distorted 8- and 10-ring openings.

The investigation by Rabo and Kasai[58] with the zeolites and salts of Table 9 showed that side reactions could take place. Although much zeolitic water was removed during slow heating from room temperature to 300°C, a little zeolitic water together with some structural –OH groups at crystal surfaces or as lattice defects such as [$\geqslant$Al HO–Si$\leqslant$] can still exist on or in the zeolite. Evidence of this was found in the evolution of some hydrogen halide when the salts were halides like NaCl, LaCl$_3$ or NaBr. With sodium halides very little hydrogen halide was evolved; rare-earth forms yielded larger quantities; and the NH$_4$-Y evolved considerable amounts. With NaCl the relevant processes are likely to be:

$$\left[\genfrac{}{}{0pt}{}{}{}\!\!\!-\!\!\!Al \quad HO-Si\!\!\!-\right] + NaCl \longrightarrow \left[\genfrac{}{}{0pt}{}{}{}\!\!\!-\!\!\!\overset{\ominus}{Al}-O\overset{Na^+}{-}Si\!\!\!-\right] + HCl$$

$$-Si-OH + NaCl \longrightarrow -SiO^-Na^+ + HCl \text{ (at crystal surfaces)}.$$

With hydrolysable salts such as LaCl$_3$ further possibilities are

$$LaCl_3 + H_2O \rightarrow La(OH)Cl_2 + HCl$$

and

$$2La^{3+} + H_2O \rightleftharpoons [La^{3+}...O^{2-}...La^{3+}] + 2H^+$$

$$H^+ + \left[\genfrac{}{}{0pt}{}{}{}\!\!\!-\!\!\!\overset{\ominus}{Al}-O-Si\!\!\!-\right] \longrightarrow \left[\genfrac{}{}{0pt}{}{}{}\!\!\!-\!\!\!Al \quad HO-Si\!\!\!-\right]$$

The cation–oxide–cation clusters can arise in the sodalite cages and so deny access of halide ions to these cages. Accordingly it will no longer be possible in these circumstances to reach the limiting uptake for sodalite cages of one anion per cage. Also the reactions leading to evolution of hydrogen halide mean that the ratio $Na^+_{guest} : Cl^-_{guest}$ is greater than one. Both these situations were encountered in the investigation.

4. Structures of Zeolite–Salt Solid Solutions

Investigations of the structures of salt-bearing zeolites have been concerned with the locations of anions and other non-framework atoms, ions and molecules, and with the framework perturbations brought about by the presence of the salts. X-ray studies include those made for the following systems: silicate- and nitrate-bearing cancrinites;[64] Ag-A, $AgNO_3$;[52,55] Na-A, $NaNO_3$[57,58] and Li-A, $LiNO_3$;[52,55] and Ba-ZK-5, $BaCl_2$ (Species P) and Ba-ZK-5, $BaBr_2$ (Species Q).[65]

Barrer and Meier[52] prepared the complex $Ag_{12}[Al_{12}Si_{12}O_{48}]AgAlO_2$. $9AgNO_3$ at a temperature a little above the melting point of $AgNO_3$ (209°C). The nine $AgNO_3$ molecules were in large cavities (26-hedra or α cages), and a superstructure was observed. All the X-ray powder diffraction lines satisfied the superlattice with a tetragonal unit cell having $a = b = 24 \cdot 6$ Å and $c = 12 \cdot 3$ Å. A structure for the $AgNO_3$ clusters was proposed in which there was a rock salt arrangement between cations and anions with the $AgNO_3$ falling into a set of five and a set of four, differing in orientation, as shown in Fig. 10. The sets in adjacent cavities were in complementary arrangements, giving rise to the correct superlattice cell dimensions. This was the only arrangement which satisfied both the type of superstructure and the available space in the large cavities. When the Ag-A, $AgNO_3$ complex was quenched rather than being cooled slowly the X-ray patterns were rather diffuse with

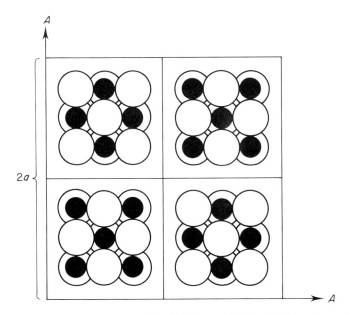

Fig. 10. The arrangement of Ag^+ (●) and NO_3^- (○) in adjacent 26-hedral cavities of zeolite Ag-A. This arrangement gives the observed superlattice spacings and cell.[52]

hardly any superstructure lines, indicating considerable disorder. The Ag-A, AgNO$_3$ complex examined by Petranovic et al.[55] also showed much disorder and failed to refine in the space group Pm3m.

The cancrinites examined had the following unit cell compositions and dimensions:[64]

$$Na_6[Al_6Si_6O_{24}]1\cdot08SiO_2, 0\cdot52Na_2O, 4\cdot05H_2O; \qquad a_0 = 12\cdot72 \pm 0\cdot02 \text{ Å}$$
$$c_0 = 5\cdot19 \pm 0\cdot02 \text{ Å}$$

$$Na_6[Al_6Si_6O_{24}]1\cdot86NaNO_3, 1\cdot44H_2O; \qquad a_0 = 12\cdot67 \pm 0\cdot02 \text{ Å}$$
$$c_0 = 5\cdot19 \pm 0\cdot02 \text{ Å}$$

The first of these compositions refined to give R(I) and R(F) equal to $0\cdot07$ and the second to R(I) $= 0\cdot13$ and R(F) $= 0\cdot12$. The co-ordinates of framework Al and Si agreed with those given by Jarchow,[66] but in the silicate-bearing cancrinite positions of oxygens and sodiums showed some differences. The non-framework silicate could be present as monomeric species such as NaH_3SiO_4 or as chain polymer, $NaHSiO_3$, in the wide channels. A cross-section of one of these channels is shown in Fig. 1. Detrital silicate could account for the experimentally observed narrow-pore character and poor capacity of other cancrinite hydrates examined as sorbents.[28] The cancrinite, $NaNO_3$ complex had only part of the $1\cdot86$ molecules of $NaNO_3$ per unit cell in the wide channels, the other part being in the cancrinite cages shown in Fig. 1. Consideration of the locations found for nitrate and water indicated that with maximum uptakes of guest species only eight sodiums per unit cell are possible of which six are required to neutralize framework charge. Therefore the maximum uptake of $NaNO_3$ would not exceed two molecules per unit cell, a limit approached in the above cancrinite nitrate.

Barrer and Villiger[57] and Petranovic et al.[55] investigated the structure of the Na-A, $NaNO_3$ complex. The former workers examined a complex having the following pseudocell composition and cubic cell edge:

$$Na_{12}[Al_{12}Si_{12}O_{48}]9\cdot3NaNO_3, 6\cdot7H_2O; \qquad a_0 = 12\cdot39 \pm 0\cdot02 \text{ Å}$$

The parent zeolite had the pseudocell content $Na_{12}[Al_{12}Si_{12}O_{48}]27H_2O$ so that one $NaNO_3$ is the equivalent of about $2\cdot2$ water molecules in filling the intracrystalline pore space and on this basis the ideal salt-filled water-free Na-A would have about 11 or 12 $NaNO_3$ molecules per unit pseudocell. The nitrate ions in the large cavities of zeolite A were in front of each of the eight 6-ring openings that lead into the eight sodalite cages surrounding each 26-hedral cage (α cage) as shown in Fig. 11(a). Petranovic *et al.*[35] studied a Na-A, $NaNO_3$ complex of pseudocell composition

$$Na_{12}[Al_{12}Si_{12}O_{48}]10NaNO_3, 6 \cdot 6H_2O; \qquad a_0 = 12 \cdot 295(1) \text{ Å}$$

and also found this location for eight of the nitrates. Three sodiums are located around the nitrate as shown in Fig. 11. Barrer and Villiger found two nitrates in the sodalite cage on the same three-fold axis as the NO_3 outside this cage. The arrangement is shown in Fig. 11(*b*).[57] Petranovic *et al.* do not refer to nitrate in the sodalite cage, but assume some near the centre of the large α cage. Both refinements were made in the space group Pm3m. It was pointed out by Barrer and Villiger that two weak lines did not fit the unit cell assumed, so that the actual structure could be more complex. These lines were neglected because the available number of powder lines would not in any case have been enough to permit one to

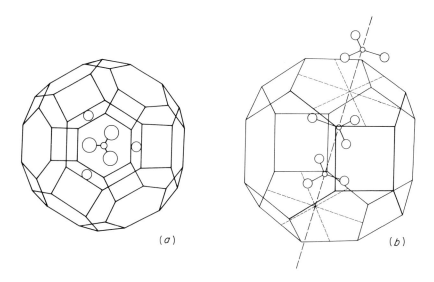

(a) $\qquad\qquad\qquad\qquad\qquad$ (b)

Fig. 11. (*a*) The arrangement of $NaNO_3$ inside the large cavity of zeolite A. The view is taken along the [111] axis and all details obscuring the drawing are omitted. If nitrate positions were all occupied there would be one in front of each of the eight 6-rings. Na^+ ions associated primarily with NO_3^- are also shown. (*b*) The arrangement of the two nitrate anions inside the sodalite cage, showing also their relation with the nitrate anions in the large cavity.[57]

The scales of (a) and (b) are not the same.

lower the symmetry. The arrangement of cations and nitrate in Li-A, $LiNO_3$ was considered by Petranovic *et al.* to be virtually the same as in Na-A, $NaNO_3$.

Three compositions were examined for the ZK-5 preparations.[65] These had cubic unit cell contents of:

Species P: $Ba_{15}[Al_{30}Si_{66}O_{192}]1 \cdot 7Ba(OH)_2, 12 \cdot 5BaCl_2, 35H_2O$

Species Q: $Ba_{15}[Al_{30}Si_{66}O_{192}]1 \cdot 6Ba(OH)_2, 11 \cdot 9BaBr_2, 30H_2O$

Extracted Species P: $(Na_2, Ba)_{15}[Al_{30}Si_{66}O_{192}]1 \cdot 7(Na_2, Ba)Cl_2, 72H_2O$

The unit cell edges were respectively $18 \cdot 65 \pm 0 \cdot 03$ Å; $18 \cdot 66 \pm 0 \cdot 03$ Å; and $18 \cdot 78 \pm 0 \cdot 03$ Å. The framework of ZK-5 (Fig. 6), which is also that of Species P and Q, is constructed from hexagonal prisms, 18-hedral cavities and finally 26-hedral cages of the same kind as in Na-A in the ratio $4:3:1$. The distributions of the non-framework atoms of P and Q in the 18-hedral cavities are shown in Fig. 12. Those of non-framework atoms in the 26-hedra of both P and Q were also determined and are shown for P in the stereoscopic pair of Fig. 13. The locations of many non-framework atoms in Species P from which most of the $BaCl_2$ had been extracted were also found both in the 18-hedra and the 26-hedra.

The structural studies described have all, by reason of the small size of the crystals, had to be made using powder data only. More detail would obviously be obtained if the crystals could be made large enough for

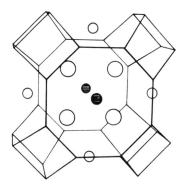 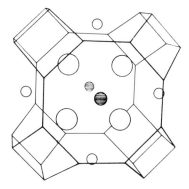

Fig. 12. Distribution of non-framework atoms in the 18-hedron of Species P (left $\bigcirc$ = Ba(1) ● = Ba(2) $\bigcirc$ = Cl(1)) and Species Q (right $\bigcirc$ = Ba(1), ⊜ = Ba(2), $\bigcirc$ = Br(1)).[65]

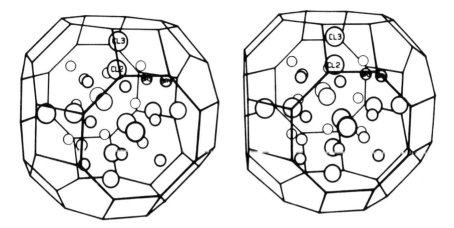

Fig. 13. Stereoscopic pair showing the distribution of located non-framework atoms in the 26-hedral cavities of Species P.[65]

single-crystal measurements. This possibility could certainly be realized for cancrinite and for zeolites A and X (Chapter 4 §7).

5. Salt Uptake from Aqueous Solution and Isotherm Formulation

Several studies have been made of the uptake of salts from aqueous solution by pre-formed zeolites.[52,37,67] Attention has also been given to the interpretation of the isotherms as well as that of isotherms for salt trapping during syntheses as considered in §2.2 and illustrated in Fig. 3.[31] Among the most extensive equilibrium studies was that of Barrer and Walker.[37] They observed that electrolytes penetrated zeolite A only slowly but that zeolites X and Y took up electrolytes much more rapidly under comparable conditions. In zeolite A the penetrant salt must traverse 8-ring windows of free diameter ~4·2 Å, while in X and Y it traverses 12-ring windows of free diameter ~7·8 Å. This difference allowed the kinetics of penetration of electrolyte into zeolite A to be investigated, and led to the adoption of different equilibration procedures for zeolite A and the faujasites.

For zeolite A the zeolite was first equilibrated with the vapour of the solution before a known weight was transferred to a tube together with exactly 4 ml of solution. The sealed tube was rotated in an air oven for some days at temperatures up to 90°C. It was finally rotated for another

day at 30°C before separating the liquid and the crystals by centrifugation. The solution was then removed to a stoppered container for
analysis. The concentration of halide ion in this mother liquor differed
from that in the parent solution. From the volume of halide solution
taken initially the total quantity of salt remaining in the solution was
calculated and so the amount of salt imbibed by the zeolite was found.
For zeolites X and Y the zeolite formed a bed resting on a sintered glass
disc across one arm of a U-shaped cell in a thermostat. The electrolyte
(5 ml) could be passed to and fro by means of a gentle pressure of
nitrogen gas and a suitable arrangement of taps and by-passes. After
equilibration, the electrolyte was driven by nitrogen pressure from the
zeolite to the arm of the cell not containing the zeolite and was then
withdrawn for analysis as for zeolite A.

From the water content of the zeolite equilibrated with the vapour of
the electrolyte solution, the salt uptake per g of anhydrous zeolite was
obtained and isotherms and kinetic curves constructed. In doing this two
extremes were considered: firstly that salt uptake did not displace
measurable amounts of zeolitic water; and secondly that water was
displaced equal in volume to the salt imbibed. The salt uptake was not
large, but there was a small difference in isotherms calculated according
to the two extremes, as shown in Fig. 14(a) for LiCl, LiBr and LiI in
zeolite Li-X.

It was possible to check the above procedures for the Na-A + NaCl

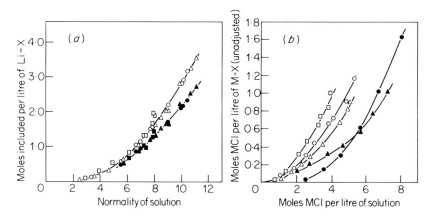

Fig. 14. (a) Isotherms for uptakes of LiCl, LiBr and LiI in Li-X. In the top
curve it is assumed that the water displaced from Li-X by the salt is
negligible; in the bottom curve the water displaced is assumed equal
in volume to that of salt included. △, ▲ = LiCl at 20°C; □, ■ = LiI at
25°C; ○, ● = LiBr at 25°C.[37] (b) Salt inclusion isotherms of chlorides at
25°C in zeolite X. □ = KCl; ○ = NaCl; △ = CaCl$_2$; ● = LiCl; ▲ = CsCl.[37]

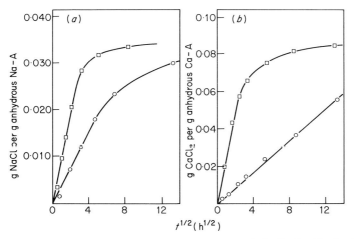

Fig. 15. The kinetics of imbibition of NaCl and CaCl₂ in Na-A. (a); □ = 87·5°C; ○ = 70°C. (b) □ = 50·5°C; ○ = 27°C.[37]

system because at 25°C the rate of inclusion was too small to measure. The zeolite equilibrated at 90°C was quickly cooled to room temperature and washed with cold distilled water until free of adhering salt. The zeolite itself was then analysed for Cl⁻. The result validated the procedure for zeolite A of the previous paragraph, and the assumption that for small uptakes of NaCl the amount of water displaced was minimal.

Figure 14(a) also shows that the uptake of the invading lithium halide in Li-X was nearly independent of the nature of the anion (Cl⁻, Br⁻ and I⁻). On the other hand, the uptakes of Li, Na, K, Cs and Ca chlorides by the corresponding Li-, Na-, K-, (Cs, Na)- and Ca-forms of zeolite X did depend upon the nature of the cation, as shown in Fig. 14(b). The zeolites X, except for the Cs-enriched form, were homoionic. A further aspect of salt uptake was that the equilibrium uptake of KCl in K-X at 25, 50 and 75°C was independent of the temperature within the accuracy of measurement.

The kinetics of imbibition of NaCl and CaCl₂ into Na-A are shown in Fig. 15. Over the early stages the amount taken up is proportional to $t^{1/2}$ (t = time), indicating diffusion control. The large temperature coefficient shows that control is not through film diffusion in electrolyte external to the crystals. If at temperatures T_1 and T_2 the times taken for equal uptakes of salt are t_1 and t_2, then the energy of activation E for intracrystalline diffusion of salt is

$$E = 2 \cdot 30 R \frac{T_1 T_2}{T_2 - T_1} \log_{10}\left(\frac{t_1}{t_2}\right)$$

The following values of E were then found:

	NaCl	CaCl$_2$	CaBr$_2$
E (kJ mol^{-1})	108 ± 8	(i) 116 ± 21	52 ± 8
		(ii) 103 ± 8	

The value for NaCl compares with 107 kJ mol^{-1} for elution of (Li, Na)Cl from (Li, Na)-A (§3.2). The smaller energy barrier for CaBr$_2$ than for CaCl$_2$ may in part arise from weaker solvation of Br$^-$ than of Cl$^-$. Some of the water of solvation must momentarily be shed when the anion passes through the 8-ring windows. Despite the difference in energy of activation between CaCl$_2$ and CaBr$_2$ the rates of uptake for similar salt concentrations at room temperature were nearly the same. This behaviour can arise from differences in the two entropies of activation.[37] At 50°C the chloride was imbibed about twice as fast as the bromide.

Salt uptake from solution into pre-formed zeolites was considered in terms of Donnan membrane equilibria.[52,37] One may write for uni-univalent salts:

	Solution	Zeolite
Components	M$^+$, X$^-$, H$_2$O	M$^+$, X$^-$, Z$^-$, H$_2$O
Concentrations	C C	C_m C_i C_r

M$^+$ is the cation, X$^-$ the anion and Z$^-$ the amount of anionic framework containing one negative charge. C_m, C_i and C_r are respectively in equivalents per dm^3 of crystal and C in equivalents per dm^3 of solution. For electrical neutrality

$$C_m = C_i + C_r \tag{1}$$

TABLE 11
Quotients $C_i(C_i + C_r)/C^2 = R$ for NaCl in Na-X and Na-A[37]

Na-X (25°C)			Na-A (30°C)		
C (mol dm^{-3})	C_i (mol dm^{-3})	R	C (mol dm^{-3})	C_i (mol dm^{-3})	R
1·039	0·073	0·600	1·055	0·037	0·357
1·819	0·163	0·442	1·583	0·069	0·297
2·495	0·305	0·446	2·576	0·265	0·438
2·971	0·442	0·463	3·544	0·403	0·356
3·828	0·654	0·422	4·077	0·541	0·367
4·086	0·715	0·407	4·680	0·687	0·357
4·857	0·910	0·375	4·687	0·695	0·361
5·265	1·159	0·406	4·992	0·777	0·358
			5·310	0·867	0·356

At equilibrium of salt between solution and crystals

$$a_i a_m = K a_\pm^2 \qquad (2)$$

where a_i and a_m are the activities X^- and M^+ in the crystal and $a_\pm$ is the mean activity of salt in the solution. In terms of concentrations and mean activity coefficients this relation becomes

$$\frac{C_i(C_i + C_r)}{C^2} = K f_\pm^2 / (f_\pm)_z^2 = R \qquad (2a)$$

where $f_\pm$ refers to salt in solution and $(f_\pm)_z$ to salt in the zeolite. If $f_\pm^2 / (f_\pm)_z^2$ is constant so must be the quotient of concentrations. Experimental values of $R = C_i(C_i + C_r)/C^2$ are given in Table 11 for NaCl in Na-X and Na-A, which show reasonable independence of C. The same was also true for NaCl in Na-Y, KCl in K-X and CsCl in (Cs, Na)-X as shown by the plots of $C_i(C_i + C_r)$ against C^2 in Fig. 16. Greater variations in $C_i(C_i + C_r)/C^2$ were observed for lithium halides in Li-X and in $C_i^2(C_i + C_r)/C^3$ for $CaCl_2$ in Ca-X and Ca-A. In these systems the ratios $f_\pm^2/(f_\pm)_z^2$ or $f_\pm^3/(f_\pm)_z^3$ were accordingly stronger functions of C.

Zeolites are solid electrolytes with high concentrations of cationic and anionic charge. Values of C_r for some of the zeolites used are given in Table 12, together with mean values of R. NaCl is sorbed more strongly in Na-X than in Na-Y and KCl more strongly in K-X than NaCl in Na-X. From Table 11 it is also seen that C_i is always much less than C, which accords with a Donnan partial exclusion of anions by the anionic charge of the framework. C_i is much below the value corresponding with filling

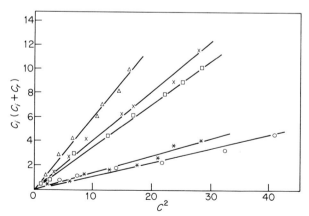

Fig. 16. Test of ideal Donnan equilibrium for salt inclusion. Plots of $C_i(C_i + C_r)$ against C^2.[37] △ = KCl in K-X; × = NaCl in Na-X; □ = NaCl in Na-A; * = NaCl in Na-Y; ○ = CsCl in (Cs, Na)-X.

the intracrystalline pore space with salt. For these dilute solutions of salt in zeolite (Table 11) all the foregoing observations are accounted for in terms of Donnan membrane effects.

When the sodalites bearing salts like $NaClO_3$, $NaClO_4$, NaCl, or NaBr, or Species P and Q respectively bearing $BaCl_2$ or $BaBr_2$, have been synthesized directly, the salt contents, as shown in §2.2, can approach much more nearly to saturation of the intracrystalline pore space than the equilibrium uptakes observed from solution in pre-formed zeolites A, X or Y. Saturation of sodalite corresponds with $C_i = 4\cdot8$ mol dm^{-3} (for 2NaCl per unit cell), and of zeolite A with $C_i = 8\cdot9$ mol dm^{-3} (for $10NaNO_3$ per pseudocell). For NaCl in Table 11 the largest concentration in zeolite Na-A was $0\cdot867$ mol dm^{-3}. It is of interest to develop an alternative formulation of isotherms in which the salt concentrations can approach the saturation values.[25] As suggested in §2.2, during crystal growth the salt may play a specific role by complexing either to form precursor aluminosilicate species in solution which then add to the growing crystal, or at the surface of the crystal in incomplete cavities which are then completed as growth proceeds.

The sodalite cage can contain one salt molecule, two molecules of alkali or four of water. A treatment, based on detailed balancing, of the competitive uptakes and releases of these three guest species between the solution and the sodalite cages before they were finally closed leads to definitive isotherm equations.[25] If N_{xyz} is the number of cages containing x water + y NaOH + z salt then the only possible compositions are

$$N_{000}, N_{100}, N_{200}, N_{300}, N_{400}, N_{010}, N_{020}, N_{110}, N_{210} \text{ and } N_{001}$$

For detailed balancing we then have the relations

$$\left.\begin{array}{ll}
4r_1N_{000} = N_{100} & 2r_1N_{010} = N_{110} \\
3r_1N_{100} = 2N_{200} & r_1N_{110} = 2N_{210} \\
2r_1N_{200} = 3N_{300} & 2r_2N_{000} = N_{010} \\
r_1N_{300} = 4N_{400} & r_2N_{010} = 2N_{020} \\
r_3N_{000} = N_{001} & r_2N_{100} = N_{110}
\end{array}\right\} \qquad (3)$$

In these relations the r_i are the products† $K_i a_i$, where for water, NaOH and salt $i = 1$, 2 and 3 respectively. The a_i are activities in the aqueous solutions and the K_i are equilibrium constants for the distribution of component i between crystals and solution. K_i is the ratio of the rate constants for the processes of entry into and removal from the incomplete sodalite cages for a molecule of guest species i. To simplify the treatment guest–guest interactions are ignored, as in Langmuir's isotherm.

† In the original treatment $K_i C_i$ was taken as the product. Substitution of C_i by a_i allows for non-ideality of the solution. Although this is preferred here, no part of the conclusions of the original treatment is changed by this preference.

TABLE 12
Salt uptake equilibria and values of C_r in some zeolites[37]

Zeolite	Cations per unit cell	Unit cell edge (Å)	$C_r =$ Equivalents of framework charge per dm^3	Salt imbibed	Mean value of $R = \dfrac{C_i(C_i + C_r)}{C^2}$
Na-X	82	24·9$_4$	8·8	NaCl	0·42
Na-Y	60	24·7$_1$	6·6	NaCl	0·13
K-X	82	25·0	8·7	KCl	0·60
(Cs, Na)-X	82	25·0	8·7	CsCl	0·11
Na-A	12	12·3	10·7	NaCl	0·36

From the above relations one may find each N_{xyz} in terms of N_{000}, the number of empty trapping centres:

$$
\left.
\begin{aligned}
N_{100} &= 4r_1 N_{000} & N_{110} &= 4r_1 r_2 N_{000} \\
N_{200} &= 6r_1^2 N_{000} & N_{210} &= 2r_1^2 r_2 N_{000} \\
N_{300} &= 4r_1^3 N_{000} & N_{010} &= 2r_2 N_{000} \\
N_{400} &= r_1^4 N_{000} & N_{020} &= r_2^2 N_{000} \\
N_{001} &= r_3 N_{000} &
\end{aligned}
\right\}
\tag{4}
$$

The total number, N, of trapping centres is the sum of the ten N_{xyz}. Equations (4) give

$$
N = N_{000}\{[(1 + r_1)^2 + r_2]^2 + r_3\}
\tag{5}
$$

One may similarly find the total numbers N_1, N_2 and N_3 respectively of water, hydroxide and salt molecules in the trapping centres as

$$
\left.
\begin{aligned}
N_1 &= 4r_1(1 + r_1)N_{000}[(1 + r_1)^2 + r_2] \\
N_2 &= 2r_2 N_{000}[(1 + r_1)^2 + r_2] \\
N_3 &= r_3 N_{000}
\end{aligned}
\right\}
\tag{6}
$$

The fractional saturations of trapping centres by these three guests (θ_1, θ_2 and θ_3) and the fractional unsaturation (θ_0) of trapping centres are given by:

$$
\left.
\begin{aligned}
\theta_0 &= (1 - \theta_1 - \theta_2 - \theta_3) = (1 + r_1)[(1 + r_1)^2 + r_2]/Q \\
\theta_1 &= \frac{N_1}{4N} = r_1(1 + r_1)[(1 + r_1)^2 + r_2]/Q \\
\theta_2 &= \frac{N_2}{2N} = r_2[(1 + r_1)^2 + r_2]/Q \\
\theta_3 &= \frac{N_3}{N} = r_3/Q
\end{aligned}
\right\}
\tag{7}
$$

where $Q = [(1+r_1)^2 + r_2]^2 + r_3$. The relations (7) can be re-written with the θ_i as the independent variables:

$$\left.\begin{array}{l} r_1 = \theta_1/\theta_0 \\ r_2 = \theta_2(\theta_0+\theta_1)/\theta_0^2 \\ r_3 = \theta_3(1-\theta_3)(\theta_0+\theta_1)^2/\theta_0^4 \end{array}\right\} \qquad (8)$$

In thermodynamic terms the distribution equilibrium of guest i between crystals and aqueous solution is given by

$$K_i = (a_i)_z/a_i$$

where the subscript "z" refers to the guest within the crystals and the a_i are activities in solution as before. Thus r_1, r_2 and r_3 can be identified as the activities of the three guest molecules in the zeolite. The solutions used in studying the uptake of salts during synthesis of sodalites (§2.2, Fig. 3) were concentrated and complex and the activities of the dissolved

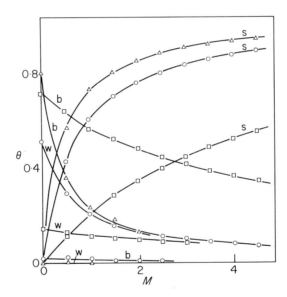

Fig. 17. Some calculated salt (s), water (w) and base (b) isotherms in sodalite.[25] Uptake (during synthesis) occurs competitively from the mixtures, but the concentrations of NaOH and of water are fixed while that of the salt varies. The assignments of the R_i and R_3' (salt) which generated these isotherms were:

	R_3' (salt)	R_2 (base)	R_1 (water)
△	99	5	0·1
○	30	0·1	1·1
□	50	10	1

species were not known. The concentrations of water and caustic soda were constant in all runs (respectively ~55·5 and 4 M), while that of the salt was nearly constant for any one run but was systematically varied from one run to the next. The unknown activities were therefore replaced by concentrations, so that the r_i are replaced by

$$r_i/f_i = R_i = K_i C_i \qquad (9)$$

where the f_i are activity coefficients. They were assumed to be constant so that R_1 and R_2 are constants and $R_3 = R_3' m_3$, where m is the molality of salt in the aqueous solution and R_3' is a constant. With these necessary approximations isotherms were calculated of θ_i against molality of the salt with several assignments of R_1, R_2 and R_3'. Some of the results, shown in Fig. 17, are to be compared with experimental isotherms like those in Fig. 3. The qualitative correspondence might be improved by other choices of R_1, R_2 and R_3' but, especially in view of the approximations, can be regarded as support for the model developed.

6. Salt Storage, Release and Decomposition

The NaCl trapped in the sodalite cages of zeolite A and faujasites (Na-X, Na-Y) conferred greater thermal stability upon the crystals, by about 50–100°C.[68] Stability differences were recorded also among salt-bearing sodalites and cancrinites according to the nature of the anions involved and the amount of trapped salts.[25] The stabilities were examined by means of X-ray powder diffraction either after programmed heating to each of a succession of temperatures, or continuously, using a heating X-ray camera. Rabo et al.[69] examined the carbonium ion type of catalytic activity of Na-Y containing NaCl trapped in sodalite cages, and usually found a diminution in activity. This decrease was attributed to elimination of acid centres by reaction with trapped salt:

$$\left[\begin{matrix} \diagdown \\ -Al \\ \diagup \end{matrix} \quad \begin{matrix} HO \diagdown \\ Si {<} \\ \diagup \end{matrix} \right] + NaCl \longrightarrow \left[\begin{matrix} \diagdown \\ -\overset{\ominus}{Al}-O-Si{<} \\ \diagup \end{matrix} \right] + Na^+ + HCl\uparrow$$

However, treatment of Na-Y with $LaCl_3$ gave a product with considerable catalytic activity, attributed to an increase in proton content of the zeolite as a result of hydrolysis of $LaCl_3$. Barrer and Villiger[57] observed that, when a number of zeolites were heated with $NaNO_2$ at 290°C or $NaNO_3$ at 330°C, there was with certain of them evolution of oxides of nitrogen and formation of sodium oxide. This reaction was not observed with

Na-X or -Y while Na-A decomposed only $NaNO_2$. With the remainder of the zeolites examined the reaction became more marked in the sequence cancrinite < mordenite < chabazite. Thus the catalysis is sensitive to the zeolite used. Exchange of Ca^{2+} by Na^+ in chabazite could account for some reaction since the $Ca(NO_2)_2$ and $Ca(NO_3)_2$ formed thereby are less stable than the corresponding sodium salts.

A number of spectroscopic studies (infrared and Raman) have been made of salt-zeolite complexes, especially those of the nitrates.[55,59,70] These show small shifts in frequencies of the absorption maxima of polyatomic ions such as NO_3^- as compared with the melt and similar small shifts in some framework frequencies. Such shifts can be attributed to specific and mutual environmental influences. A quantitative investigation has been made of the thermal decomposition of $NaClO_2$, $NaClO_3$ and $NaClO_4$ trapped in sodalite and of sodium azide trapped in cancrinite, in which reaction was followed by measuring the oxygen or nitrogen evolution as a function of temperature and time.[23] Figure 18(a), (b) and (c) respectively show at a series of temperatures the kinetics of release of oxygen from $NaClO_4$ (18·1% by weight) and $NaClO_3$ (17·3%) by weight trapped in sodalite, and of nitrogen from NaN_3 (5·5% by weight) in cancrinite. The latter reaction is clearly more complex than can be explained in terms of simple diffusion control. Each process has a large positive temperature coefficient and some apparent activation energies, evaluated from the times required at each of the two different temperatures to reach the same fraction of decomposition, are given in Table 13 for $NaClO_4$- and $NaClO_3$-sodalite. These energies appear to increase with Q_t/Q_∞ for $NaClO_3$, as could be expected if at least two consecutive reactions were involved, the second one with a higher activation energy than the first. Suggested steps were:

$$2NaClO_3 \rightarrow NaClO_2 + NaClO_4$$
$$NaClO_2 \rightarrow NaCl + O_2$$
$$NaClO_4 \rightarrow NaClO_2 + O_2 \rightarrow NaCl + 2O_2$$

The values of E for oxygen evolution from both salt-bearing sodalites are large, that for $NaClO_4$-sodalite being the greater. The decomposition steps for this complex may be those in the last of the above equations.

So far no large-scale use has been made of the many possible salt-bearing zeolites and felspathoids. The trapped salts are securely packaged on a molecular scale by the inert tectosilicate frameworks and may thereby be protected from attack by reactive species. Potentially unstable salts prone in bulk to vigorously exothermal or explosive decompositions and reactions could be rendered safe by the packing. Thus, bulk $NaClO_3$, often used as a weed killer, can be a dangerously strong oxidant in some circumstances and is banned for general sale in certain countries for this reason. Packaged in a zeolite or felspathoid it is protected from such contacts but could still be released slowly after application of the chlorate-bearing zeolite, by displacement by water or weathering of the packaging tectosilicate framework. Rapid and safe release of oxysalts as oxidizing agents in wet chemical processes can be readily achieved at the moment required by adding acid to the mixture containing the previously inert packaged zeolite-oxysalt complex. The acid quickly decomposes the aluminosilicate framework,

Release of salts is subject to molecule sieving controls in the same way

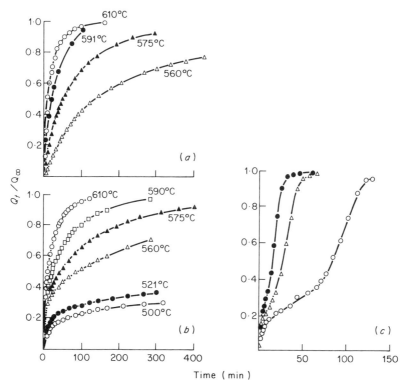

Fig. 18. (a) The evolution of oxygen at different temperatures from NaClO$_4$-sodalite containing 18·1% by weight of NaClO$_4$. (b) Evolution of oxygen from NaClO$_3$-sodalite containing 17·3% by weight of NaClO$_3$. (c) Evolution of nitrogen from NaN$_3$-cancrinite containing 5·5% by weight of NaN$_3$. In (c), $\bigcirc = 550°C$; $\triangle = 575°C$; and $\bullet = 600°C$.[23] In (a), (b) and (c), Q_t = gas evolved at time t and Q_∞ = gas evolved when reaction is complete.

TABLE 13

Apparent activation energies E (kJ mol^{-1})[23]

	NaClO$_4$-sodalite				NaClO$_3$-sodalite		
$\dfrac{Q_t}{Q_\infty}$	T_1 (K)	T_2 (K)	E	$\dfrac{Q_t}{Q_\infty}$	T_1 (K)	T_2 (K)	E
0·4	848	833	376	0·2	794	773	220
0·6	848	833	398	0·4	848	833	252
	864	848	352	0·6	848	833	312
0·8	883	848	288		863	848	310
					883	863	258
				0·8	883	848	318

as for non-ionic species. The only way to release salts from sodalite is to decompose the sodalite; from their complexes with Na-A, NaCl and NaNO$_3$ can be released slowly in water (§3.2); while from complexes with wide-pore zeolites (Na-X, Na-Y), water removes salts rapidly. Accordingly, rate of release by water of packaged salts in zeolites can be controlled according to the use envisaged. In weed killer uses of NaClO$_3$ a notable flexibility may thus be possible with no danger of explosion or fires from its misuse. The same is true of release rates in wet chemistry for oxidants packaged in variously open frameworks (chabazite, zeolite A, ZSM-5, offretite, mordenite, mazzite, zeolites X and Y). The yields of oxygen molecules from sodalite and zeolite A bearing oxysalts are as follows:

$$
\begin{array}{llll}
\text{Na}_6[\text{Al}_6\text{Si}_6\text{O}_{24}]2\text{NaX} & \text{X} = \text{ClO}_2 & \text{O}_2 = 43\ \text{cm}^3 \text{ at s.t.p. per g reagent} \\
\quad \text{(sodalite)} & \text{X} = \text{ClO}_3 & \text{O}_2 = 62 \\
& \text{X} = \text{ClO}_4 & \text{O}_2 = 81 \\
\text{Na}_{12}[\text{Al}_{12}\text{Si}_{12}\text{O}_{48}] & \text{X} = \text{ClO}_2 & \text{O}_2 = 86 \\
\quad \text{(zeolite A)} & \text{X} = \text{ClO}_3 & \text{O}_2 = 121 \\
& \text{X} = \text{ClO}_4 & \text{O}_2 = 152
\end{array}
$$

The oxidizing potential of such materials is thus considerable. Controlled release of H$_2$ can also be envisaged by the action of water on zeolite containing borohydride, for example Al(BH$_4$)$_3$, or zeolite or felspathoid containing sodium. Outgassed sodalite hydrate can be saturated with Na vapour to give a dark product in which the packaged Na is protected from potential reactants too large to pass through the 6-rings.[71] Even water reacts only very slowly with the packaged Na, but if rapid release of H$_2$ is required acid quickly breaks down the aluminosilicate packaging. At high temperatures the elementary Na can be distilled out of the sodalite in a controlled way. Controlled release of acids is also possible, for example by treating zeolites bearing $AlCl_3$, PCl_3, $SOCl_2$ with water. Here the danger of lattice breakdown, characteristic of more aluminous zeolites, can be avoided by using silica-rich packaging (mordenite; zeolites ZSM-5 and -11 and their silica end members silicalites 1 and 2; or the silica end member of faujasite[72]).

The ideas developed above may be considered as examples of some of the possibilities of reagent packaging and storage in tectosilicate frameworks. The account given in this chapter indicates that there can be a diversity among zeolite-salt complexes which parallels that found with covalent guest molecules. Considerable developments in the chemistry and physics of salt–zeolite solid electrolytes may be made in the future.

References

1. E. Galli, G. Gottardi and D. Pongiluppi, *Neues Jb. Mineral. Mh.* (1979) **1**, 1.
1a. R.M. Barrer, "Zeolites and Clay Minerals as Sorbents and Molecular Sieves", p. 45. Academic Press, London and New York, 1978.
2. H.J. Emeleus and J.S. Anderson, "Modern Aspects of Inorganic Chemistry", p. 212. Routledge, London, 1938.
3. J.S. Prener and R. Ward, *J. Amer. Chem. Soc.* (1950) **72**, 2780.
4. F.M. Jaeger, *Trans. Faraday Soc.* (1929) **25**, 320.
5. C.A. Kumins and A.E. Gessler, *Ind. Eng. Chem.* (1953) **45**, 567.
6. K. Leschewski, *Z. Angew. Chemie* (1935) **48**, 533.
7. K. Leschewski and E. Podschus, *Ber. Deut. Chem. Gesell.* (1935) **68**, 1872.
8. K. Leschewski, H. Möller and E. Podschus, *Z. Anorg. Allg. Chemie* (1934) **220**, 317.
9. R.M. Barrer and J.S. Raitt, *J. Chem. Soc.* (1954) 4641.
10. Y. Kubo and G. Yamaguchi, *Bull. Chem. Soc. Japan* (1969) **42**, 1897.
11. G. Yamaguchi and Y. Kubo, *Bull. Chem. Soc. Japan* (1968) **41**, 2641.
12. G. Yamaguchi and Y. Kubo, *Bull. Chem. Soc. Japan* (1968) **41**, 2645.
13. W. Eitel, "The Physical Chemistry of the Silicates", p. 896. University of Chicago Press, 1954.
14. W.A. Lambertson, *J. Amer. Ceram. Soc.* (1952) **35**, 161.
15. J. Lemberg, *Z. Deut. Geol. Gesell.* (1875) **28**, 601.
16. J. Morozewicz, *Tsch. Mineral. Petr. Mitt.* (1898) **18**, 128.
17. Z. Weyberg, *Centr. Mineral. Geol.* (1905) 717.
18. J. Morozewicz, *Tsch. Mineral. Petr. Mitt.* (1899) **18**, 20.
19. J. Lemberg, *Z. Deut. Geol. Gesell.* (1883) **33**, 579.
20. St.J. Thugutt, *Z. Anorg. Chem.* (1892) **2**, 65.
21. C. Friedel and G. Friedel, *Bull. Soc. Mineral. Fr.* (1890) **13**, 182 and 239.
22. E. Flint, W. Clarke, E.S. Newman, L. Shartsis, D. Bishop and L.S. Wells, *J. Res. Nat. Bur. Stds* (1945) **36**, 63.
23. R.M. Barrer, E.A. Daniels and G.A. Madigan, *J. Chem. Soc. Dalton* (1976) 1805.
24. R.M. Barrer, J.F. Cole and H. Sticher, *J. Chem. Soc. A* (1968) 2475.

25. R.M. Barrer and J.F. Cole, *J. Chem. Soc. A* (1970) 1516.
26. J. Lemberg, *Z. Deut. Geol. Gesell.* (1887) **39,** 559.
27. R.M. Barrer and J.D. Falconer, *Proc. Roy. Soc.* (1956) **236A,** 227.
28. R.M. Barrer and D.E.W. Vaughan, *J. Phys. Chem. Solids* (1971) **32,** 731.
29. R.M. Barrer, "Zeolites and Clay Minerals as Sorbents and Molecular Sieves", Table 4, p. 36. Academic Press, London and New York, 1978.
30. R.M. Barrer and E.A.D. White, *J. Chem. Soc.* (1952) 1561.
31. R.M. Barrer and E.F. Freund, *J. Chem. Soc. Dalton* (1974) 1049.
32. R.M. Barrer and D.J. Marshall, *J. Chem. Soc.* (1965) 6621.
33. R.M. Barrer, *J. Chem. Soc.* (1948) 127.
34. R.M. Barrer and C. Marcilly, *J. Chem. Soc. A* (1970) 2735.
35. R.M. Barrer and D.W. Riley, *J. Chem. Soc.* (1948) 133.
36. W.M. Meier and D.H. Olson, "Atlas of Zeolite Structure Types", p. 47. Structure Commission of the International Zeolite Association, 1978.
37. R.M. Barrer and A.J. Walker, *Trans. Faraday Soc.* (1964) **60,** 171.
38. R.M. Barrer, "Zeolites and Clay Minerals as Sorbents and Molecular Sieves", Ch. 7, §§6, 7. Academic Press, London and New York, 1978.
39. M. Grandjean, *Bull. Soc. Fr. Mineral.* (1910) **33,** 5.
40. R.M. Barrer and M. Woodhead, *Trans. Faraday Soc.* (1948) **44,** 1001.
41. R.M. Barrer and J.L. Whiteman, *J. Chem. Soc. A* (1967) 19.
42. R.M. Barrer and S. Wasilewski, unpublished.
43. V.S. Komerov, L.P. Shirinskaya and N.P. Bokhan, *Zhu. Fiz. Khim.* (1976) **50,** 2464.
44. C.R. Castor, 1963, U.S.P. 3,329,482.
45. T.R.P. Gibb, *J. Chem. Ed.* (1948) **25,** 577.
46. F.W. Clarke and G. Steiger, *Amer. J. Sci.* (1899) **8,** 245.
47. F.W. Clarke and G. Steiger, *Amer. J. Sci.* (1900) **9,** 117.
48. C.M. Callahan and M.A. Kay, *J. Inorg. Nucl. Chem.* (1966) **28,** 233 and 2743.
49. M. Liquornik and Y. Marcus, *J. Phys. Chem.* (1968) **72,** 2885 and 4074.
50. M. Liquornik and J.W. Irvine, *Inorg. Chem.* (1970) **9,** 1330.
51. R.M. Barrer and A.G. Kanellopoulos, *J. Chem. Soc. A* (1970) 775.
52. R.M. Barrer and W.M. Meier, *J. Chem. Soc.* (1958) 299.
53. N.A. Petranovic and M.V. Susic, *J. Inorg. Nucl. Chem.* (1969) **31,** 551.
54. M.V. Susic, N.A. Petranovic and D.A. Mioc, *J. Inorg. Nucl. Chem.* (1971) **33,** 2667.
55. N. Petranovic, U. Mioc, M. Susic, R. Dimitrijevic and I. Krstanovic, *J. Chem. Soc. Faraday 1* (1981) **77,** 379.
56. N. Petranovic and M. Susic, *J. Inorg. Nucl. Chem.* (1974) **36,** 1381.
57. R.M. Barrer and H. Villiger, *Zeit. Krist.* (1975) **142,** 82.
58. J.A. Rabo and P.H. Kasai, in "Progress in Solid State Chemistry" (Ed. J.O. McCaldin and G. Somorjai) Vol. 9, p. 1. Pergamon Press, Oxford, 1975.
59. A.T. Petfield and R.P. Cooney, *Aust. J. Chem.* (1980) **33,** 659.
60. A. Araya and A. Dyer, *Zeolites* (1981) **1,** 35.
61. G.W. Skeels, unpublished. See "Zeolite Chemistry and Catalysis", (Ed. J.A. Rabo) p. 342. American Chemical Society Monograph 171, 1976.
62. J.A. Rabo, in "Zeolite Chemistry and Catalysis" (Ed. J.A. Rabo), p. 344. American Chemical Society Monograph 171, 1976.
63. G.V. Antoshin, Kh.M. Minachev, E. N. Sevastjanov and D.A. Kondratjev, *Russ. J. Phys. Chem.* (1970) **44,** 1491.

64. R.M. Barrer, J.F. Cole and H. Villiger, *J. Chem. Soc. A* (1970) 1523.
65. R.M. Barrer and D.J. Robinson, *Zeit. Krist.* (1972) **135,** 374.
66. O. Jarchow, *Zeit. Krist.* (1965) **122,** 407.
67. W.A. Platek and J.A. Marinsky, *J. Phys. Chem.* (1961) **65,** 2118.
68. Ref. (62), p. 347.
69. J.A. Rabo, M.L. Poutsma and G.W. Skeels, *in* "Proceedings of the 5th International Congress on Catalysis", p. 1353. North Holland, Amsterdam, 1972.
70. K.R. Loos and J.F. Cole, *in* "Proceedings of the 3rd International Conference on Zeolites", p. 230. (Ed. J.B. Uytterhoeven). Leuven University Press, Leuven, 1973.
71. R.M. Barrer and J.F. Cole, *J. Phys. Chem. Solids* (1968) **29,** 1755.
72. H.K. Beyer and I. Belenykaja, *in* "Catalysis by Zeolites", p. 203. (Ed. B. Imelik *et al.*) Elsevier, Amsterdam, 1980.

Subject Index

Substance Index

Adularia, 3, 79, 98, 202, 239–40, 274
Afghanite, 307–8
Akermanite, 301
Albite, 3, 7, 45, 49, 81, 84, 90, 92–3, 95, 190, 195, 200, 203, 274, **278–281**, 295, 311
  Ga-, (Ga, Ge)-, Ge-, 95
Allanite, 96
Allophane, 79
Aluminates, (Ba, Na)-, 210
  Ca-, 193, 198
  Na-, 271
Aluminate sodalites 270, 281
Aluminogermanates, 1, 2
Aluminophosphates, 294–5, 298–300
Alunite, 79
Amblygonite, 96
Amicite, 18
Analcime (analcite), 1, 3–7, 9, 18, 21, 48, 81, 84, 91–2, 98, 124–5, 138–9, 158, 162, 164, 168–9, 173, 179, 181, 189–91, 195–6, **199**, 201–2, 208, 218–220, 227–8, 230–1, 233–4, 237–9, 242–4, 255–6, 275, 284–5, 290, 294–8, 312, 315, 322
  Be-, 283–4
  Ca-, 143, 198
  NH$_4$-, 162, 191
  Rb-, 18, 191
  Sr-, 193
Anorthite, 48, 81, 90, 95, 193, 198,
252, 272, 275, 282, 310
  Ga-, (Ga, Ge)- and Ge-, substitution in, 95
  hexagonal dimorph of, 162, 193, 198, 202, 239
Antigorite, Co-, Ni-, 84
Apatite, 49, 79, 96
Axinite, 109

Banalsite, 12
Barium silicate hydrate, 173
Barrerite, 17, 18, 227
Bauxite, 45
Bayerite, 94, 192, 197
Beidellite, 48, 79, 81, 83–4
Benitoite, 109
Beryl, 109, 138
Bicchulite, 262, 271, 281
Bikitaite, 8, 12, 18, 21, 201–2
Blende, 49, 92
Boehmite, 82, 83, 162, 191–3, 198, 202, 244.
Boralites, 287–9
Borohydrides, 324, 326.
Borosilicates, Ba-, 289–90
Brewsterite, 18, 21
Brucite, 79

Calcite, 79

355

Geology Museum:

Zeolites shown by: Susanna Van Rose

Contact Mr Alan Jobbins,
589-3444
ext 250